Phenolic Compounds in Foods and Natural Health Products

ACS SYMPOSIUM SERIES **909**

Phenolic Compounds in Foods and Natural Health Products

Fereidoon Shahidi, Editor
Memorial University of Newfoundland

Chi-Tang Ho, Editor
Rutgers, The State University of New Jersey

Sponsored by the
ACS Division of Agricultural and Food Chemistry, Inc.

American Chemical Society, Washington, DC

Library of Congress Cataloging-in-Publication Data

Phenolic compounds in foods and natural health products / Fereidoon Shahidi, Chi-Tang Ho, editor.

    p. cm.—(ACS symposium series ; 909)

    Includes bibliographical references and index.

    ISBN 0-8412-3891-X (alk. paper)

    1. Phenols—Physiological effect—Congresses. 2. Functional foods—Congresses.

    I. Shahidi, Fereidoon, 1951– II. Ho, Chi-Tang, 1944– III. Series.

QP801.P4P455   2005
612′.0157—dc22   2005041203

The paper used in this publication meets the minimum requirements of American National Standard for Information Sciences—Permanence of Paper for Printed Library Materials, ANSI Z39.48–1984.

Copyright © 2005 American Chemical Society

Distributed by Oxford University Press

All Rights Reserved. Reprographic copying beyond that permitted by Sections 107 or 108 of the U.S. Copyright Act is allowed for internal use only, provided that a per-chapter fee of $30.00 plus $0.75 per page is paid to the Copyright Clearance Center, Inc., 222 Rosewood Drive, Danvers, MA 01923, USA. Republication or reproduction for sale of pages in this book is permitted only under license from ACS. Direct these and other permission requests to ACS Copyright Office, Publications Division, 1155 16th Street, N.W., Washington, DC 20036.

The citation of trade names and/or names of manufacturers in this publication is not to be construed as an endorsement or as approval by ACS of the commercial products or services referenced herein; nor should the mere reference herein to any drawing, specification, chemical process, or other data be regarded as a license or as a conveyance of any right or permission to the holder, reader, or any other person or corporation, to manufacture, reproduce, use, or sell any patented invention or copyrighted work that may in any way be related thereto. Registered names, trademarks, etc., used in this publication, even without specific indication thereof, are not to be considered unprotected by law.

PRINTED IN THE UNITED STATES OF AMERICA

Foreword

The ACS Symposium Series was first published in 1974 to provide a mechanism for publishing symposia quickly in book form. The purpose of the series is to publish timely, comprehensive books developed from ACS sponsored symposia based on current scientific research. Occasionally, books are developed from symposia sponsored by other organizations when the topic is of keen interest to the chemistry audience.

Before agreeing to publish a book, the proposed table of contents is reviewed for appropriate and comprehensive coverage and for interest to the audience. Some papers may be excluded to better focus the book; others may be added to provide comprehensiveness. When appropriate, overview or introductory chapters are added. Drafts of chapters are peer-reviewed prior to final acceptance or rejection, and manuscripts are prepared in camera-ready format.

As a rule, only original research papers and original review papers are included in the volumes. Verbatim reproductions of previously published papers are not accepted.

ACS Books Department

Contents

Preface xi

1. Phenolics in Food and Natural Health Products: An Overview 1
   Fereidoon Shahidi and Chi-Tang Ho

Chemistry and Analysis

2. Phenolic Acid Composition of Wheat Bran 10
   Kequan Zhou, John W. Parry, and Liangli (Lucy) Yu

3. Structure and Analysis of Flavonolignans from *Silybum marianum* 19
   David Y-W Lee and Yanze Liu

4. Antioxidant Activity of Sesame Fractions 33
   Fereidoon Shahidi and Chandrika M. Liyana-Pathirana

5. Effect of Lactic Acid Fermentation on Quercetin Composition and Antioxidative Properties of *Toona sinensis* Leaves 46
   Tzou-Chi Huang, Hseng-Kuang Hsu, Hui-Yin Fu, and Chi-Tang Ho

6. Antioxidant Capacity of Phenolics from Canola Hulls as Affected by Different Solvents 57
   M. Naczk, R. Amarowicz, R. Zadernowski, and F. Shahidi

7. Antioxidant Activity of Polyphenolics from a Bearberry-Leaf (*Arctostaphylos uva-ursi* L. Sprengel) Extract in Meat Systems 67
   Ronald B. Pegg, Ryszard Amarowicz, and Marian Naczk

8. Beans: A Source of Natural Antioxidants 83
   Terrence Madhujith and Fereidoon Shahidi

9. **Antioxidant and Antibacterial Properties of Extracts of Green Tea Polyphenols**..................94
   Ryszard Amarowicz, Ronald B. Pegg, Gary A. Dykes, Agnieszka Troszynska, and Fereidoon Shahidi

10. **Radical Scavenging Properties of Cold-Pressed Edible Seed Oils**..................107
    John W. Parry, Kequan Zhou, and Liangli (Lucy) Yu

11. **Honeybush Tea: Chemical and Pharmacological Analyses**..................118
    Mingfu Wang, Rodolfo Juliani, James E. Simon, Albert Ekanem, Chia-Pei Liang, and Chi-Tang Ho

12. **Combined Inhibitory Effects of Catechins with Fe^{3+} on the Formation of Potent Off-Odorants from Citral**..................129
    Toshio Ueno, Hideki Masuda, and Chi-Tang Ho

13. **Influence of Flavonoids on the Thermal Generation of Aroma Compounds**..................143
    Devin G. Peterson and Vandana M. Totlani

14. **Capsaicinoid Oxidation by Peroxidases: Kinetic, Structural, and Physiological Considerations**..................161
    Douglas C. Goodwin, Kimberley A. Laband, and Kristen M. Hertwig

Synthesis, Production, and Mechanism of Action

15. **Artepillin C Isoprenomics: Facile Total Synthesis and Discovery of Amphiphilic Antioxidant**..................176
    Yoshihiro Uto, Shutaro Ae, Hideko Nagasawa, and Hitoshi Hori

16. **Production of Theaflavins and Theasinensins during Tea Fermentation**..................188
    Takashi Tanaka, Chie Mine, Sayaka Watarumi, Yosuke Matsuo, and Isao Kouno

17. **New Momentum on the Action Mechanisms of Black Tea Polyphenols, the Theaflavins**..................197
    Jen-Kun Lin, Min-Hsiung Pan, Yu-Chih Liang, Shoui-Yn Lin-Shiau, and Chi-Tang Ho

Health Effects

18. **Biotransformation and Bioavailability of Tea Polyphenols: Implications for Cancer Prevention Research**..........212
    Joshua D. Lambert, Jungil Hong, Mao-Jung Lee, Shengmin Sang, Xiaofeng Meng, Hong Lu, and Chung S. Yang

19. **Prevention of Cancer by Dietary Phytochemicals**..........225
    Zigang Dong and Ann M. Bode

20. **Effect of Black Tea Theaflavins and Related Benzotropolone Derivatives on 12-*O*-Tetradecanoylphorbol-13-acetate-Induced Mouse Ear Inflammation and Inflammatory Mediators**..........242
    Divya Ramji, Shengmin Sang, Yue Liu, Robert T. Rosen, Geetha Ghai, Chi-Tang Ho, Chung S. Yang, and Mou-Tuan Huang

21. **Antioxidant and Antitumor Promoting Activities of Apple Phenolics**..........254
    Ki Won Lee, Hyong Joo Lee, and Chang Yong Lee

22. **Cranberry Phenolics: Effects on Oxidative Processes, Neuron Cell Death, and Tumor Cell Growth**..........271
    Catherine C. Neto, Marva I. Sweeney-Nixon, Toni L. Lamoureaux, Frankie Solomon, Miwako Kondo, and Shawna L. MacKinnon

Indexes

Author Index..........285

Subject Index..........287

Preface

Phenolic and polyphenolic compounds in foods and natural health products represent one of the most widely distributed plant secondary metabolites with varied structured characteristics, properties and health effects. Many of the beneficial health effects of plant phenolics are due to their antioxidant activity and their role in balancing the deleterious effects of oxidants. However, this is not always the case. The health promotion activity of phenolics is in the control of certain types of cancer, cardiovascular disease, autoimmune disorders, inflammatory diseases and the process of ageing. Thus, consumption of food ingredients and natural health products capable of neutralizing reactive oxygen and other oxidant species in the body is of much interest.

This volume reports the latest developments in the phenolic and polyphenolic areas of research and discusses their chemistry, analysis, mechanism of action, production, and health effects of different products and phenolics thereof. We are grateful to the authors of different chapters for their state-of-the-art contributions that made publications of this book possible.

Fereidoon Shahidi
Department of Biochemistry
Memorial University of Newfoundland
St. John's, Newfoundland A1B 3X9
Canada

Chi-Tang Ho
Department of Food Science
Rutgers, The State University of New Jersey
New Brunswick, NJ 08901–8520

Chapter 1

Phenolics in Food and Natural Health Products: An Overview

Fereidoon Shahidi[1] and Chi-Tang Ho[2]

[1]Department of Biochemistry, Memorial University of Newfoundland,
St. John's, Newfoundland A1B 3X9, Canada
[2]Department of Food Science, Rutgers, The State University of New Jersey,
65 Dudley Road, New Brunswick, NJ 08901–8520

Phenolic and polyphenolic compounds occur in food as secondary metabolites. Different classes of phenolics are often encountered in foods and these exert beneficial health effects related to their antioxidant activity, chemopreventive effect and anticarcinogenic potential, among others. These activities are often governed by the structural characteristics of the bioactive components present. In addition, processing of foods might affect the chemical and activity of the compounds involved.

Introduction

Phenolic compounds in foods and natural health products originate from one of the main classes of secondary metabolites in plants derived from phenylalanine or tyrosine (*1*). The term 'phenolic' or 'polyphenol' can be defined as a substance which possesses an aromatic ring bearing one or more hydroxyl substituents, including functional derivatives (esters, methyl ethers, glycosides etc.) (*2*). Most phenolics have two or more hydroxyl groups and are bioactive substances occurring widely in foods.

Occurrence of Phenolics

The phenolics which occur commonly in foods and natural health products may be classified into simple phenols, hydroxybenzoic acid and hydroxycinnamic acid derivatives, flavonoids, stilbenes, lignans and hydrolyzable as well as condensed tannins.

The Simple Phenols and Phenolic Acids

The simple phenols include monophenols such as *p*-cresol isolated from several fruits and diphenols. Diphenols such as hydroquinones are probably the most widespread simple phenols (*3*).

There are two classes of phenolic acids, hydroxybenzoic acids and hydroxycinnamic acids. Except in certain red fruits and onions, the content of hydroxybenzoic acids in edible plants is usually very low (*1*). Tea can be an important source of gallic acid which is 3,4,5-trihydroxybenzoic acid (*4*). Gallic acid usually occur in plants in souble form as catechin esters, quinic acid ester or hydrolyzable tannins.

The hydroxycinnamic acids and their derivatives are more common than are the hydroxybenzoic acids in foods. Hydroxycinnamic acids and their derivatives are almost exclusively derived from *p*-coumaric acid, caffeic acid, and ferulic acid whereas sinapic acid is, in general, less encountered. The occurrence of hydroxycinnamic acids in food has been reviewed by Herrmann (*5*). Caffeic acid and its esterified derivatives are the most abundant hydroxycinnamic acids in a variety of fruits. On the other hand, ferulic acid and its derivatives are the most abundant hydroxycinnamic acids found in cereal grain (*6*).

Flavonoids

Flavonoids are ubiquitous in plants; almost all plant tissues are able to synthesize flavonoids. There are also a wide variety of types – at least 2000 naturally occurring flavonoids. Flavonoids are present in edible fruits, leafy vegetables, roots, tubers bulbs, herbs, spices, legumes, tea, coffee, and red wine (*1,2,7*). They can be classified into seven groups: flavones, flavanones, flavonols, flavanonols, isoflavones, flavanols (catechins), and anthocyanidins. The structures of these flavonoids are given in Figure 1. Examples of common flavonoids in foods are listed in Table I. In general, the leaves, flowers, and fruits or the plant contain flavonoid glycosides; woody tissues contain aglycones, and seeds may contain both.

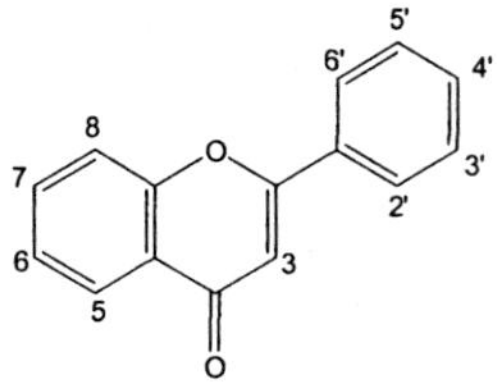

Flavone 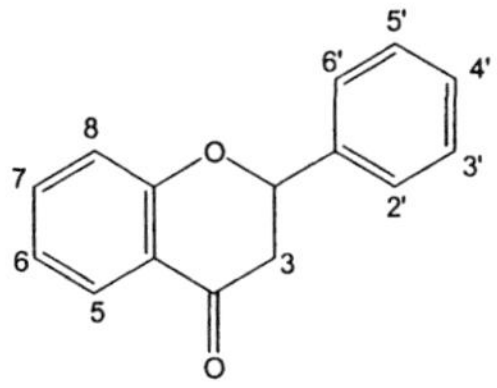   Flavanone 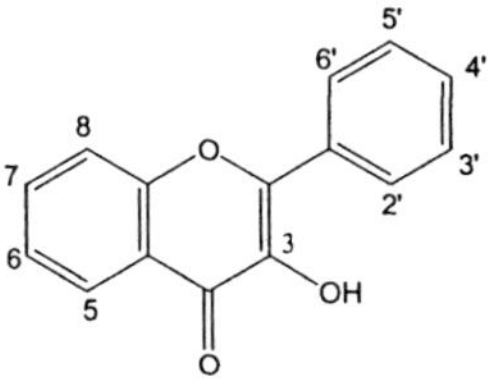   Flavonol

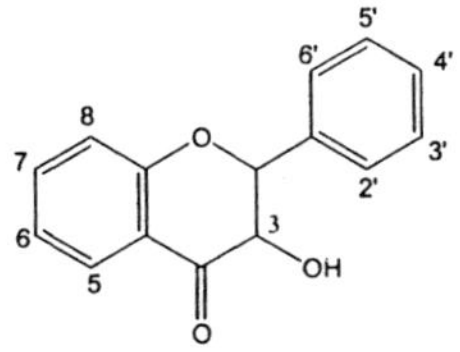

Flavanonol 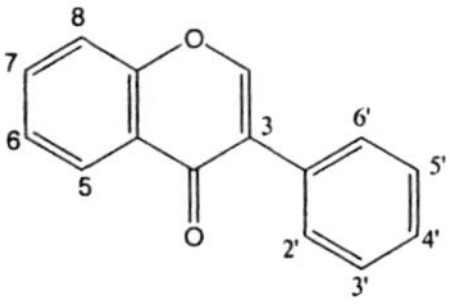   Isoflavone

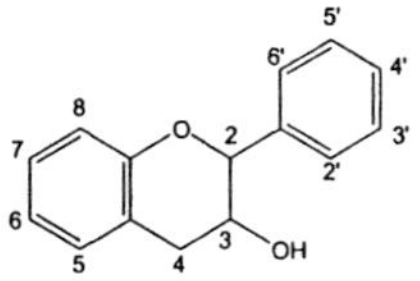

Flavanol (catechin) 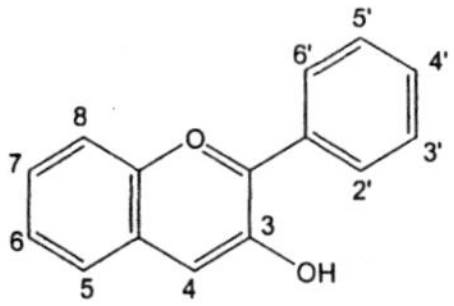   Anthrocyanidin

Figure 1. Sturctures of flavonoids.

Table I. Different Classes of Flavonoids, Their Substitution patterns and Dietary Sources

Class	Name	Substitution	Dietary Source
Flavone	Apigenin	5,7 –OH	Parsley, celery
	Rutin	5,7,3',4' - OH, 3-O-rutinose,	Buckwheat, citrus
	Tangeretin	4,5,6,7,4' - OCH$_3$	Citrus
Flavanone	Naringin	5,4' - OH	Citrus
	Naringenin	5,7,4' - OH	Orange peel
Flavonol	Kaempferol	3,5,7,4' - OH	Broccoli, tea
	Quercetin	3,5,7,3',4' - OH	Onion, broccoli, apples, berries
Flavononol	Taxifolin	3,5,7,3',4' -OH	fruits
Isoflavone	Genistein	5,7,4' - OH	Soybean
	Daidzein	4' - OH, 7–O-glucose	Soybean
	Puerarin	7,4' - OH, 8 – C-glucose	Kudzu
Flavanol (catechin)	(-)-Epicatechin	3,5,7, 3',4' -OH	Tea
	(-)-epigallocatechin	3,5,7, 3',4' ,5'-OH	Tea
	(-)-epigallocatechin gallate	5,7, 3',4',5'-OH, 3 – gallate	Tea
Anthocyanidin	Cyanidin	3,5,7,3',4' - OH	Cherry, strawberry
	Delphinidin	3,5,7, 3',4' ,5'-OH	Dark fruits

As a result of their ubiquity in plants, flavonoids are an integral part of the human diet. It is estimated (8) that the average American's daily intake of flavonoids is close to 1 g per person. However, this estimated daily intake of flavonoids may be too high when compared to the results of the Zutphen Elderly Study by Hertog et al. (9). They measured the content of selected flavonoids, namely quercetin, kaempferol, myricetin, apigenin, and luteolin in various foods most commonly consumed in The Netherlands. By assessing the flavonoid intake of 805 men aged 65-84 years, they found that the mean baseline flavonoid intake was 25.9 mg daily. Hertog et al. (*10*) also found that the major sources of flavonoid intake were tea (61%), onion (13%), and apple (10%). Certain later studies provided more precise individual data concerning the intake of various classes of flavonoids. For example, the consumpstion of flavonols for American has been estimated at ≈20-25 mg/d (*11*).

Stilbenes

Stilbenes are phenolic compounds which contain two benzene rings separated by an ethane bridge. They are widely distributed in higher plants and their main physiological roles relate to their action as phytoalexins and growth regulators (*12*).

Stilbenes had not caught the attention of food and nutritionists until one of its family members, resveratrol (3,5,4'-trihydroxystilbene) was reported to demonstrate a preventing effect on cancer (*13,14*).

Lignans

Lignans are dimers of phenylpropanoid units linked by the central carbons of their side chains. In plants, lignans and their higher oligomers act as defensive substances (*1*). Lignan-rich plant products were found to be active ingredients in the treatment of disease in Chinese folk medicine. Unfortunately, many of the active ingredients of these plant products have not been scientifically tested as therapeutic agents (*15*). However, flax and sesame lignans have been considered as important components with health benefits (*16,17*).

Tannins

Depending on their structures, tannins are defined as hydrolyzable or condensed. Hydrolyzable tannins are glycosylated gallic acids (*1*). Condensed tannins also known as proanthocyanidins and are linear polymers of flavan-3-ol (catechin and gallocatechin), and flavan-3,4-diol units. The consecutive units of condensed tannins are linked through the interflavonoid bond between C-4 and C-8 or C-6 (*18*). Tannins occur widely in different foods and are often concentrated in the skin of fruits and seed coats, among others.

Antioxidant Properties of Phenolics

Antioxidants are added to fats and oils, or foods containing fats, to prevent the formation of various off-flavors and other undesirable compounds that result from the oxidation of lipids. The most widely used synthetic antioxidants, BHA (butylated hydroxyanisole), BHT (butylated hydroxytoluene) and TBHQ (tertiary-butylhydroquinone) are all phenolics.

Antioxidants not only protect the quality of lipid-containing foods, but in living systems dietary antioxidants may also be effective in protecting *in vivo* lipid peroxidation and combating carcinogenesis, heart disease and aging. Although dietary essential antioxidant nutrients such as tocopherols and ascorbic acid have the ability to inhibit lipid peroxidation *in vivo*, numerous other nonessential antioxidants are consumed daily, often at concentrations far exceeding those of nutrient antioxidant (*19*). Flavonoids are the most important among these nonessential dietary antioxidants.

Flavonoids are known antioxidants and act as free radical acceptors and chain breakers. Many of the flavonoids have shown marked antioxidant characteristics. Flavonols are known to chelate metal ions at the 3-hydroxy-4-keto group, 5-hydroxy-4-keto group, or both. An ortho-diphenolic group on the B-ring can also demonstrate metal chelating activity (*20*). A recent review compared the antioxidant activity of a large number of food phenolics (*21*).

Phenolics and Disease Prevention

Almost all phenolics possess several common biological and chemical properties: (a) antioxidant activity, (b) the ability to scavenge active oxygen species, (c) the ability to scavenge electrophiles, (d) the ability to inhibit nitrosation, (e) the ability to chelate metals, (f) the potential to producing hydrogen peroxide in the presence of certain metals, and (g) the capability to modulate certain cellular enzyme activities. Rencetly, phenolics, particularly flavonoids have been intensively investigated because of these broad potential pharmacological activities (*22*). It appears that diets in rich in phenolics may protect against cardiovascular diseases, neurodegenerative disorders, and some forms of cancer (*23,24*).

Several phenolics have been recognized as active chemopreventive agents. Epigallocatechin 3-gallate (EGCG) from green tea and theaflavins from black tea exert strong inhibitory effects on diverse cellular events asocociated with multistage carcinogenesis. Several chapters in this volume provide recent studies on tea polyphenols and cancer. Curcumin, a yellow ingredient from tumeric (*Curcuma longae* L.) has also been extensively investigated for its cancer preventive potential (*25*). A similar compound, 6-gingerol grom ginger

(Zingiber officinale Roscoe) also has antitumer promotional effects. The molecular mechanisms explaining the anticancer effects of 6-gingerol has been discussed in detail in this volume. Many of the beneficial health effects are known to be linked to the antioxidant activity of the components involved although this is not always the case.

References

1. Shahidi, F.; Naczk, M. *Phenolics in Food and Nutraceuticals*. CRC Press, Boca Raton, FL **2004**.
2. Harborne, J.B. In *Methods in Plant Biochemistry, Vol. 1: Plant Phenolics*; Harborne, J.B., Ed.; Academic Press: London, UK, 1989, pp. 1-28.
3. Van Sumere, C. F. In *Methods in Plant Biochemistry, Vol. 1: Plant Phenolics*; Harborne, J.B., Ed.; Academic Press: London, UK, 1989, pp. 29-73.
4. Tomas-Barberan, F. A.; Clifford, M. N. *J. Sci. Food Agric.* **2000**, *80*, 1024-1032.
5. Herrmann, K. *CRC Crit. Rev. Food Sci. Nutr.* **1989**, *28*, 315-347.
6. Monach, C.; Scalbert, A.; Morand, C. R.; Jiménez, L. *Am. J. Clin. Nutr.* **2004**, *79*, 727-747.
7. Ho. C.-T. In Food Factors for Cancer Prevention; Ohigashi, H.; Osawa, T.; Terao, J.; Watanabe, S.; Yoshikawa, T., Eds.; Spinger: Tokyo, Japan, 1993, pp. 593-597.
8. Kuhnau, J. *World Rev. Nutr. Diet.* **1976**, *24*, 117-191.
9. Hertog, M. G. L.; Feskens, E. J. M.; Hollman, P. C. H.; Katan, M. B.; Kromhout, D. *Lancet* **1993a**, *342*, 1007-1011.
10. Hertog, M. G. L.; Hollman, P. C. H.; Katan, M. B.; Kromhout, D. *Nutr. Cancer* **1993b**, *20*, 21-29.
11. Manach, C.; Scalbert, A.; Morand, C.; Ramesy, C.; Jimenez, L. *Am. J. Clin. Nutr.* **2004**, *79*, 727-741.
12. Gotham, J. In *Methods in Plant Biochemistry, Vol. 1: Plant Phenolics;* Harborne, J.B., Ed.; Academic Press: London, UK, 1989, pp. 159-196.
13. Jang, M.; Cai, L.; Udeani, G. O.; Slowing, K. V.; Thomas, C. F.; Beecher, C. W. W.; Fong, H. H. S.; Farnsworth, N. R.; Kinghorn, A. D.; Mehta, R. G.; Moon, R. C.; Pezzuto, J. M. *Science* **1997**, *275*, 218-220.
14. Savouret, J. F.; Quesne, *M. Biomed. Pharmacother.* **2002**, *22*, 1111-1117.
15. Ayres, D. C.; Loike, J. D. *Lingans: Chemcial, Biological and Clinical Properites;* Cambridge Univeristy Press: Cambridge, UK, 1990.
16. Kang, M-H.; Naito, M., Tsujihara, N.; Osawa, T. *J. Nutr.* **1998**, *128*, 1018-1022.

17. Spence, J.D.; Thornton, T.; Muir, A.D.; Westcott, N.D. *J. Am. College Nutr.* **2003**, *22*, 494-502.
18. Hemingway, R. W. In *Chemistry and Significance of Condensed Tannin*, Hemingway, R. W.; Karchesy, J.J., Eds., Plenum Press: New York, 1989, pp. 83-98.
19. Decker, E. A. *Nutr. Rev.* **1995**, *10*, 210-219.
20. Ho, C-T. In *Food Factors for Cancer Prevention*. Ohigashi, H.; Osawa, T.; Terao, J.; Walanabe, S.; Yoshikawa, T., Eds.; Springer: Tokyo, Japan. **1993**, pp. 593-597.
21. Kim, D. O.; Lee, C. Y. *CRC Crit. Rev. Food Sci. Nutr.* **2004**, *44*, 233-273.
22. Huang, M. T.; Ferraro, T. In *Phenolic Compounds in Food and Their Effects on Health II: Antioxidants & Cancer Prevention*; Huang, M. T.; Ho, C.-T.; Lee, C. Y., Eds., American Chemical Society: Washington, D.C., 1882, pp. 8-34.
23. Hertog, M. G. L.; Hollman, P. C. H.; Katan, M. B. *J. Agric. Food Chem.* **1992**, *40*, 2379-2383.
24. Block, G. *Nutr. Rev.* **1992**, *50*, 207-.
25. Lin, J. K. In *Phytochemicals: Mechanisms of Action*, Meskin, M. S.; Bidlack, W. R.; Davies, A. J.; Lewis, D. S.; Randolph, R. K., Eds., CRC Press: Boca Raton, FL, 2004, pp. 79-108.

Chemistry and Analysis

Chapter 2

Phenolic Acid Composition of Wheat Bran

Kequan Zhou, John W. Parry, and Liangli (Lucy) Yu

Department of Nutrition and Food Science, 0112 Skinner Building,
University of Maryland, College Park, MD 20742

Bran samples of wheat grown in Colorado were extracted with aqueous acetone followed by alkaline hydrolysis. After neutralization with HCl, the total free phenolic acids were extracted by ethyl acetate/ethyl ether (1:1, v/v), and analyzed by a reverse phase HPLC and LC-MS. The major phenolic acids detected in the wheat bran extracts were ferulic, syringic, p-coumaric, p-hydroxybenzoic, and vanillic acids. The results also showed that both wheat variety and the growing condition may influence the phenolic acid composition in wheat bran.

Reactive oxygen species, such as hydroxyl radicals and peroxyl radicals, may attack DNA, bioactive proteins, membrane lipids, and carbohydrates and cause damage, which may result in cell injury and death, and consequently the development of several aging-related chronic diseases including cancer and heart disease (*1*). Antioxidants may prevent these physiologically important molecules from oxidative damages, and reduce the risk of aging-related diseases (*2,3*). In addition to their health benefits, antioxidants may also suppress lipid oxidation in food products and improve food quality, stability and safety. Natural antioxidants are in high demand because of consumer preference and the potential safety concerns of synthetic antioxidants such as butylated hydroxytoluene (BHT), which is currently used as an antioxidative food additive.

There has been a continuous effort to develop novel edible natural antioxidants for disease prevention and health promotion, as well as for food preservation.

Previous research has shown that wheat and wheat based food products have significant antioxidative activities and may serve as dietary sources of natural antioxidants (*2-10*). Wheat antioxidants were shown to directly react with and quench free radicals, chelate transition metals, suppress lipid peroxidation in fish and soy oils, and prevent liposome peroxidation (*3,4,6-8*). Phenolics were present in several varieties of wheat and are believed to contribute to the overall antioxidant properties of wheat. A group of phenolic acids have been detected in wheat extracts. In 1982, Sosulski and others (*11*) reported that ferulic, syringic, and vanillic acids were the most prevalent phenolic acids present in the flour of Neepawa wheat, with ferulic acid accounting for 89.1% of the total. Later in 1992, Onyeneho and Hettiarachchy (*7*) detected ferulic, vanillic, p-coumaric, caffeic, chlorogenic, gentisic, syringic and *p*-hydroxybenzoic acids in bran extracts of durum wheat (*Triticum durum*). Among these phenolic acids, ferulic, vanillic, and *p*-coumaric acids were present in the greatest amounts. In addition, Hatcher and Kruger (*12*) detected six phenolic acids in extracts of five varieties of Canadian wheat. These included sinapic, ferulic, and vanillic acids, along with minor amounts of coumaric, caffeic and syringic acids. Previous studies have also shown that antioxidative properties of wheat might vary among wheat cultivars, and may be significantly altered by growing conditions and the possible interactions between genotype and growing condition (*3,4,6*). It is of interest whether growing condition and wheat cultivar may influence the contents of phenolic acids in wheat. Therefore, the present study was conducted to examine and compare the phenolic acid composition of bran extracts from Lakin, Venago, and Enhancer wheat grown at Burlington and Walsh in Colorado

Materials and Methods

Materials.

Bran samples of Lakin, Venago, and Enhancer wheat from Burlington and Walsh in Colorado were provided by Dr. Scott Haley in the Department of Soil and Crop Science at Colorado State University, Fort Collins, Colorado. All other chemicals and solvents were of the highest commercial grade and used without further purification.

Extraction and testing sample preparation

4 grams of each bran sample was ground and extracted for 15 hours with 40 mL of 50% acetone under nitrogen at ambient temperature. After removing acetone using a rotary evaporator at 35°C, the extracts were hydrolyzed with 4N NaOH for 4 hours at 55°C under nitrogen, acidified using 6N HCl, and extracted with ethyl ether/ethyl acetate (1:1, v/v) according to the procedure described previously (*13*). The ethyl ether/ethyl acetate was evaporated at 25°C using a nitrogen evaporator, and the solid residue was re-dissolved in methanol, filtered through a 0.45 μm membrane filter, and kept in dark under nitrogen until HPLC analysis.

HPLC analysis

Phenolic acid composition in the methanol solution was analyzed by HPLC using a Phenomenex C18 column (250 mm × 4.6 mm). Phenolic acids were separated using a linear gradient elution program with a mobile phase containing solvent A (acetic acid/H_2O, 2:98, v/v) and solvent B (acetic acid/acetonitrile/H_2O, 2:30:68, v/v/v). The solvent gradient was programmed from 10 to 100% B in 42 min with a flow rate of 1.5 mL/min (*14*). Identification of phenolic acids was accomplished by HPLC-MS and comparing the retention time of peaks in wheat samples to that of the standard compounds.

Statistic analysis

Data were reported as mean ± SD for triplicate determinations. Analysis of variance and least significant difference tests (SPSS for Windows, Version Rel. 10.0.5., 1999, SPSS Inc., Chicago, IL) were conducted to identify differences among means. Statistical significance was declared at $P<0.05$.

Results and Discussion

Phenolic acid composition in the hydrolysate of bran extracts was examined. Five phenolic acids, including p-hydroxy benzoic, coumaric, ferulic, syringic, and vanillic acids, were detected in both hard winter red and white varieties grown in Colorado (Figure 1A-B). Among these phenolic acids, ferulic acid had the greatest concentration in all three tested hard winter wheat varieties, and followed by syringic acid (Figure 1B), regardless of wheat genotype. This is in agreement to the observation of Sosulski and others (*11*) and Onyeneho and

Hettiarachchy (7) that ferulic acid was the predominant phenolic acid in wheat samples. Ferulic acid accounted 65.3-74.4% and 65.4-73.8% of total phenolic acids on per weight basis in hard winter red and white wheat bran samples, respectively.

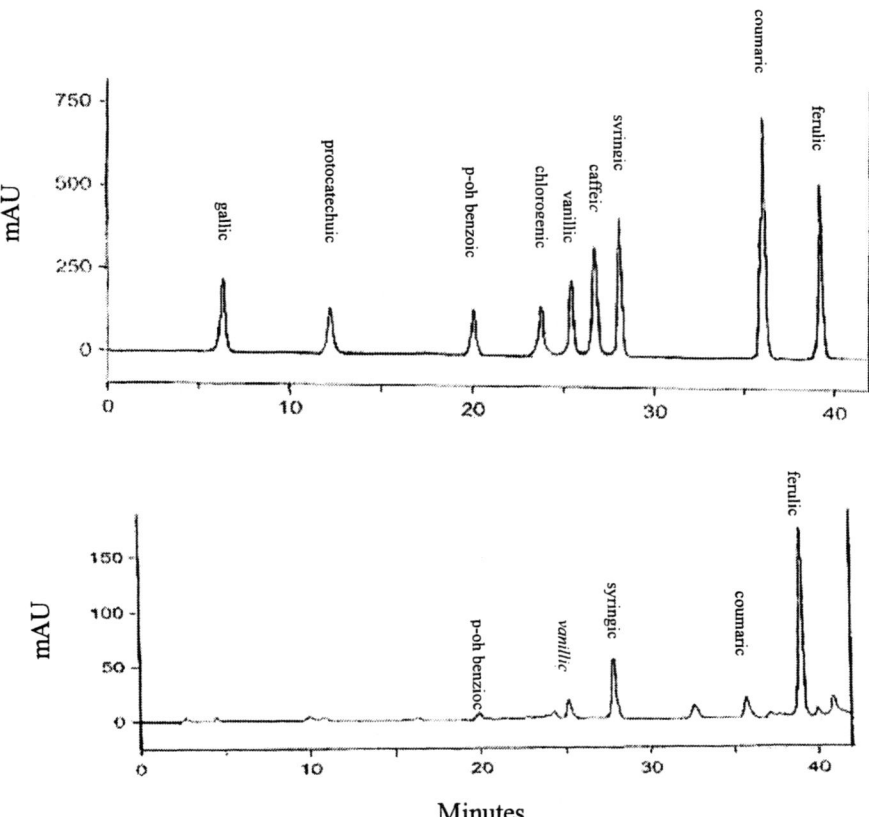

Figure 1. HPLC chromatograms of standard phenolic acids and the hydrolysate of bran extract. A: represents the HPLC of the standard phenolic acids, while B: is the HPLC result of the hydrolate of bran extract. Gallic, protocatechuic, p-OH benzoic, chlorogenic, vanillic, caffeic, syringic, coumaric, and ferulic stand for gallic, protocatechuic, p-OH benzoic, chlorogenic, vanillic, caffeic, syringic, coumaric and ferulic acids, respectively.

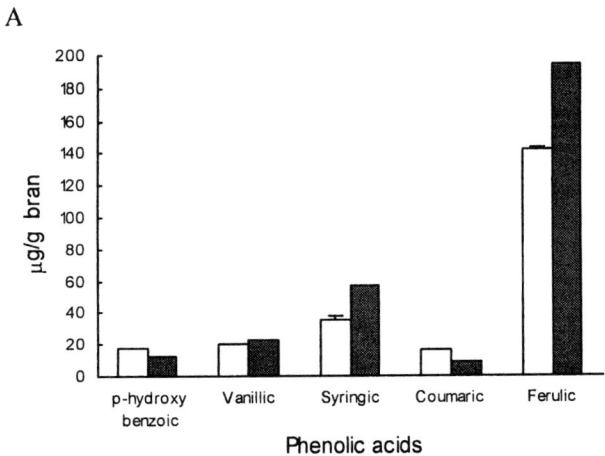

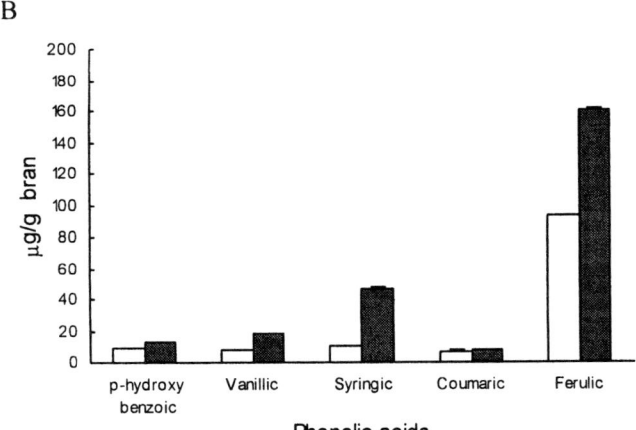

Figure 2. Comparison of white and red wheat for their phenolic acid contents. A: represents the phenolic acid composition of Lakin and Enhancer wheat grown at Burlington, while B represents that of Lakin and Enhancer wheat grown at Walsh Colorado, respectively. Lakin is hard white winter wheat, and Enhancer is hard red winter wheat. The clear column stands for Lakin extract, while the solid column represents Enhancer extract, respectively. Vertical bar on the top of each column is the standard deviation (n = 3).

Lakin wheat, a hard white winter variety, was compared with Enhancer wheat, a hard red winter variety, for their phenolic acid compositions using bran samples from wheat grown at Burlington and Walsh testing locations in Colorado. Significant differences were observed in all of their phenolic acid compositions, except for the vanillic acid content in the two wheat brans from the Burlington location and the coumaric acid concentration in that from the Walsh location, suggesting that white and red wheat may differ in phenolic acid composition. Bran extracts prepared from Enhancer wheat grown at both locations had greater concentrations of total phenolics, ferulic, and syringic acids than that in Lakin bran extracts (Figure 2) under the experimental conditions. More varieties of both white and red wheat grown at different locations are required to compare and make a conclusion how red and white wheat differ in their phenolic acid composition in general.

It is interesting to compare the two same colored wheat varieties for their phenolic acid compositions. Both Enhancer and Venago are hard red winter wheat. Bran extracts were prepared from Enhancer and Venago bran from Burlington and Walsh, and analyzed for their phenolic acid compositions. The results showed that the two red wheat extracts differed in their contents of total phenolic, ferulic, syringic, and vanillic acid contents (Figure 3). Enhancer bran had greater total and individual phenolic acid contents than Venago bran, regardless of growing location, suggesting the dependence of phenolic acid composition on wheat genotype. In addition, bran extracts of both varieties from Burlington exhibited greater total and individual phenolic acid contents than those grown at Walsh, indicating the potential effect of growing conditions on the phenolic acid content of wheat. These are in agreement with previous observations that both genotype and growing conditions may influence the antioxidant properties and total phenolic contents of wheat (*3,4 6 - 8*).

An additional experiment was conducted to further evaluate the effect of growing condition on phenolic acid composition of wheat using Lakin (white wheat) and Enhancer (red wheat) bran samples. Figure 4A represents the phenolic acid composition of Lakin bran, and Figure 4B represents that of Enhancer bran from Burlington and Walsh, respectively. Lakin bran from Burlington exhibited greater total and all five individual phenolic acid contents than that from Walsh (Figure 4A). Enhancer bran from Burlington also had greater total, ferulic, syringic acid contents than that from Walsh, but the bran samples from the two locations had no difference in their coumaric and p-hydroxy benzoic acid contents (Figure 4B). These results suggest that growing condition may alter the total and individual phenolic acid contents in wheat, and the interaction between genotype and environment may also play a role in phenolic acid composition of wheat.

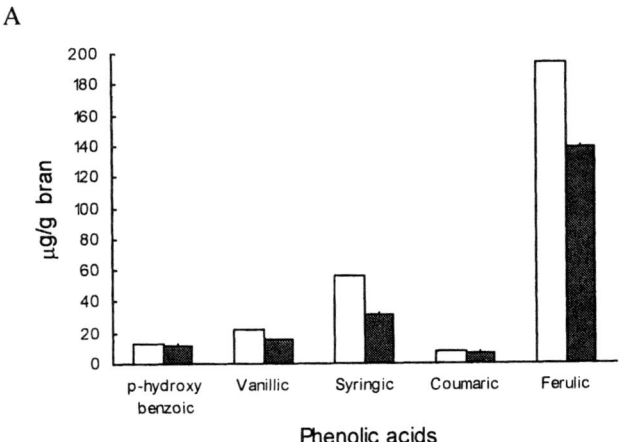

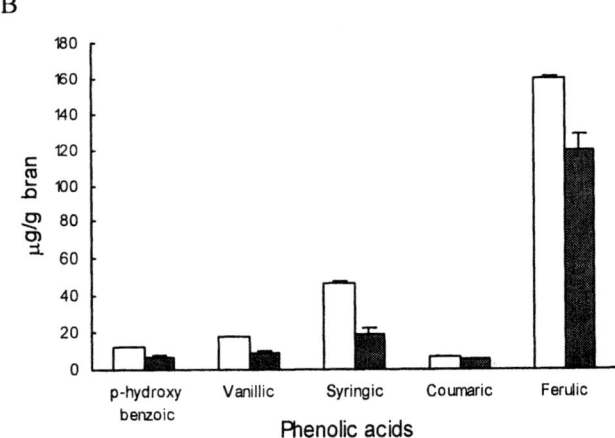

Figure 3. Phenolic acid compositions of two red wheat bran extracts. A: represents the phenolic acid composition of Enhancer and Venago wheat grown at Burlington, while B represents that of Enhancer and Venago wheat grown at Walsh Colorado, respectively. Both Enhancer and Venago are hard red winter wheat. The clear column stands for Enhancer extract, while the solid column represents Venago extract, respectively. Vertical bar on the top of each column is the standard deviation (n = 3).

A

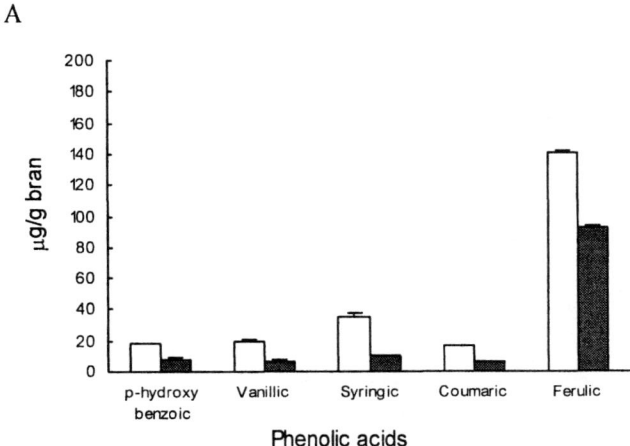

B

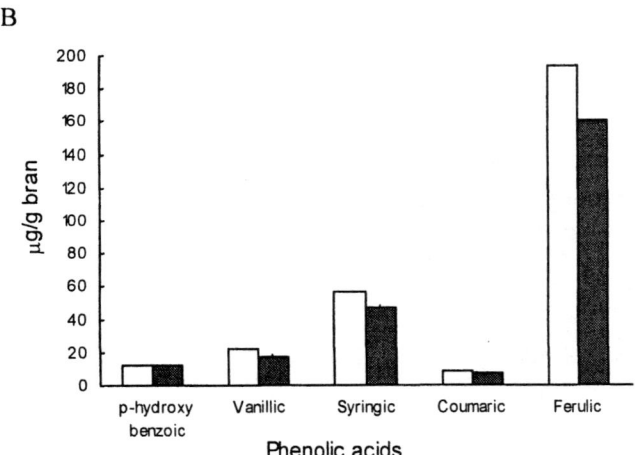

Figure 4. Effect of growing condition on phenolic acid composition of wheat. A: represents the phenolic acid composition of Lakin wheat grown at Burlington and Walsh, while B represents that of Enhancer wheat grown at Burlington and Walsh, Colorado. Lakin is hard white winter wheat, and Enhancer is hard red winter wheat. The clear column stands for wheat grown at Burlington, while the solid column represents that from Walsh, Colorado, respectively. Vertical bar on the top of each column is the standard deviation (n = 3).

Acknowledgement

The author would like to thank Dr. Scott Haley in the Department of Soil and Crop Science, Colorado State University for providing wheat bran samples.

References

1. Halliwell, B.; Gutteridge, J. M. C.; Cross, C. E. *J. Lab. Clin. Med.* **1992**, *119*, 598-620.
2. Baublis, A. J.; Clydesdale, E. M.; Decker, E. A. *Cereal Foods World.* **2000**, *45*, 71-74.
3. Yu, L.; Haley, S.; Perret, J.; Harris, M.; Wilson, J.; Qian, M. *J. Agric. Food Chem.* **2002a**, *50*, 1619-1624.
4. Yu, L.; Haley, S.; Perret, J.; Harris, M. *Food Chem.* **2002b**, *78*, 457-461.
5. Yu, L.; Davy, B.; Wilson, J.; Melby, C. L. *J. Food Sci.* **2002c**, *67*, 2600-2603.
6. Yu, L.; Perret, J.; Harris, M.; Wilson, J.; Haley, S. *J. Agric. Food Chem.* **2003**, *51*, 1566-1570.
7. Onyeneho, S. N.; Hettiarachchy, N. S. *J. Agric. Food Chem.* **1992**, *40*, 1496-1500.
8. Zielinski, H.; Kozlowska, H. *J. Agric. Food Chem.* **2000**, *48*, 2008-2016.
9. Miller, H. E.; Rigelhof, F.; Marquart, L.; Prakash, A.; Kanter, M. *J. American College of Nutrition.* **2000**, *19*, 312S-319S.
10. Abdel-Aal, E.; Hucl, P. *J. Agric. Food Chem.* **2003**, *51*, 2174-2180.
11. Sosulski, F.; Krygier, K.; Hogge, L. *J. Agric. Food Chem.* **1982**, *30*, 337-340.
12. Hatcher, D. W.; Kruger, J. E. *Cereal Chem.* **1997**, *74*, 337-343.
13. Krygier, K.; Sosulski, F.; Hogge, L. *J. Agric. Food Chem.* **1982**, *30*, 330-334.
14. Tüzen, M.; Özdemir, M. *Turk. J. Chem.* **2003**, *27*, 49-54.

Chapter 3

Structure and Analysis of Flavonolignans from *Silybum marianum*

David Y-W Lee and Yanze Liu

Bio-Organic and Natural Product Laboratory, McLean Hospital / Harvard Medical School, Belmont, MA 02478

The structures and stereochemistry of silybin A, silybin B, isosilybin A, and isosilybin B, diastereoisomers of silybin and isosilybin, were characterized, together with other known flavonolignans isolated from the seeds of *Silybum marianum* so far. The absolute stereochemistry of each diastereoisomer was unambiguously established through the X-ray crystallographic study of isosilybin A in combination with optical rotation data. Three new flavonolignans, namely, 2,3-*cis*-silybin A, 2,3-*cis*-silybin B, and neusilychristin were separated for the first time. The fingerprints of HPLC and ^{1}H NMR of silymarin, standardized extract from the seeds of *Silybum marianum* were also reported.

Milk thistle is an ancient medicinal plant used to treat a range of liver and gallbladder disorders, including hepatitis, cirrhosis, and to protect the liver against poisoning from wild mushroom, alcohol, chemical, and environmental toxins. The scientific name for milk thistle is *Silybum marianum*. Dioscorides, a first century Greek physician who served Roman army, gave the name "Silybum" to edible thistles (*1*) and "marianum" derives from the legend that the white veins running through the plant leaves were symbolized by a drop of the Virgin Mary's milk (*1-3*). In Germany where the milk thistle is often named after the Virgin Mary such as Marian thistle, Mary thistle, St. Mary's thistle, Lady's thistle, Holythistle, sow thistle, thistle of the blessed virgin, and Christ's crown.

Medicinal use of milk thistle as a liver-protecting herb dates back to the earliest Greek references to the plant. Pliny the Elder (A.D. 23-79), the first century Roman physician mentioned that the juice of the plant, mixed with honey, is excellent for "carrying off bile." This is perhaps the first written reference to the use of milk thistle for liver-related conditions. After a thousand years of historical use, it was mentioned in an important medieval German manuscript, the *Physica* of Hildegarde of Bingen published in 1553. Still used in the 18th century, Culpepper (*4*) noted that it is effectual "to open the obstructions of the liver and spleen, and thereby is good against the jaundice." Another German physician from the 19th century, Johannes Gottfried Rademacher developed a tincture made from milk thistle seeds for chronic liver disease, acute hepatitis, and jaundice. In the early 20th century, a German scientist began to look into the clinical significance of milk thistle. Investigations into the liver-protecting properties of the plant were intensified around 1960 and the active components of the seeds (Figure 1) were isolated as silymarin by Dr. Wagner at the University of Munich in 1968 (*5*). Later, improved chemical separation revealed that silymarin was not a single component but a mixture of flavonolignans. The primary components include silybin (**1**), isosilybin (**2**), and silychristin (**3**). Collectively, these flavonolignans are found in concentrations of 4 to 6 % in the ripe seeds. European milk thistle products are currently standardized to 80% silymarin. The chemical structure of silybin (**1**) was first established by Pelter and Hänsel in 1975 via degradation (*6*) and synthesis of dehydrosilybin (*7*). It was further confirmed by biomimetic synthesis by Merlini et al (*8, 9*). However, biomimetic synthesis gave a mixture of regioisomers: silybin and isosilybin in 57:43 ratio respectively. Isosilybin (**2**), a regioisomer of silybin, was first isolated by preparative TLC and reported by Arnore et al in 1979 (*10*). It is also reported that both silybin (**1**) and isosilybin (**2**) were mixtures of diastereoisomer as evidenced by ^{1}H NMR (*10*), HPLC (*11-13*), X-ray crystallography (*14*), and MECC (*15*) analyses. Tanaka et al (*16*) regioselectively synthesized silybin in 63% overall yield. Unfortunately, separation of individual diastereoisomer of silybin and isosilybin had not been achieved for 30 years until recently the successful separation of silybin A (**11**), silybin B (**12**), isosilybin A (**9**) and isosilybin B (**10**) by HPLC. Afterwards, the absolute configuration of each diastereoisomer was established by X-ray analysis

Figure 1. Seeds of Silybum marianum (L.) Gaertn.

of the single crystal of isosilybin A in combination with ORD measurements (*17*).

Stereochemistry at C-2 and C-3 of Phenolic Flavonolignans

Naturally occurring dihydroquercitin, known as (+)-taxifolin, has four stereoisomers, i.e. 2*R*, 3*R*-dihydroquercitin (**5a**), 2*S*, 3*S*-dihydroquercitin (**5b**), 2*R*, 3*S*-dihydroquercitin (**5c**), and 2*S*, 3*R*-dihydroquercitin (**5d**). The difference among **5a, 5b** and **5c, 5d** can be easily identified by ^{1}H NMR coupling, in which the former pair of isomers have *trans* configuration with a coupling constant of *ca* $J_{2,3}$= 11.0 Hz, whereas the later pair have *cis* configuration with a coupling constant of *ca* $J_{2,3}$= 2.0 Hz. The difference between 2*R*, 3*R* and 2*S*, 3*S*, and 2*R*, 3*S* and 2*S*, 3*R* can be observed from their corresponding optical rotation values as shown in Table I.

The conformation of 2*R*, 3*R* dihydroquercetin was confirmed by Arnone A et al (*9*) *via* biomimetic synthesis. 2*R*, 3*R*-dihydroquercetin (**5a**) and coniferyl alcohol (**6**) were reacted in benzene/acetone to give a pair of diastereoisomer in 78% yield (Figure 2). Simple recrystallization gave pure silybin (**1**). TLC, HPLC, NMR, IR, and CD spectra are identical with those of the natural silybin with a 2*R*, 3*R* configuration.

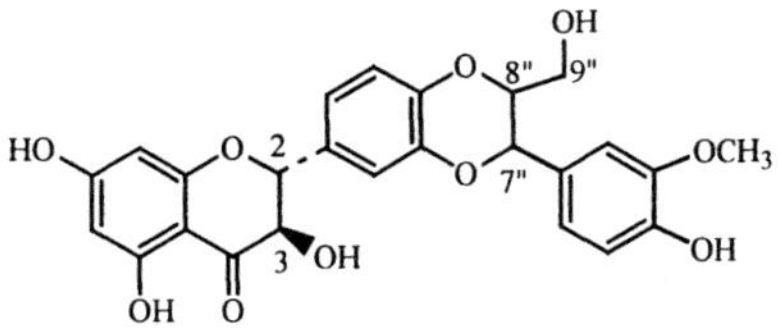

Silybin **1**

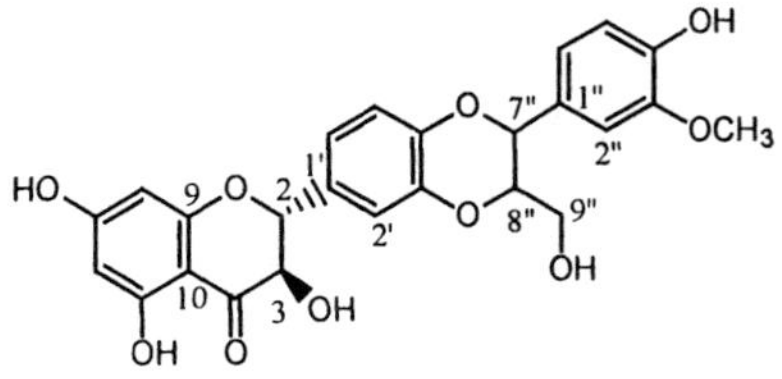

Isosilybin **2**

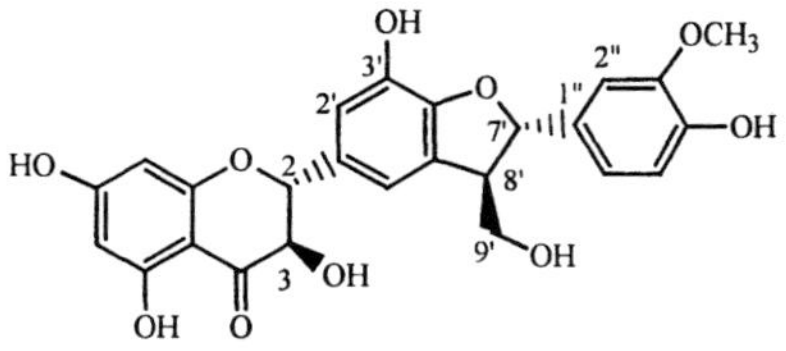

Silychristin **3**

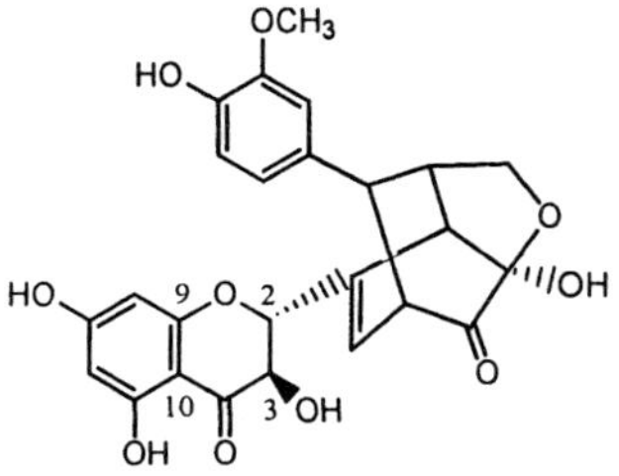

Silydianin **4**

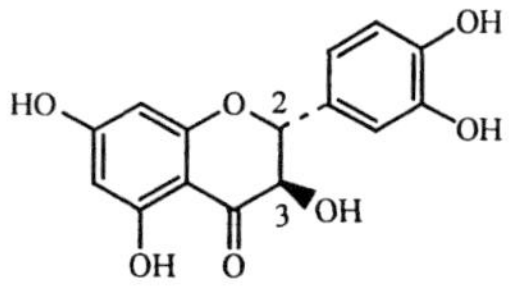

(+)-Taxifolin **5a**

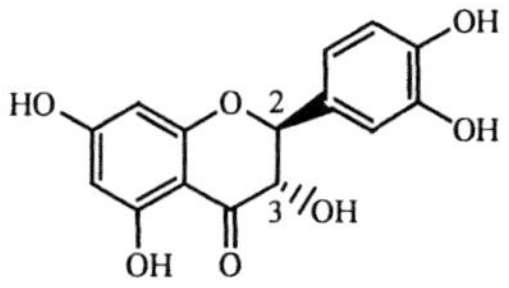

2S, 3S-dihydroquercetin **5b**

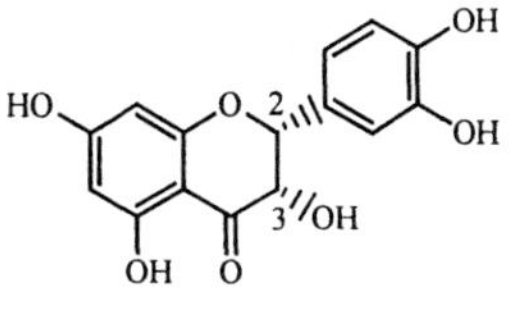

2R, 3S-dihydroquercetin **5c**

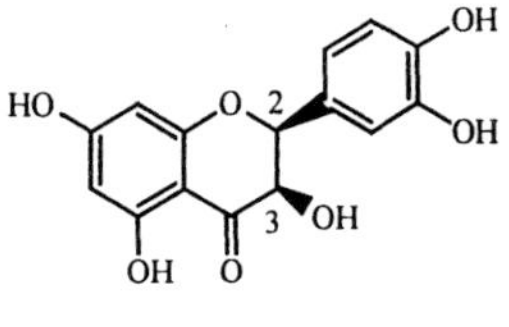

2S, 3R-dihydroquercetin **5d**

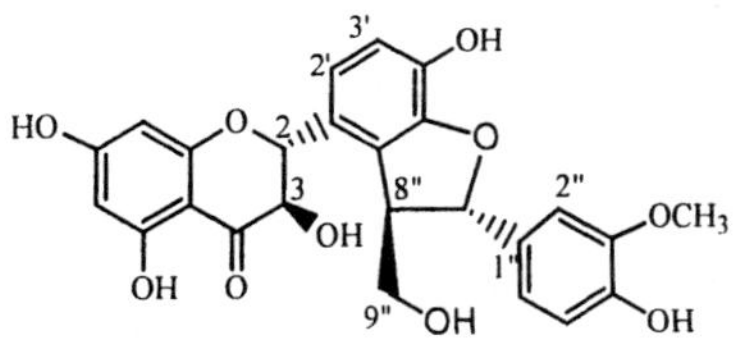

Isosilychristin **7**

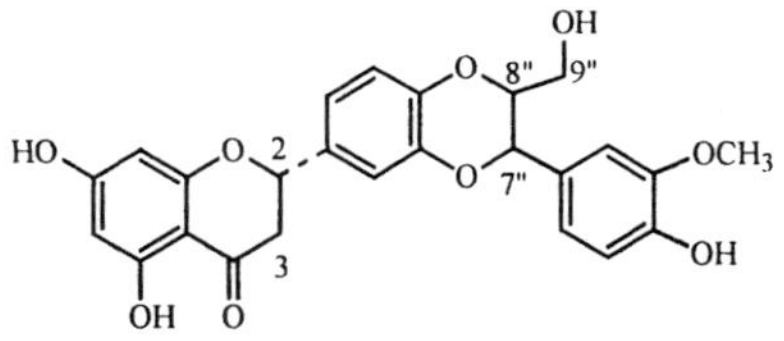

Silandrin **8**

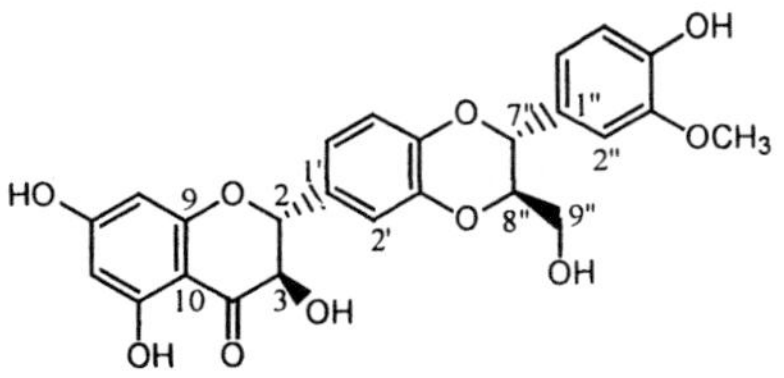

Isosilybin A **9**

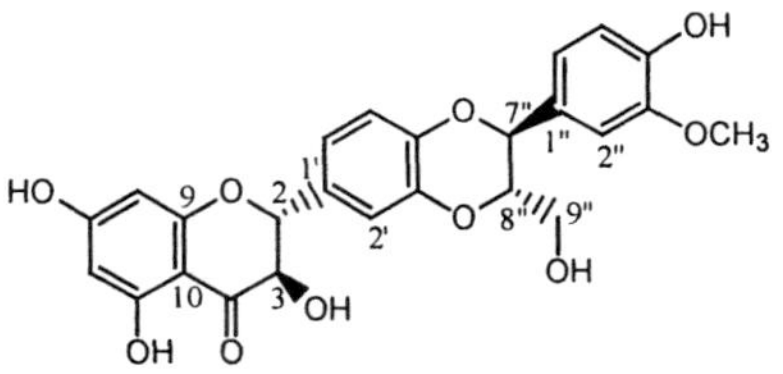

Isosilybin B **10**

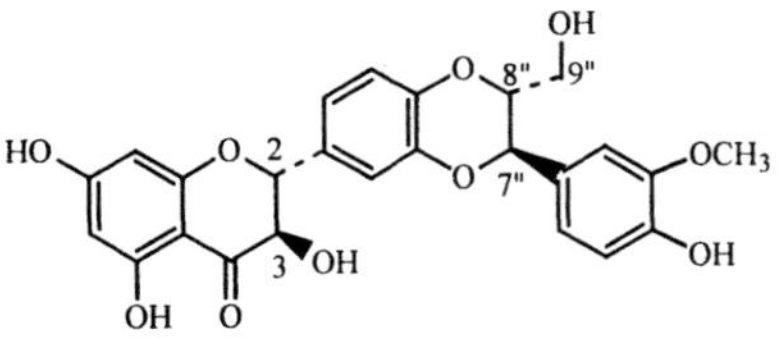

Silybin A **11**

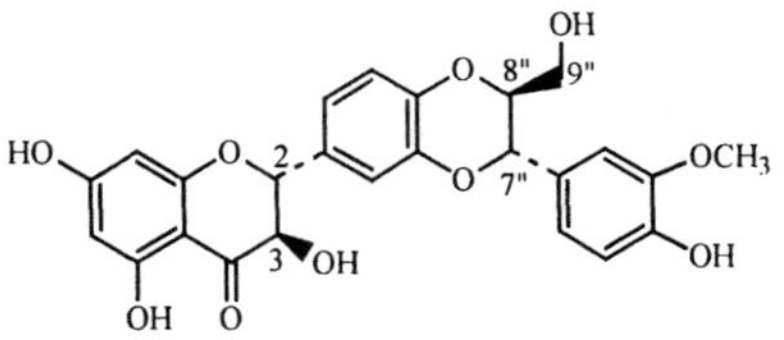

Silybin B **12**

Table I. Characterization of Isomeric Dihydroquercetin

Dihydroquercetin	$[\alpha]_D$	^{1}HNMR ($J_{2,3}$, Hz)
2R, 3R (**5a**)	+4.6	10-12
2S, 3S (**5b**)	- 5.6	10-12
2R, 3S (**5c**)	(-)	1-3
2S, 3R (**5d**)	(+)	1-3

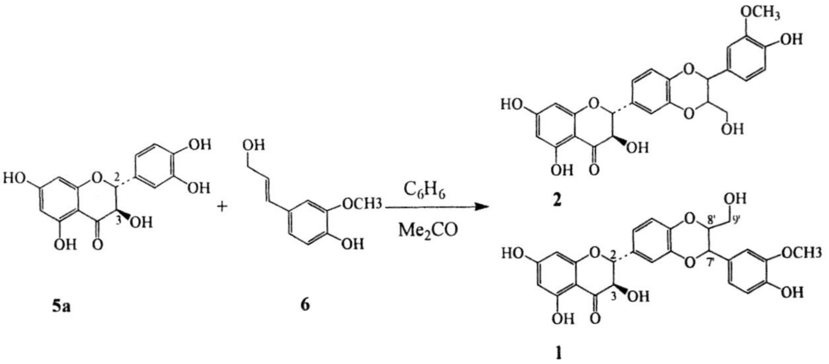

Figure 2. Biomimetic synthesis of diostereoisomeric silybin and isosilybin.

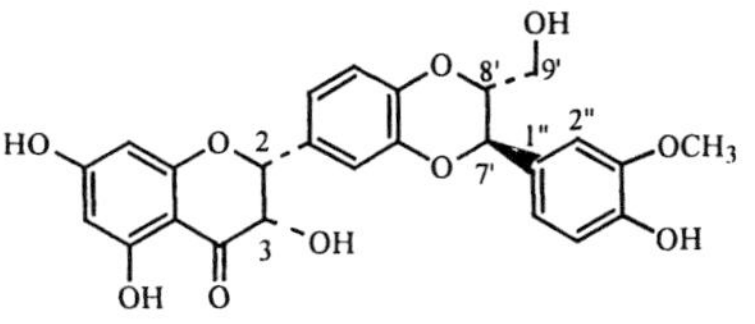

2,3-cis-Silybin A (**13a**)

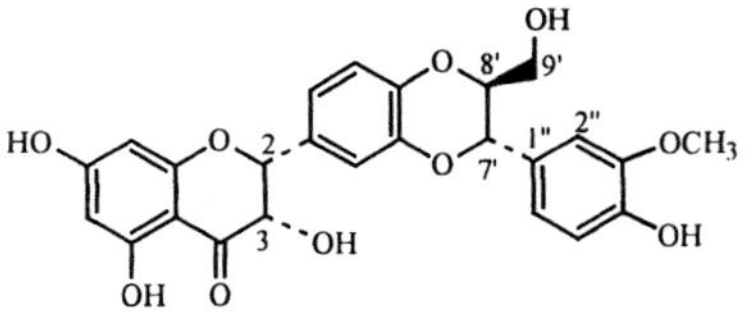

2,3-cis-Silybin B (**14a**)

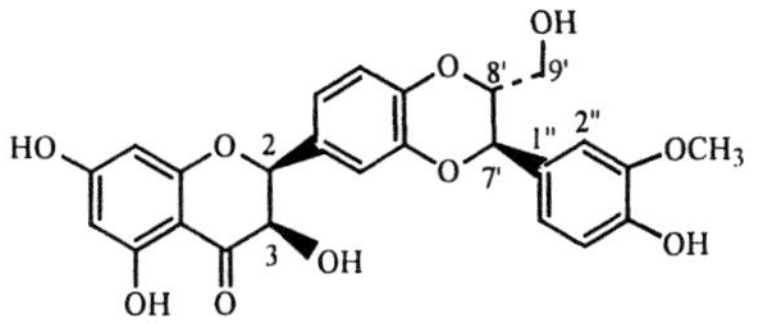

2,3-cis-Silybin A (**13b**)

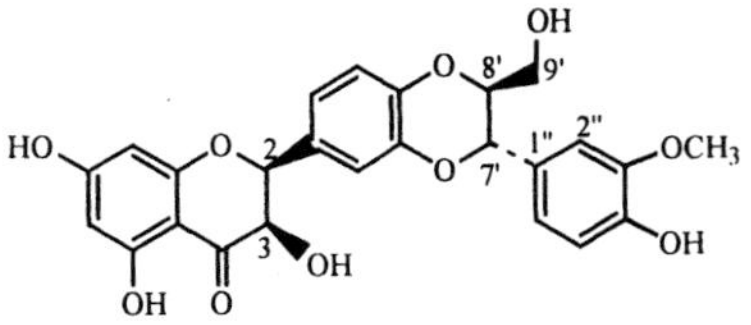

2,3-cis-Silybin B (**14b**)

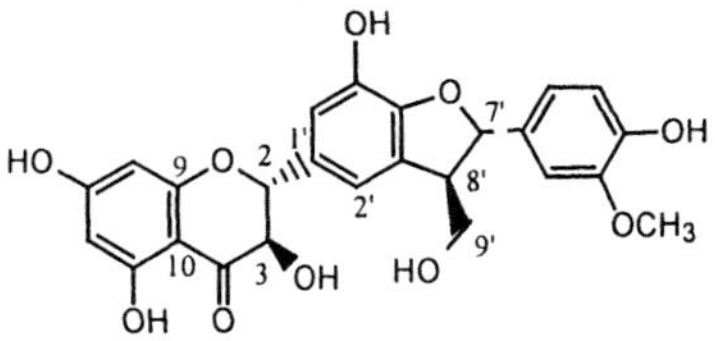

Neusilychristin (**15**)

In addition to silybin (**1**), isosilybin (**2**), silychristin (**3**), and silydianin (**4**), other minor flavonolignans, such as isosilychristin (**7**) (*18*), silandrin A (**8**), (*19*) were also isolated from the same resource.

Stereochemistry at C-7' and C-8' of Phenolic Flavonolignans

As described above, the basic structure of silybin and isosilybin were determined in 1975 (*6, 7*) and 1979 (*10*), respectively. 2*R*, 3*R* conformation of silybin and isosilybin were established through biomimetic synthesis in 1980 [9]. But the stereochemistry at C-7' and C-8' of dioxane ring remained unknown for near 30 years until recently we obtained the single crystal (Figure 3) of isosilybin A and successfully conducted an X-ray crystallographic analysis (*17*). The flavonol part with 2*R*, 3*R* conformation in the structure of isosilybin A (**9**), the first one corresponding to two peaks of isosilybin (**2**) in HPLC, could be used as internal standard. So, it is clearly to know that C-7' and C-8' are *R* conformation based on this internal standard (Figure 4). Isosilybin B (**10**) has same configuration at C-7' and C-8' with isosilybin B (**9**), but the optical rotations were opposite. Therefore, the stereochemistry at C-7' and C-8' of isosilybin B were established as 7'*S* and 8'*S* unambiguously.

By comparing the optical rotations of silybin A (**11**) and silybin B (**12**) with those of isosilybin A (**9**) and isosilybin B (**10**), it was relatively easy to identify silybin A (**11**) and silybin B (**12**). The optical rotation data and stereochemistry of four isomers are shown in Table II.

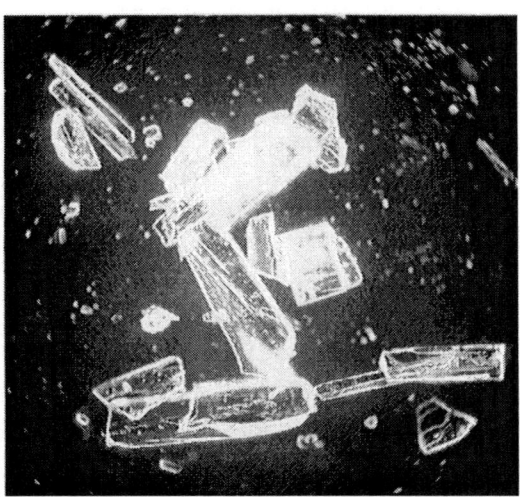

Figure 3. Single crystal of isosilybin A viewed under microscope.

Table II. Optical Rotation Data and Stereochemistry of Compounds 9-12

Compound	c (%)	solvent	$[\alpha]_D$	stereochemistry
9	0.27	acetone	+48.5	2R, 3R, 7'R, 8'R
10	0.31	acetone	-23.6	2R, 3R, 7'S, 8'S
11	0.21	acetone	+20.0	2R, 3R, 7'R, 8'R
12	0.28	acetone	-1.07	2R, 3R, 7'S, 8'S

Kim et al (*20*) first reported the stereochemistry of silybin A (**11**), silybin B (**12**), isosilybin A (**9**), and isosilybin B (**10**) by CD spectral method. Unfortunately, the data presented were ambiguous particularly with regard to the stereochemistry. Recently, we isolated silychristin (**3**), silydianin (**4**), (+)-taxifolin (**5a**), silybin A (**11**), silybin B (**12**), isosilybin A (**9**), and isosilybin B (**10**) along with three new flavonolignans by preparative HPLC. According to ^{1}H NMR, ^{13}C NMR and MS analysis, the structures of three new flavonolignans were assigned as 2,3-*cis*-silybin A (**13**), 2,3-*cis*-silybin B (**14**), and neusilychristin (**15**) (*21*). 2,3-*cis* and 7', 8'-*trans* conformations were clearly shown in ^{1}H NMR spectrum. Another regioisomer of silychristin was also separated and assigned as **15** by comparing its ^{1}H and ^{13}C NMR spectra with those of silychristin (**3**) (*22*) and isosilychristin (**7**) (*18*). Interestingly, silychristin and neusilychristin have same ^{1}H NMR coupling pattern, while their HPLC behavior are quite different, and isosilychristin has an *ortho* coupling constant of $J_{2',3'}$.

The difference of ^{1}H and ^{13}C NMR spectra between silybin A, silybin B, isosilybin A, and isosilybin B is too small to separate them, fortunately, HMBC spectrum provides unique correlations to identify silybin (**1**) and isosilybin (**2**). The HMBC correlation of isosilybin A (**9**) was illustrated on Figure 5. Other methods of analysis including newly established HPLC and ^{1}H NMR fingerprints will be discussed in next section.

Analytical Methods for Phenolic Flavonolignans

As discussed above, the phenolic flavonolignans found in silymarin are structurally related isomers with identical molecular formula and physical properties. Therefore, the separation and identification of each isomer presented the greatest challenge.

27

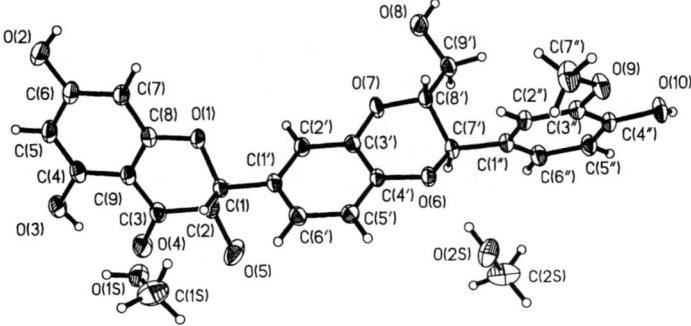

*Figure 4. X-ray structure of isosilybin A (**9**) bis-methanolate.*

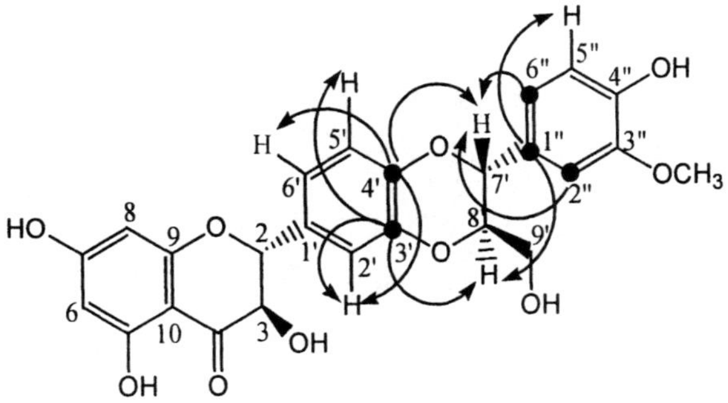

*Figure 5. Key HMBC correlation of isosilybin A (**9**).*

HPLC

RP-HPLC is a convenient analytical method for qualitative and quantitative analysis of phenolic flavonolignans. Tittel et al (*11, 12*) first reported the HPLC analysis using naphthol-1 as internal standard. Taxifolin (dihydroquercetin, **5a**), silychristin (**3**), silydianin (**4**), silybin (**1**), and two unknown peaks were identified. During our systematic evaluation of silymarin, we optimized the HPLC conditions and identified every flavonolignan peaks shown in Figure 6. HPLC was carried out with a Waters HPLC system (1525 binary HPLC pump and Waters 2487 dual wavelength detector) using an ODS-A column (4.6 mm i.d. x 150 mm) with a solvent system consisting of methanol-water (45:55) at a flow rate of 1.0 mL/min. The UV detection was monitored at 254 nm. Based on our study, we found that HPLC is a highly sensitive method for quantitative and qualitative analysis of silymarin, especially in the identification of isomeric compounds such as silybin A (**11**), silybin B (**12**), isosilybin A, (**9**) and isosilybin B (**10**).

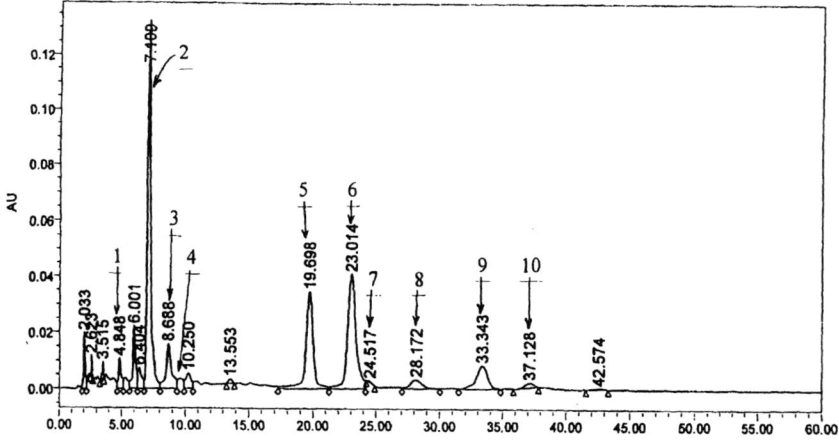

Figure 6. HPLC fingerprint spectrum of silymarin (NPI). 1. (+)-taxifolin (5a), 2. silychristin (3), 3. neusilychristin (15), 4. silydianin (4), 5. silybin A (11), 6. silybin B (12), 7. 2,3-cis-silybin A (13), 8. 2,3-cis-silybin B (14), 9. isosilybin A (9), 10. isosilybin B (10)

MECC

Micellar electrokinetic capillary chromatographic (MECC) method for the separation and determination of silybin (**1**), isosilybin (**2**), silychristin (**3**), and silydianin (**4**) was developed by Ding et al (*15*). The diastereoisomers of silybin and isosilybin, silychristin, silydianin, and taxifolin were well separated, but other peaks (23-27 min) were poorly resolved. Germany silymarin (Germany, Madans AG) contains higher percent of silydianin in comparison with Chinese silymarin. This may result from different growing environment or different method of extraction.

NMR

It is well known that NMR including various 2D techniques is a very powerful tool for structure determination and shows applicable value for recognizing organic mixture or crude extract in pharmacognosy (*23*). The ^{1}H NMR of these stereoisomers, such as silybin A (**11**), silybin B (**12**), isosilybin A (**9**), isosilybin B (**10**), and silychristin (**3**) have very similar ^{1}H NMR spectra. It is almost impossible to assign silybin A and silybin B, isosilybin A and isosilybin B by ^{1}H NMR spectrum alone. Interestingly, these isomers have significant difference in their HPLC retention times. We investigated a total of five different NMR solvents along with D_2O exchange. ^{1}H NMR spectrum taken in acetone-d_6 +D_2O displayed a very specific characterization of total phenolic flavonolignans contained in silymarin. It is conceivable that NMR spectrum could be used as fingerprint for identification and quality control. Figure 7 shows eight characteristic areas (A-H) within the fingerprint. The specific signals in each area reveal the presence of a particular compound: A: Aromatic protons of silybin A (**11**), silybin B (**12**), isosilybin A (**9**), isosilybin B (**10**), silychristin (**3**), and silydianin (**4**); B: H-6, H-8 on the A-ring of flavonolignans including silybin A (**11**), silybin B (**12**), isosilybin A (**9**), isosilybin B (**10**), silychristin (**3**), and silydianin (**4**); C: H-7' of silychristin (**3**); D: H-2 of silybin A (**11**), silybin B (**12**), isosilybin A (**9**), isosilybin B (**10**), and silychristin (**3**); H-7' of silybin A (**11**), silybin B (**12**), isosilybin A (**9**), and isosilybin B (**10**); E: H-3 of silybin A (**11**), silybin B (**12**), isosilybin A (**9**), isosilybin B (**10**), and silychristin (**3**); F: H-8' of silybin A (**11**), silybin B (**12**), isosilybin A (**9**), and isosilybin B (**10**); G: OCH$_3$ of silybin A (**11**), silybin B (**12**), isosilybin A (**9**), isosilybin B (**10**), and silychristin (**3**); H: 2 H-9' of silybin A (**11**), silybin B (**12**), isosilybin A (**9**), and isosilybin B (**10**); H-8' of silychristin (**3**).

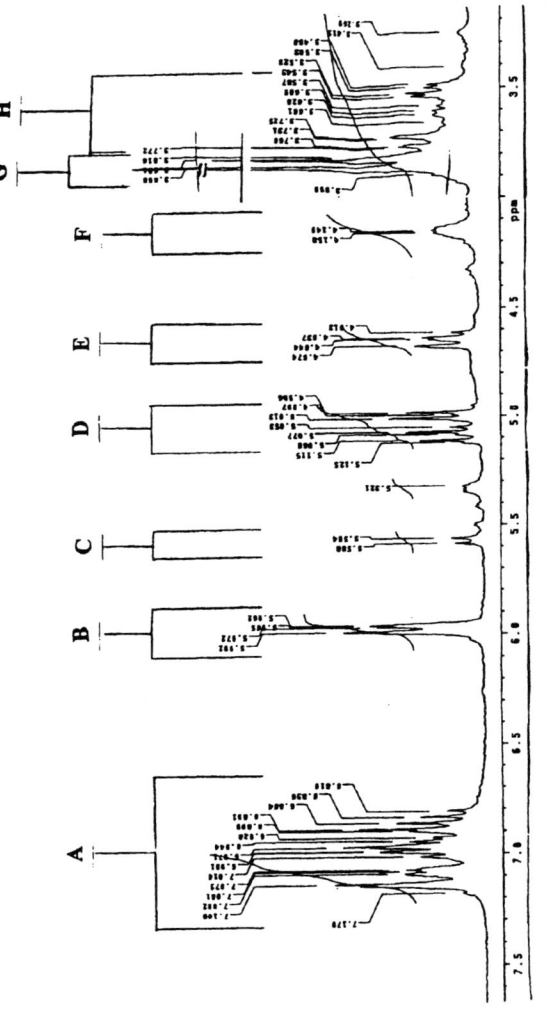

Figure 7. ^{1}H NMR fingerprint spectrum of silymarin (300 MHz, Me$_2$CO-d$_6$).

X-Ray

X-ray crystallography is still the most powerful technique for determination of chemical structures. By adding certain heavy atom or a chemical group with known chiral carbon as internal standard, it can provide unambiguous information on absolute stereochemistry. Lotter et al (*14*) reported the X-ray crystallographic study of silybin (**1**), but it was derived from a mixture of diastereoisomers with 2*R*, 3*R*, 7'*R*, 8'*R* and 2*R*, 3*R*, 7'*S*, 8'*S* stereochemistry. Our experiment showed that silybin A (**11**) and silybin B (**12**) were difficulty to form single crystal under various solvent conditions. Fortunately, when isosilybin A (**9**) was dissolved in aqueous methanol and slowly evaporation of solvent, it gave beautiful single crystals with the shape of chrysanthemum. X-ray crystallography of A (**9**) in combination with optical rotation measurements (Table II) of silybin A (**11**), silybin B (**12**), isosilybin A (**9**), and isosilybin B (**10**) resulted in the assignments of absolute stereochemistry.

Summary

Silybin A (**11**), silybin B (**12**), isosilybin A (**9**), and isosilybin B (**10**) were successfully separated by preparative HPLC and their stereochemistry were established by X-ray crystallography of isosilybin A (**9**) in combination with optical rotation data. Silychristin (**3**) and silydianin (**4**) were also found as main flavonolignans in silymarin. Neusilychristin (**15**), 2,3-*cis*-silybin A (**13**), and 2,3-*cis*-silybin B (**14**) were identified as minor components. RP-HPLC is a very convenient, sensitive and specific method for qualitative and quantitatively determination of each isomer including silybin A (**11**), silybin B (**12**), isosilybin A (**9**), and isosilybin B (**10**). HPLC trace can be used as fingerprint in quality control. ^{1}H NMR of silymarin in acetone-d_6 + D_2O or DMSO-d_6 + D_2O can also be used as fingerprint for a comprehensive analysis of total flavonolignans in silymarin. Further studies about the structure-activity relationship are in progress.

References

1. Brown, D. *Herbal Research Update.* **1993**, 23-26.
2. Foster, S. Milk thistle: *Silybum marianum.* Austin (TX): American Botanical Council, **1990**.
3. Luper, S. *Altern. Med. Rev.* **1998**, 3, 410-421.
4. Culpepper, N. *The English Physician Enlarged.* Dublin: H. Colbert, **1787**.
5. Wagner, H.; Hörhammer, L.; Münster, R. *Arzneim.-Forsch.* **1968**, 18, 688.
6. Pelter, A.; Hänsel, R. *Chem. Ber.* **1975**, *108,* 790-802.
7. Pelter, A.; Hänsel, R. *Chem. Ber.* **1975**, *108,* 1482-1501.

8. Merlini, L.; Zanarotti, A. *J. Chem. Sos., Chem. Commun.* **1979**, 695.
9. Merlini, L.; Zanarotti, A.; Pelter, A.; Rochefort, M.P.; Hänsel, R. *J. Chem. Soc.,Perkin Trans.* 1 **1980**, 775-778.
10. Arnone, A.; Merlini, L.; Zanarotti, A. *J. Chem. Soc., Chem. Commun.* **1979**, 696-697.
11. Tittel, G.; Wagner, H. *J. Chromatogr.* **1977**, *135*, 499-501.
12. Tittel, G.; Wagner, H. *J. Chromatogr.* **1978**, *153*, 227-232.
13. Ding, TM, Tian, SJ, Zhang, ZX. *Chinese Journal of Pharmaceutical Analysis*, **1999**; *19*: 304-307.
14. Lotter, H.; Wagner, H. *Z. Naturforsch.* **1983**, *38c*, 339-341.
15. Ding, T.; Tian, S.; Zhang, Z.; Shi, Y.; Sun, Z. *Acta Pharm. Sin.* **2000**, *35*, 778-781.
16. Tanaka, H.; Shibata, M.; Ohira, K.; Ito, K. *Chem. Pharm. Bull.* **1985**, *33*, 1419-1423.
17. Lee, David YW; Liu, Yanze. *Journal of Natural Product.* **2003**, *66*, 1171-1174
18. Kaloga, Macki. *Zeitschrift fuer Naturforschung, Teil B: Anorganische Chemie,Organische Chemie* **1981**, *36B* (2), 262-265.
19. Szilagi, I.; Tetenyi, P.; Antus, S.; Seligmann, O.; Chari, V. M.; Seitz, M.; Wagner, H. *Planta Medica* **1981**, *43* (2), 121-7.
20. Kim, Nam-Cheol; Graf, Tyler N.; Sparacino, Charles M.; Wani, Mansukh C.; Wall, Monroe E. *Organic & Biomolecular Chemistry* **2003**, *1* (10), 1684-1689.
21. Lee, David Y. W.; Liu, Yanze. *Abstracts of Papers, 226th ACS National Meeting*, New York, NY, United States, September 7-11, **2003**, AGFD-057.
22. Wagner, Hildebert; Chari, V. Mohan; Seitz, Maria; Riess-Maurer, Ingrid. *Tetrahedron Letters* **1978**, (4), 381-4.
23. Qin HL, Shang YJ, Zhao W, Zhao TZ. *Chinese Traditional and Herbal Drugs* **2000**; *31*: 881-884.

Chapter 4

Antioxidant Activity of Sesame Fractions

Fereidoon Shahidi[1,2] and Chandrika M. Liyana-Pathirana[1]

Departments of [1]Biology and [2]Biochemistry, Memorial University of Newfoundland, St. John's, Newfoundland A1B 3X9, Canada

The total phenolic content (TPC), total antioxidant activity (TAA), free radical scavenging capacity, metal chelation capacity and inhibition of copper-mediated oxidation of low density lipoprotein (LDL) cholesterol of ethanolic extracts of black and white sesame seeds and their hull fractions demonstrated that both the seeds and hulls of black sesame possessed greater antioxidant potential than those of white sesame. The TPC and TAA were highest for black sesame hulls while white sesame seeds were lowest. The results obtained for free radical scavenging, metal chelation and inhibition of LDL oxidation corresponded well with total phenolic content and total antioxidant activity. All the extracts exhibited dose-dependent activity in each antioxidant assay. However, among all the different fractions examined, the black sesame hulls exhibited considerable antioxidant properties and may serve as an excellent source of natural antioxidants for the food and nutraceutical industries.

Currently there is much interest in the phytochemicals such as phenylpropanoids and polyphenolic compounds because of their potential health benefits related to their antioxidant, anti-inflammatory and anti-aggregatory properties (1,2). In general, increased consumption of foods of plant origin has been associated with a reduced risk of a variety of chronic diseases (3,4). This has, in fact, been attributed to the presence of phytochemicals such as vitamins, polyphenols and carotenoids, among others that may possess antioxidant and free radical scavenging properties that play a significant role in the etiology of chronic diseases via modulating oxidative damage to cells and biological molecules (5). Different methods have been developed to measure the efficiency of dietary antioxidants either as pure compounds or in food/plant extracts (6). These methods focus on different antioxidant mechanisms including scavenging of superoxide anion and hydroxyl radicals, reduction of lipid peroxyl radicals, inhibition of lipid peroxidation or chelation of metal ions, among others (6). Most assays involve a pro-oxidant, which is usually a free radical and an oxidizable substrate. Thus, the pro-oxidant induces oxidative damage to the substrate that may be inhibited in the presence of an antioxidant. The pro-oxidants concerned in these methods, in general, are of pathologic importance. Hence, the existence of various harmful pro-oxidants such as $O_2^{-\bullet}$, H_2O_2, $ROO^{\bullet}$ and $^{\bullet}OH$ *in vivo* makes antioxidants crucial for the maintenance of a healthy life (7). Thus an antioxidant may efficiently reduce a pro-oxidant subsequently giving rise to products with no or low toxicity (8). In particular, polyphenolic compounds of higher plants may act as antioxidants contributing to anticarcinogenic or cardioprotective actions (2).

The reactive species, in general, possess the ability to alter chemically, all major biomolecules such as lipids, proteins and nucleic acids with subsequent changes in structure and function leading to various pathologic conditions and/or diseases (9). Dietary antioxidants may exert protection in the body against these reactive species thereby preventing the incidence of these diseases (10). Recently much effort has been paid in the preparation of antioxidants from natural sources by extraction, purification and fractionation. Plants synthesize the well known antioxidants such as tocopherols, ascorbic acid and carotenoids. In addition, plants synthesize substantial amounts of phenolic and polyphenolic compounds. Plants use these chemicals to protect themselves against oxidative damage by inhibiting or quenching free radicals and reactive species (11). It has been reported that phenolic and polyphenolic compounds in fruits and vegetables are better antioxidants than other common antioxidants such as vitamin C and E and contribute more than 80% to the total antioxidant activity (12,13).

Sesame (*Sesamum indicum* L.) is an important crop which provides an excellent source of edible oil (14). Sesame is cultivated on a worldwide basis for both oil and protein where the seed is composed of 55% lipid and 20% protein. Sesame seed hulls contain large amounts of oxalic acid and fiber (15). In general, seed hulls play a major role in physical and chemical defense of the seed

(*16*) and serve a good source of antioxidants (*16-18*). Chang et al. (*19*) have demonstrated that sesame hulls possess considerable antioxidant activity, in part due to the presence of phenolic compounds. Kuo et al. (*16*) reported that seeds of the medicinal plant adlay had a moderate antioxidant activity and their hulls exhibited greater antioxidant capacity than other parts of the adlay seed such as testa, bran, and polished seed (*20*). In oat, most of the phenolic compounds were located in the seed coat/aleurone/sub-aleurone layers. Thus, there was a decreasing concentration of phenolics toward the interior of the seed (*20*). This study was designed to determine total phenolic content of whole seed and the hull fractions of white and black sesame seeds and to compare the antioxidant activity using different antioxidative assays.

Materials and Methods

Materials

The samples of black sesame as such were obtained from a supermarket in Singapore while white sesame seeds were from Egypt. Their hulls were separated by a combined mechanical and aspiration method. The chemicals used were obtained from Sigma Chemical Co. (St. Louis, MO) or Aldrich Chemical Co. (Milwaukee, WI). Solvents used in this study were ACS-grade or better and were purchased from Fischer Scientific Co. (Nepean, ON).

Methods

Preparation of samples

Sesame seeds were ground in an electric coffee grinder for 10 min. Ground samples were then defatted by blending with hexane (1:5 w/v, 5 min X 3) in a Waring blender at ambient temperature. The resulting slurry was filtered under suction and the residue was air dried for 12 h. The dried defatted meal was stored in vacuum packed polyethylene pouches and kept at -20 °C prior to analysis. Sesame hulls were also defatted in the same manner prior to analysis.

Preparation of crude phenolic extracts

Phenolic constituents were extracted from both sesame meal and hulls under reflux condition in a thermostated water bath. Phenolic compounds of sesame samples (6 g) were extracted into 80% aqueous ethanol (100 mL) at 70 °C for 30 min. The resulting slurries were centrifuged for 5 min at 4000 X g. The supernatants were collected and the residues re-extracted under the same conditions. The solvent from the combined supernatants was removed under vacuum at 40 °C; the resulting concentrated solutions were lyophilized for 72 h at -49 °C and 46 X 10^{-3} mbar.

Determination of total phenolic content of sesame extracts

The content of total phenolics was determined according to a modified version of the procedure described by Singleton and Rossi (*21*). Extracts were dissolved in methanol to obtain a 1 mg/mL solution. Folin-Ciocalteu's reagent (0.5 mL) was added to centrifuge tubes containing 0.5 mL of the extracts. Contents were mixed and 1 mL of a saturated solution of sodium carbonate was added to each tube followed by adjusting the volume to 10 mL with distilled water. The contents in the tubes were thoroughly mixed by vortexing. Tubes were allowed to stand at ambient temperature for 45 min until the characteristic blue color appeared and then centrifuged for 5 min at 4000 X g. Absorbance of the supernatants was recorded at 725 nm. The content of total phenolics in each extract was reported as mg catechin equivalents per gram of extract.

Measurement of total antioxidant activity by Trolox equivalent antioxidant capacity (TEAC) assay

Total antioxidant activity was determined according to the procedure described by van den Berg *et al.* (*22*). The extracts and reagents were prepared in a 100 mM phosphate buffered saline (pH 7.4, 150 mM NaCl) solution (PBS buffer). A solution of 2,2'-azinobis-(3-ethylbenzthiazoline-6-sulfonate) radical anion (ABTS$^{\bullet+}$) was prepared by mixing 2.5 mM 2,2'-azobis-(2-methylpropionamidine) dihydrochloride (AAPH) with 2.0 mM ABTS^{2+} in a 1:1 (v/v) ratio, and heating at 60 °C for 12 min. The radical solution was stored at room temperature and protected from light. A standard curve was prepared using different concentrations of Trolox. The reduction in absorbance of the ABTS$^{\bullet-}$ solution (1960 µL) at different concentrations of Trolox (40 µL) over a 6 min period was measured and plotted. TEAC values of the extracts (1 mg/mL) were determined in the same way and expressed as µM Trolox equivalents.

Scavenging of 1,1-diphenyl-2-picrylhydrazyl (DPPH) radical

The method described by Kitts *et al.* (*23*) was used to assess the DPPH radical scavenging capacity of sesame extracts. A 100 µM DPPH solution in 95% ethanol was mixed with various amounts of sesame extracts (5, 10, 20, 40 µg/mL) and vortexed thoroughly. The mixture was allowed to stand at ambient temperature for 30 min. The absorbance was measured at 519 nm. The scavenging percentage was calculated according to the equation:

Scavenging % = {($Abs_{control}$ − Abs_{sample}) / $Abs_{control}$} X 100.

Determination of iron (II) chelating capacity

Solutions of ferrous sulfate (400 µM), extracts/ standard were prepared in a 10 mM hexamine-HCl buffer containing 10 mM KCl (pH 5.0). One milliliter of ferrous sulfate was mixed with 1 mL of extracts/ standard followed by the addition of 0.1 mL of a 1 mM solution of tetramethylmurexide prepared in the same buffer. The final concentration of extracts/ standard in the assay medium was 50 or 100 ppm based on catechin equivalents. Absorbance of the reaction mixture was recorded at 460 and 530 nm and the ratio of A_{460} to A_{530} calculated. A standard curve of absorbance ratio versus free iron (II) was prepared. The difference between the total iron (II) and the free iron (II) indicates the concentration of chelated iron (II). Iron (II) chelating capacities of extracts/ standard were calculated using the following equation.

Iron (II) chelating capacity, % = {Concentration of chelated iron (II)/ Concentration of total iron (II)} X 100

Inhibition of copper-mediated human low density lipoprotein (LDL) cholesterol

Human LDL cholesterol was dialyzed in 10 mM PBS (pH 7.4, 150 mM NaCl) at 4 °C in the dark for 24 h. Human LDL (0.2 mg protein/mL) was mixed with different amounts of sesame extracts (25-100 ppm phenolics). Catechin was used as the reference antioxidant compound. Reaction was initiated by adding a solution of $CuSO_4$ (10 µM); samples were then incubated at 37 °C for 22 h. The formation of conjugated dienes was measured at 234 nm as described by Hu and Kitts (*24*).

Statistical analysis

All experiments were carried out in triplicate and the significance of differences among mean values determined at $p<0.05$ using one way ANOVA followed by Tukey's multiple range test. The type of relationship between variables were determined by simple regression analysis. All percentage data were transformed to arcsin values and tested for normality before being subjected to an appropriate statistical analysis.

Results and Discussion

The total phenolic content (TPC) of extracts of black and white sesame seeds and their hulls are shown in Figure 1. The yield of phenolic compounds was highest in black sesame hulls (146.6 ± 4.1 mg catechin equivalents/g crude extract) while this was lowest in white sesame seeds (10.6 ± 1.6 mg catechin equivalents/g crude extract). The yield of crude phenolics was approximately 5 times higher in the hull of black sesame than that of white sesame. Phenolic and polyphenolic compounds are naturally present in essentially all plant material including foods of plant origin and hence exist ubiquitously in fruits, vegetables, cereal grains, and nuts, among others (*25*). Among different plants, several oilseeds and their by-products have been investigated for their phenolic compounds (*26*). Naczk and Shahidi (*27*) have shown that hulls of canola seeds contain a considerably high proportion of phenolic compounds. Using histochemical studies it has been shown that the highest concentration of phenolic compounds is located in the outer layers of a seed (*28*). The hulls of sesame, in particular black sesame, possessed a higher phenolic content compared to the whole sesame seeds reflecting a dilution effect as the endosperm contains a very low amount of phenolics compared to those of the hulls.

The antioxidant activity (TAA) of sesame extracts was studied using the Trolox equivalent antioxidant capacity (TEAC) assay (*22*). The TEAC assay is based on the inhibition of $ABTS^{\bullet+}$, that has a characteristic long wavelength absorption spectrum showing maxima at 660, 734 and 820 nm, by the presence of antioxidants (*29*). Total antioxidant activity of sesame extracts, expressed as Trolox equivalents, is also shown in Figure 1. The concentration of sesame extracts used in the determination of TAA was 1 mg/mL. The TAA was highest in the hulls of black sesame followed by black sesame seeds, white sesame hulls and white sesame seeds. The greater antioxidant activity of black sesame hulls may be related to presence of high amounts of phenolics. A good correlation ($R^2 = 0.99$) existed between TPC and TAA indicating the main influence of phenolic compounds for the observed antioxidant activity. Antioxidant activity of phenolic compounds may be attributed to the presence of hydroxyl groups although this is not the only factor that determines their potency as antioxidants (*16*).

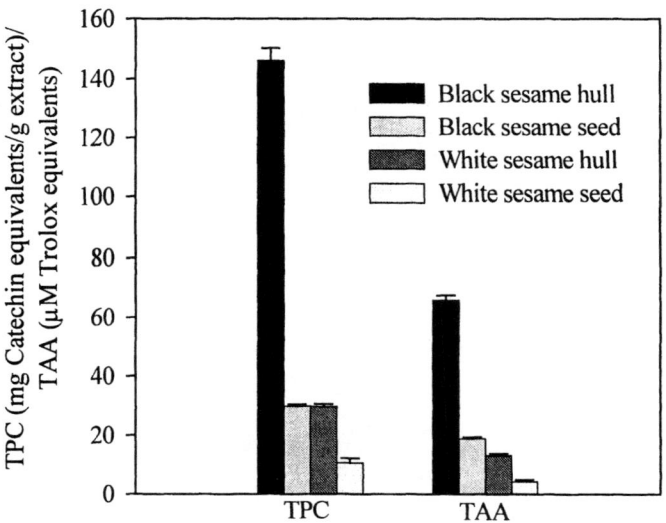

Figure 1. Total phenolics content (TPC; mg catechin equivalents per g ethanolic extracts) and total antioxidant activity (TAA; μM Trolox equivalents) of black and white sesame seeds and hulls

Free radical scavenging is the accepted mechanism for antioxidant action inhibiting lipid oxidation. The scavenging of the stable organic free radical DPPH may be used in the evaluation of antioxidant activities of various compounds (*30*). The DPPH radical scavenging method has been used extensively to predict the antioxidant activity owing to the relatively short time required for their analysis (*30*). Hence, antioxidant potential of sesame extracts was evaluated using DPPH radical. The percentage DPPH scavenging capacity by sesame extracts compared to that of catechin is shown in Figure 2. Results demonstrates a concentration-dependent scavenging of DPPH radical where black sesame hulls exhibited a significantly ($p<0.05$) higher activity than all the other extracts. At higher concentrations black sesame hulls showed similar ($p>0.05$) scavenging activity to that of catechin. However, this was significantly ($p<0.05$) different at lower concentrations.

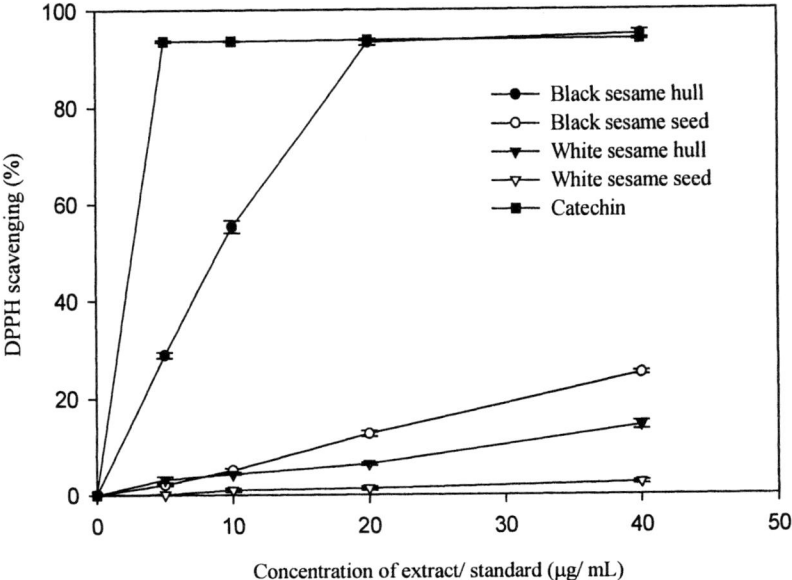

Figure 2. Scavenging capacity of DPPH radical by extracts of black and white sesame seeds and their hulls

In general, free radical scavenging activity of an antioxidant is related to its ability to donate a hydrogen atom and stabilization of the resulting antioxidant radical by electron delocalization (*31*). It has been suggested that the reducing ability of polyphenols may seem to be an important factor dictating the antioxidant and free radical scavenging capacity of these compounds (*6*).

The chelating effect of sesame extracts in comparison to that of catechin is shown in Figure 3. Different sesame fractions demonstrated significantly different chelating properties at the two concentration levels examined in this study. Black sesame hulls exhibited the highest chelating activity at both 50 and 100 ppm level. At both levels catechin showed strong iron (II) chelation ability.

Polyphenolic compounds, that occur commonly in fruits, vegetables, tea, wine and spices are diverse in their chemical structure and characteristics and exhibited a wide range of effects both *in vivo* and *in vitro*. Some of these biological effects include antioxidant and metal-chelating activities (*32*). Green tea and rosemary extracts are known to reduce the oxidation of food products thereby extending their shelf-life possibly due to their antioxidative potential attributed to their ability to scavenge free radicals and to chelate metal ions (*33,34*). It has been reported that high iron status or iron overload is positively correlated with coronary heart disease risk (*35*). Hence, increased intake of flavonoids, for instance, may maintain a relatively low iron status thereby reducing the risk of iron overload (*36*). Since there is a reduction of nonheme-iron absorption in the presence of tea and rosemary extracts, it has been suggested that iron chelation provides one of the mechanisms of antioxidant action *in vivo* (*36*). Thus, once the metal ions are chelated this renders them inactive to participate in free radical generating reactions that may otherwise bring about damage to biomolecules (*37,38*). In general, chelating agents are known to stabilize pro-oxidative transition metals by complexation where unshared pairs of electrons in the molecular structure of chelator promote the complexation (*39*). Citric acid and its salts, phosphates and salts of ethylenediaminetetraacetic acid (EDTA) are the most commonly used chelators. In addition, plant phenolic compounds may function as good metal chelators (*40*). Results revealed that sesame extracts especially black sesame hulls chelated iron (II) efficiently at 100 ppm level.

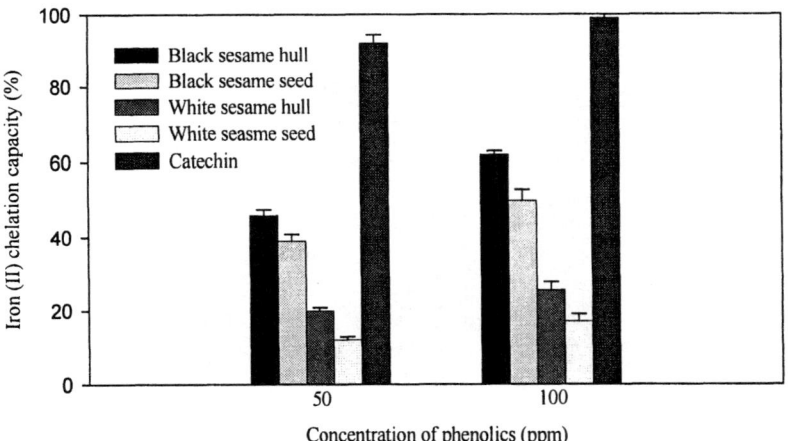

Figure 3. Ferrous ion chalation capacity by extracts of black and white sesame seeds and hulls

The protection provided by fruits and vegetables against chronic diseases has been attributed to the presence of various antioxidants including phenolics and polyphenolics in these foods (*41*). In our study, sesame phenolics, especially those of black sesame hulls, inhibited human LDL oxidation under *in vitro* conditions hence exhibiting a protective effect against experimentally-induced oxidative stress (Figure 4). The sesame extracts inhibited LDL oxidation in a dose-dependent manner. Addition of 100 ppm phenolics inhibited 57.3 - 99% of lipid oxidation of LDL while inhibition was 24.4 -o 59.3% at 25 ppm.

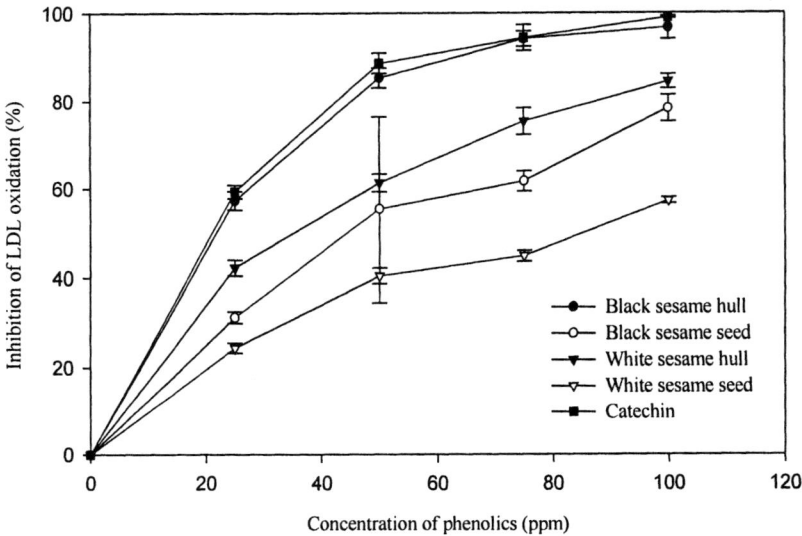

Figure 4. Inhibition of copper-induced oxidation of human LDL by black and white sesame seeds and hulls

The effect of catechin and black sesame hull extracts against LDL oxidation was not significantly (p<0.05) different at each concentration. Catechin prevents the consumption of α-tocopherol in the LDL particles (2). Cupric ions (Cu^{2+}) are the most widely studied transition metal with respect to LDL oxidation in *in vitro* model systems. It has been suggested that inhibition of copper-mediated LDL oxidation may occur by free radical scavenging and/or copper ion chelation (42). The ability of an antioxidant to inhibit copper-induced LDL oxidation may also be attributed to efficient removal of copper from the surface of LDL (42). Certain phenolic compounds have been suggested to act as peroxyl radical scavengers thereby increasing the resistance of LDL to oxidation. These phenolic compounds may function as chain-breaking antioxidants in inhibiting LDL oxidation (43). The copper-mediated oxidation of tryptophan residues in LDL-apolipoprotein B has also reported to be a major cause in initiating lipid oxidation in LDL molecules (44). A unique antioxidant mechanism that involves the blockage of copper binding sites of tryptophan residues on apolipoprotein B has been described (*45*). This may be achieved by binding of antioxidants to apolipoprotein B on the LDL molecule.

In general, LDL is the major cholesterol carrier in the blood. It has been well established that an elevated level of LDL in plasma is associated with an increased risk of atherosclerosis and cardiovascular disease (*46*). However, LDL may not form atherosclerotic plaques in its native form. It has now been recognized that oxidative modification of LDL is the key factor in pathogenesis of atherosclerosis. This may finally lead to develop plaque in the arteries and subsequently coronary heart disease (*47,48*). Hence, dietary antioxidants that inhibit LDL oxidation may be important in preventing these diseases (*48*).

The overall order of antioxidant activity among different sesame fractions was found to be black sesame hulls > black sesame seeds > white sesame hulls and > white sesame seeds. Under oxidative stress cellular and extracellular macromolecules such as proteins, lipids and nucleic acids may suffer from oxidative damage causing tissue injury (*49*). Therefore, it is important to maintain a balance between antioxidants and oxidants in living organisms and increased intake of dietary antioxidants may help in maintaining an adequate antioxidant status (*50*). In fact dietary antioxidants such as polyphenolic compounds have been shown to be a major dietary factor with a myriad of protective effects (*51*).

References

1. Rice-Evans, C.A.; Miller, N.J.; Bolwell, P.G.; Bramley, P.M.; Pridham, J.B. *Free Rad. Res.* **1995**, *22*, 375-383.
2. Salah, N.; Miller, N.J.; Paganga, G.; Tijburg, L.; Bowell, G.P.; Rice-Evans, C. *Arch. Biochem. Biophys.* **1995**, *322*, 339-346.
3. Temple, N.J. *Nutr. Res.* **2000**, *2*, 449-459.
4. Halliwell, B. *Nutr. Rev.* **1997**, *55*, S44-S52.
5. Jimenez-Escrig, A.; Dragsted, L.O.; Daneshvar, B.; Pulido, R.; Saura-Calixto, F. *J. Agric. Food Chem.* **2003**, *51*, 5540-5545.
6. Pulido, R.; Bravo, L.; Saura-Calixto, F. *J. Agric. Food Chem.* **2000**, *48*, 3396-3402.
7. Prior, R.L.; Cao, G. *Free Rad. Biol. Med.* **1999**, *27*, 1173-1181.
8. Prior, R.L.; Cao, G. *Hort. Sci.* **2000**, *35*, 588-592.
9. Castelluccio, C.; Bolwell, P.G.; Gerrish, C.; Rice-Evans, C. *Biochem. J.* **1996**, *316*, 691-694.
10. Deshpande, S.S. In *Food Antioxidants: Technological, toxicological and health perspectives*; Madhavi, D.L.; Deshpande, S.S.; Salunkhe, D. Eds.; Marcel Dekker Inc.; New Your, **1996**

11. Larson, R.A. *Phytochemistry* **1988**, *27*, 969-978
12. Cao, G.; Sofic, E.; Prior, R.L. *J. Agric. Food Chem.* **1996**, *44*, 3426-3431.
13. Wang, H.; Cao, G.; Prior, R.L. *J. Agric. Food Chem.* **1996**, *44*, 701-705.
14. Namiki, M. *Food rev. Int.* **1995**, *11*, 281-329.
15. Abou-Gharbia, H.A.; Shahidi, F.; Shehata, A..A.Y.; Youssef, M.M. *J. Am. Oil Chem. Soc.* **1997**, *74*, 215-221.
16. Kuo, C-C.; Shih, M-C.; Kuo, Y-H.; Chiang, W. *J. Agric. Food Chem.* **2001**, *49*, 1564-1570.
17. Asamarai, A.; Abdulwahab, M.; Addis, P.B.; Epley, R.J.; Krick, T.P. *J. Agric. Food Chem.* **1996**, *44*, 126-130.
18. Lin, Y.L.; Tsai, S.H.; Lin-Shiau, S.Y., Ho, C-T.; Lin, J.K. *Eur. J. Pharmacol.* **1999**, *367*, 379-388.
19. Chang, L.-W.; Yen, W.-J.; Huang, S.c.; Duh, P-D. *Food Chem.* **2002**, *78*, 347-354.
20. Emmons, C.L.; Peterson, D.M.; Paul, G.L. *J. Agric. Food Chem.* **1999**, *47*, 4894-4898.
21. Singleton, V.L.; Rossi, J.A. Colorimetry of total phenolics with phophomolybdic-phosphotungstic acid reagents. *Am. J. Enol. Vitic.* **1965**, *16*, 144-158.
22. van den Berg, R.; Haenen, G.R.M.M.; van den Berg, H.; Bast, A. *Food Chem.* **1999**, *66*, 511-517.
23. Kitts, D.D.; Wijewickreme, A.N.; Hu, C. *Mol. Cell. Biochem.* **2000**, *203*, 1-10.
24. Hu, C.; Kitts, D.D. Mol. Cell. Biochem. 2001, *218*, 147-155.
25. Arnous, A.; Makris, D.P.; Kefalas, P. *J. Food Comp. Anal.* **2002**, *15*, 655-666.
26. Wettasinghe, M.; Shahidi, F.; Amarowicz, R. *J. Agric. Food Chem.* **2002**, *50*, 1267-1271.
27. Naczk, M.; Shahidi, F. *Food Chem.* **1989**, *34*, 159-164.
28. Fulcher, R. In *Oats chemistry and technology*; Webster, F. ed.; American Association for Cereal Chemists; St. Paul, MN, **1986**; pp 47-74
29. Prior, R.L.; Cao, G. *Free Rad. Biol. Med.* **1999**, *27*, 1173-1181.
30. Chen, Y.; Wang, M.; Rosen, R.T.; Ho, C-T. *J. Agric. Food Chem.* **1999**, *47*, 2226-2228.
31. Andreasen, M.; Landbo, A-K.; Christensen, L.P.; Hansen, A.; Meyer, A.S. *J. Agric. Food Chem.* **2001**, *49*, 4090-4096.
32. Cook, N.C.; Samman, S. *J. Nutr. Biochem.* **1996**, *7*, 66-76.
33. Guo, Q.; Zhao, B.; Li, M.; Shen, S.; Xin, W. *Biochem. Biophys. Acta* **1996**, *1304*, 210-222.

34. Basaga, H.; Tekkaya, C.; Acikel, F. *Lebensmittel-Wissenschaft Technol.* **1997**, *30*, 105-108.
35. Salonen, J.T.; Nyyssonen, K.; Korpela, H.; Tuomilehto, J.; Seppanen, R.; Salonen, R. *Circulation* **1992**, *86*, 803-811.
36. Samman, S.; Sandstrom, B.; Toft, M.B.; Bukhave, K.; Jensen, M.; Sorensen, S.; Hansen, M. *Am. J. Clin. Nutr.* **2001**, *73*, 607-612.
37. Morel, I.; Lescoat, G.; Cillard, P.; Cillard, J. *Methods Enzymol.* **1994**, *243*, 437-443.
38. Miller, N.J.; Castelluccio, C.; Tijburg, L.; Rice-Evans, C. *FEBS Lett.* **1996**, *392*, 40-44.
39. Dziezak, J.D. *Food Technol.* **1986**, *40*, 94-102
40. van Acker, S.A.B.E.; van den Berg, D-J.; Tromp, M.N.J.L.; Griffioen, D.H.; van Bennekom, W.P.; van der Vijgh, W.J.F.; Bast, A. *Free Rad. Biol. Med.* **1996**, *20*, 331-342.
41. Cao, G.; Sofic, E.; Prior, R.L. *Free Rad. Biol. Med.* **1997**, *22*, 749-760.
42. Decker, E.A.; Ivanov, V.; Zhu, B-Z.; Frei, B. *J. Agric. Food Chem.* **2001**, *49*, 511-516.
43. Castelluccio, C.; Paganga, G.; Melikian, N.; Bolwell, G.P.; Pridham, J.; Sampson, J.; Rice-Evans, C. *FEBS Lett.* **1995**, *368*, 188-192.
44. Giessauf, A.; Steiner, E.; Esterbauer, H. *Biochem. Biophys. Acta* **1995**, *1256*, 221-232.
45. Meyer, A.S.; Donovan, J.L.; Pearson, D.A.; Waterhouse, A.L.; Frankel, E.N. *J. Agric. Food Chem.* **1998**, *46*, 1783-1787.
46. Ross, R. *Nature* **1993**, *362*, 801-809.
47. Steinberg, D. *Atheroscl. Rev.* 1992, *18*, 1-6.
48. Esterbauer, H.; Gebicki, J.; Puhl, H.; Jurgens, G. *Free Rad. Biol. Med.* **1992**, *13*, 341-390.
49. Halliwell, B.; Chirico, S. *Am. J. Clin. Nutr.* **1993**, *57*, 715-725.
50. Halliwell, B.; Murcia, M.A.; Chirico, S.; Aruoma, O.I. *Crit. Rev. Food Sci. Nutr.* **1995**, 35, 7-20.
51. Hertog, M.G.L. *Proc. Nitr. Soc.* **1996**, *55*, 385-397.

Chapter 5

Effect of Lactic Acid Fermentation on Quercetin Composition and Antioxidative Properties of *Toona sinensis* Leaves

Tzou-Chi Huang[1], Hseng-Kuang Hsu[2], Hui-Yin Fu[3], and Chi-Tang Ho[4]

[1] Department of Food Science, National Pingtung University of Science and Technology, 912, Pingtung, Taiwan
[2] Department of Physiology, Kaohsiung Medical University, Kaohsiung, Taiwan
[3] Department of Food Sanitation, Tajen Institute of Technology, 912, Pingtung, Taiwan
[4] Department of Food Science, Rutgers, The State University of New Jersey, 65 Dudley Road, New Brunswick, NJ 08901–8520

Flavonoids of fresh and lactic acid bacteria fermented *Toona sinensis* (*Cedrela sinensis* A. Juss) leaves were extracted and analyzed using HPLC. Rutin was characterized as the major quercetin derivative in fresh *Toona sinensis* leaf, whereas aglycone concentration increased after lactic acid bacteria fermentation. DPPH radical scavenging activity was used to evaluate the antioxidant activity. The β-glucosidases in both *Toona sinensis* and lactic acid bacteria were detected during fermentation resulting in a release of free hydroxyl groups from the quercetin molecule. The increased hydroxyl groups could be attributed to the increase of free radical scavenging activities of the fermented *Toona sinensis* leaves. The increased absorption of quercetin in rat could be attributed to the released quercetin-3-glucoside from rutin in fermented *Toona sinensis* leaves.

Introduction

In traditional Chinese medicine, leaf extract of *Toona sinensis* has many biological functions, especially its therapeutic effect on abdominal tumors. The leaves and stems of *Toona sinensis* Roemor have long been used for the treatment of enteritis, dysentery and itch in Oriental medicine (*1*). Evidence has shown that the components of *Toona sinensis* exert potent anti-inflammatory and analgestic actions, and inhibit boil growth *in vivo* (*2*). Recently crude extract from leaves of *Toona sinensis* was found to effectively block the cell cycle progression of human lung cancer cells by inhibiting the expression of cyclin D1 and E in A549 cells. Additionally, incubation of the extract led to activation of caspase-3-like proteases and apoptotic cell death. However, the extract did not show any significant cytotoxic effect on cultured human foreskin fibroblasts or MRC-5 human lung fibroblasts (*3*).

Few studies have been carried out on the chemical composition of *Toona sinensis*. Limonoids, sterols, sesquiterpenes, triterpenoids and flavonoids are reported to be the major components in *Toona* (Meliaceae). While investigating *Cedrela sinensis,* cedrellin, 2,6,10,15-phytatetraene-14-ol, 7-α-obacunyl acetate, 6-acetoxyobacunol acetate, 7-α-acetoxydihydronomilin, 2,6,10-phytatriene-1,14,15-triol and phytol were identified (*4*). A number of compounds, including retinoid, vitamins B and C, *o*-coumaric acid, kaempferol, methyl gallate, quercetin, afzelin, quercitrin, isoquercitrin and rutin have been isolated from the leaves of *Cedrela sinensis* (*5*). Park *et al.* (*6*) reported that methyl gallate, quercitrin, bis-(*p*-hydroxyphenyl) ether, adenosine, isoquercitrin, rutin, (+)-catechin and (-)-epicatechin were isolated and characterized from rachis of *Cedrela sinensis* A. Juss.

Fresh *Toona sinensis* leaf samples, whole or chopped, were ensilaged in an airtight mason jar. Spontaneous fermentation started to occur after two days of storage at ambient temperature (25±3 °C). A sharp decline in *Toona sinensis* silage pH from 6.4 to 4.5, followed by the continuing fermentation to a final pH of 4.2 for the whole leaf samples was observed, as shown in Figure 1. The change of pH value in the fermented chopped *Toona sinensis* leaf followed a similar tendency, with faster acid production than that of the whole leaf. This result indicates that sugars in the chopped sample were released and utilized more efficiently than in the whole leaf and lead to faster decline of pH value.

Organic acids identified by HPLC in solid-fermented *Toona sinensis* include lactic and acetic acids (Table I). The concentration of lactic acid increased

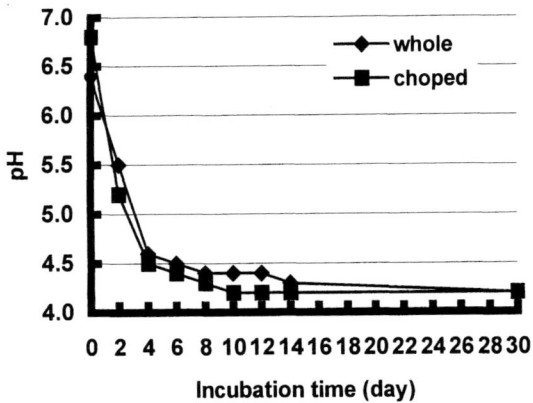

Figure 1. Changes of pH in naturally fermented Toona sinensis leaf silage.

quickly with increasing fermentation time until a value of 63±5 g/Kg dry matter was reached after 4 weeks. In addition to lactic acid, a significant amount of acetic acid (28±5 g/Kg dry matter) was produced as well, after 30 days incubation. During the first two weeks, significantly higher amounts of lactic acid (41±3 g/Kg dry matter) and acetic acid (21±2 g/Kg dry matter) were observed in the chopped *Toona sinensis* leaves. There was no significant (p<0.05) difference in either lactic or acetic acid production between whole and chopped *Toona sinensis* leaves after 2 weeks fermentation, and, no variation was observed after 4 weeks fermentation.

Triplicate samples of solid-fermented *Toona sinensis* (10 g) were homogenized with 90 mL sterile peptone physiological saline solution. Lactic acid bacteria (LAB) strains were initially selected on the basis of Gram staining, morphology, catalase test (with 3% H_2O_2) and oxidase test. Strains initially determined as LAB were further identified with a Biolog MicroPlate test panel (Biolog, Inc., USA). This provides a standardized micromethod, based on 95 biochemical tests, to assess the ability of a microorganism to utilize or oxidize various carbon sources, e.g. carbohydrates, organic acids, amino acids, etc. (7). A positive utilization reaction was indicated by the development of a purple color in the relevant well. The pattern of purple wells was then keyed into a Biolog MicroLog computer program to identify the isolate. At least 27 strains were isolated from solid-fermented *Toona sinensis* leaf samples and characterized as LAB. The isolated LAB consisted of five genera, *Lactobacillus*

(16 strains), *Leuconostoc* (5 strains), *Lactococcus* (3 strains), *Streptoococcus* (2 strains) and *Enterococcus* (1 strain). The number of *Lactobacillus* strains (70%) of the isolates suggests that *Lactobacilli* predominated in the solid-fermented *Toona sinensis*. Among them, *Lactobacillus plantarum* was predominant followed by *Leuconostoc mensenteroides* in the indigenous flora in solid-fermented *Toona sinensis* leaf samples fermented for 1 week, as shown in Table II. Our results were in good accordance with results in lactic acid fermentation of carrot, cabbage, beet and onion vegetable mixture (8) and whole-crop wheat and corn (9), in which *Lactobacillus plantarum* is predominant. Weinberg and Muck (10) reported that single strain inoculants, usually containing *Lactobacillus plantarum*, were used for wheat and corn silage because these species are fast and efficient producers of lactic acid.

Table I. Concentration of Organic Acids in Naturally Fermented *Toona sinensis* Leaf Silage

Period (week)	g Kg^{-1} dry matter			
	lactic acid		acetic acid	
	whole	chopped	whole	chopped
1	11 ± 3^a	13 ± 2^b	6 ± 2^a	7 ± 3^b
2	34 ± 2^a	41 ± 3^b	17 ± 4^a	21 ± 2^b
3	57 ± 5^a	57 ± 5^a	21 ± 4^a	24 ± 4^a
4	63 ± 5^a	62 ± 3^a	28 ± 5^a	31 ± 4^a

Within a raw, means followed by the same letter did not differ significantly ($p<0.05$) in Duncan's multiple range test.

Numbers of LAB were determined on the selective media MRS agar (Merck, Darmstadt, Germany) with glucose as a source of energy. Appropriate dilutions were plated on MRS agars and incubated at 30 °C for 48 h. The *Leuconostoc mensenteroides* growth curve differed from that of *Lactobacillus plantarum* by a rapid viability decline after four days of incubation (Figure 2). This rapid mortality could be attributed to weak resistance of *Leuconostoc mensenteroides* to the low pH (11). Naturally existing *Lactobacillus plantarum* can effectively grow in, as well as acidify, *Toona sinensis* leaf. Sauerkraut production relies typically on a sequential microbial process that involves *Leuconostoc* species for the first group and *Lactobacillus* and *Pediococcus* for the second. In the first phase of fermentation, *Leuconostoc* principally uses glucose and fructose for the production of lactic and acetic acids, which is accompanied by a rapid decrease of pH. During the second phase, homofermentative bacteria continue the fermentation process to a final pH of 3.5

to 3.8. (*12*). *Lactobacillus plantarum* is often isolated from silage and meets almost all the criteria for a successful ensilaging process (*13*).

Table II. Identification and Characterization of the Major LAB Strains Isolated from Solid-fermented *Toona sinensis* Leaves

Test	L. plantarum	L. mensenteroides
Gram-positive	+	+
Oxidase	−	−
Catalase	−	−
Gas from glucose	+	+
Arginine dihydrolase	+	−
Cell morphology	Rod-shaped	Spherical
Fermentation of:		
Glucose	+	+
Fructose	+	+
Galactose	+	+
Mannose	+	+
Lactose	+	−
Maltose	+	+
Sucrose	+	+
Trehalose	+	+
Mannitol	+	+
Sorbitol	−	−
Identification by Biolog Microplate test panel	L. plantarum	L. mensenteroides

The 2,2-diphenyl-1-picrylhydrazyl radical (DPPH) method was used to investigate the scavenging activity of partially purified methanolic extract from fermented *Toona sinensis* leaf samples. The data for DPPH free radical scavenging activity, measured as scavenging percentage at room temperature after addition of various fermented *Toona sinensis* leaf extracts, are presented in Table III. DPPH free radical scavenging activity of *Toona sinensis* leaf fermented for various time periods increased with increasing incubation time from to 51 % (0 week) to 52 % (2 weeks) and 54 % (4 weeks) and in a dose dependent manner from 5 to 15 μg/mL. As compared with that of authentic quercetin and rutin, slightly lower DPPH scavenging capacities were observed for the fermented samples, possibly due to the impurities. DPPH has been used to evaluate the potential of quecetin and glucosylated derivatives as antioxidants (*14*). Quercetin and its glycoside (rutin) were shown to have 50 % radical scavenging ability in the concentration range of 1 to 3 μM. The biological activity of quercetin is predicted to be highly dependent on the availability of hydroxyl group in the molecule. The antioxidative activity of phenolic acid increases with increasing the number of hydroxyl (OH) groups (*15*). The high proportion of quercetin and quercetin-3-glucoside with a high level of free hydroxyl groups may explain to the higher antioxidative activity of the leaves fermented for 2 weeks than that of unfermented samples.

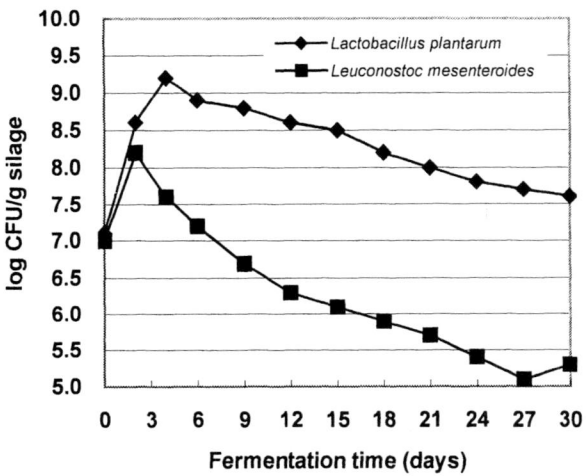

Figure 2. Changes in lactic acid bacteria counts and total bacterial counts in naturally fermented Toona sinensis leaf.

Table III. Changes of Antioxidative Activity of Extracts from Fermented *Toona sinensis* Leaves

	DPPH		
Extract (μg/mL)	*5*	*10*	*15*
Rutin	74 a	82 a	92 a
Quercetin	76 a	84 a	100 a
Toona sinensis extract *Fresh*	51 b	63 b	72 b
Fermented 2 Ws	52 b	64 b	74 b
Fermented 4 Ws	54 c	66 c	78 c

Values are means of triplicate measurements

Within the same column, means with no letter in common are significantly different ($p<0.05$)

Note: DPPH free radical scavenging capacity of Quercetin (15 μg /mL) is 100 %.

The concentration and distribution of quercetins in solid-fermented *Toona sinensis* samples were evaluated by high-performance liquid chromatography and photodiode array detection following the method of Rhodes and Price (*16*). The levels of individual quercetin were determined for the partially purified methanol extracts from solid-fermented *Toona sinensis* samples. There is no significant variation (p<0.05) among the three tested samples, fresh and fermented for 2 and 4 weeks. Changes in the composition of the quercetin conjugates due to lactic acid fermentation are shown in Table IV. Rutin is the major quecetin in fresh leaves (Table IV). Fermentation processes may lead to some extend of enzymatic autolysis in the solid-fermented *Toona sinensis* samples. For the *Toona sinensis* leaves fermented for 2 weeks, at least some of rutin was converted into quercetin-3-glucoside. Aglycone quercetin is dominant in the *Toona sinensis* leaves fermented for 4 weeks.

Table IV. Changes of Quercetin Glycosides in *Toona sinensis* Leaves during Silaging

Toona sinensis	Concentration (μg/g dry weight) (%)			
	Rutin	quercetin-3-glucoside	quercetin	total
Fresh	1583^a (78)	275^a (14)	165^a (8)	2023^a
Fermented	1070^a (53)	794^b (39)	152^a (8)	2016^a for 2 Ws
Fermented	182^b (10)	293^a (15)	1439^b (75)	914^a for 4 Ws

rutin = quercetin-3-diglucoside, Qmg = quercetin-4'-monoglucoside;

Data indicate is means of triplicate trials;

Within the same column, means with no letter in common are significantly different (p<0.05).

Rutin (Quercetin-3-*O*-β-rutinoside) has a β-glucosidic linkage at the 3-position in the A ring of the corresponding quercetin aglycons. β-Glucosidase (1,4-β-D-glucosidase glucohydrolase; EC 3.2.1.21) from *Lactobacillus casei* tends to hydrolyze compounds containing β-glucosidic linkages by splitting off the terminal β-glucose residue (*17*). High levels of β-glucosidase activity have been shown in wheat soon after the germination stage (*18*). Significant 6-Br-2-naphthyl-β-D-glucopyranoside hydrolysis activity in some strains of *Lactobacillus plantarum* (*19*) and *Leuconostoc mensenteroides* (*20*) have been detected. *Lactobacillus plantarum* seemed to cleave glycosides by a two-step (sequential) reaction (*21*). The conversion of rutin to quercetin-3'-monoglucoside and subsequently quercetin aglycone could be attributed to β-glucosidase activity either from *Toona sinensis* leaves or the dominant lactic acid bacteria, *Lactobaillus plantarum* or *Leuconostoc mensenteroides* in the fermented silage.

The extent of antioxidant potential *in vivo* for quercetin derivatives is dependent on metabolism, absorption and excretion within the body after ingestion. The bound sugar moiety on *O*-glycoside of flavonoids is known to influence their bioavalability (*22*). In a rat model system, four hours after the beginning of a meal, the quercetin metabolites (β-glucuronidase/sulfatase hydrolysable) present in plasma are highest when quercetin (20 mg) was supplied as quercetin 3-glucoside (33.2 μM) as compared with pure quercetin (11.7 μM) and rutin (about 3 μM). These data suggest that 3-*O*-glucosylation improves the absorption of quercetin in the small intestine, whereas the binding of a rhamnose or of a glucose-rhamnose moiety to the aglycone markedly depressed its absorption (*23*).

Although quercetin and rutin showed a significant linoleic acid peroxidation and DPPH scavenging capacity, it is not a quercetin aglycone but quercetin glucuronides that may participate in the antioxidant defense in blood circulation when quercetin-containing foods are consumed (*24*). Quercetin was transformed into circulating glucurono-sulfo conjugates of isorhamnetin, quercetin and tamarixetin and exhibited a total antioxidant status markedly higher than that of control rats (*25*). Quercetin is recovered in human plasma as conjugated derivatives in healthy volunteers after consumption of a meal rich in plant products. Quercetin glucuronide significantly delay the Cu^{2+}-induced oxidation of human LDL (*26*). Moon *et al.* (*27*) reported that quercetin 3-β-D-glucuronide is the metabolite in rat plasma after oral administration of quercetin that possesses antioxidative activity. Thus the absorption and transformation rate of quercetins in fermented *Toona sinensis* leaf were estimated using a rat model.

Thirty six male Spragure-Dawley rats, 220~240 g, were obtained from the National Institute of Experimental Animal, Taiwan. The animals were randomly divided into 6 groups of 6 rats. All groups were maintained in an animal facility at 25 °C under 12 h dark/light cycling, with free access to tap water and modified MF diet: starch, 53.0%; casein, 24.1%; cellulose powder, 4.2%; mineral mixture, 7.1%; vitamin mixture, 2.1%; oil (beef tallow), 4.6%; sucrose, 5.0% for 14 d before the experiment. Authentic quercetin and rutin (0.2% respectively in diet) were dissolved in 2.0 mL of propylene glycol and then mixed with the modified MF diet. About thirty grams of dried Toona leaves (unfermented, fermented for 2 weeks and fermented for 4 weeks, respectively) were mixed with modified MF diet to obtain a diet with quercetin equivalent concentration around 0.2%. Control (n=6) were administered only modified MF diet. No side effects were observed during the experimental time period. Each rat from all groups was anesthetized with diethyl ether 30 min after administration. Then the abdomen wall was opened and blood was collected from abdominal aorta into heparinized tubes. Plasma was obtained by centrifugation at 4 °C (100 x g, 20 min) and stored at -80 °C until use.

Eight hours after ingestion of quercetin containing diets, rats were sacrificed, plasmas were acidified (to pH 4.5) with 0.1 volume of 0.58 mg/L acetic acid solution. Solutions were treated for 30 min at 37 °C in the presence of β-glucuronidase/sulfatase. As expected, in rat plasma, quercetin, isorhamnetin, and minor amounts of tamarixetin were observed. Fresh Toona leaves showed a very low absorption, similar to that of authentic rutin, as described by Morand *et al* (*23*). Toona leaf containing diet (after 4 weeks fermentation) showed higher absorption, parallel to that of authentic quercetin. However, for the diet containing Toona leaf fermented for 2 weeks, the absorption rate of quercetins was found to be doubled the level for Toona leaf fermented for 4 weeks, as shown in Figure 3. The result indicates that after a Toona leaf meal, quercetin 3-β-D-glucoside was absorbed and conjugated into quercetin 3-β-D-glucuronide. This is in good agreement with the report of Spencer *et al* (*28*). The first metabolic step of quercetin is to be glucuronidation occurring in intestinal mucosa and the next step is sulfation in liver and methylation in liver and kidney.

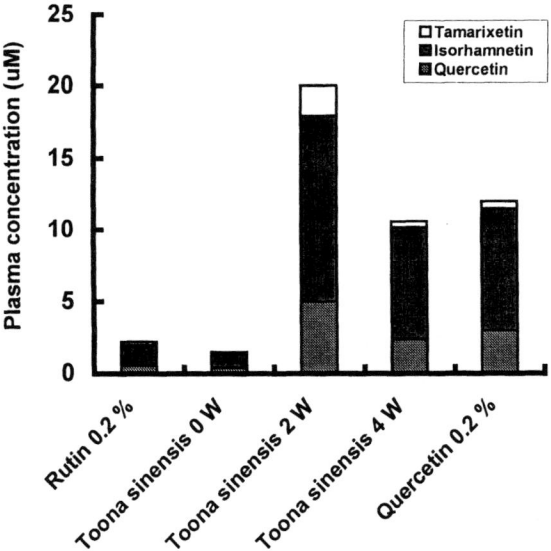

Figure 3. Quercetin and quercetin derivative concentrations in rat plasma 8 hours after ingestion of an experimental meal.

Another rat model showed that total plasma quercetin level increased from 13 µM (4 h) to 25 µM (12 h) and declined slightly as shown in Figure 4. Quercetin 3-β-D-glucuronide and other glucuronides seem to be formed from dietary quercetin in the form of glucosides by metabolic conversion in intestinal absorptive cells and remained as circulating metabolites in plasma for at least 12 hours. The antioxidative properties and biological activities of quercetins have been studied extensively *in vitro*. Their biological activities after oral administration of quercetin containing diets, are determined by availability and biotransformation. The demonstration of LDL antioxidative activities of quercetin glucuronides increases the effectiveness of quercetin as a chemopreventive agent. The maintenance of adequate levels of quercetin glucuronides/sulfates may be required for optimal antioxidative status throughout the life span. Further study of the antioxidative effect of quercetin glucuronides in plasma against LDL oxidation is needed to clarify the role of dietary rutin.

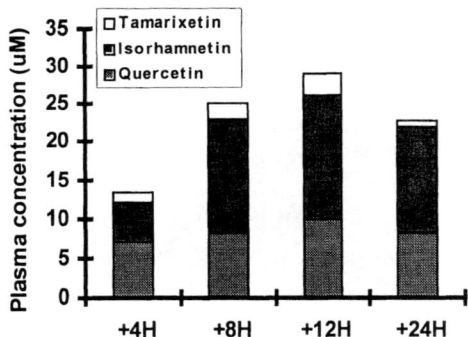

Figure 4. Evolution of the plasma concentration of quercetin metabolites in rats receiving a single meal containing dried fermented Toona sinensis leaves with 0.2 % quercetin equivalents.

References

1. Lee, T.B. Illustrated Flora of Korea. Hyangmunsa, Seoul, 1995, p. 506.
2. Park, J.C.; Yu, Y.B.; Lee, J.H. *Yakhak Hoechi* **1993**, *37*, 306-310.
3. Chang, H.C.; Hung, W.C.; Huang, M.S.; Hsu, H.K. *Am. J. Chin. Med.* **2001**, *3*, 307-314.
4. Luo, X.D.; Wu, S.H.; Ma, Y.B.; Wu, D.G. *Fitoterapia* **2000**, *71*, 492-496.
5. Park, J.C.; Yu, Y.B.; Lee, J.H.; Kim, N.J. Han'guk Yonguang Siklyong Hakhoechi. **1994**, *23*, 671-674.

6. Park, J.C.; Yu, Y.B.; Lee, J,H.; Choi, J.S.; Ok, K.D. *Kor. J. Phamacogn.* **1996**, *27*, 219-223.
7. Bochner B. *Nature* **1989**, *339*, 157-158.
8. Gardner, N.J.; Savard, T.; Obermeier, P.; Caldwell, G.; Champagne, C.P. *Int. J. Food Microbiol.* **2001**, *64*, 261-275.
9. Weinberg, Z.G.; Ashbell, G.; Hen, Y.; Azrieli, A.; Szakacs, G.; Filya, I. *J. Indus. Microbiol. Biotech.* **2002**, *28*, 7-11.
10. Weinberg, Z.G.; Muck, R.E. *FEMS Microbiol. Reviews* **1996**, *19*, 53-68.
11. McDonald, L.C.; Fleming, H.P.; Hassan, H.M. *Appl. Environ. Microbiol.* **1990**, *56*, 2120-2124.
12. Font de Valdez, G.; de Giori, G.S.; Garro, M.; Mozzi, F.; Oliver, G. *Microbiol. Alim. Nutr.* **1990**, *8*, 175-179.
13. Spoelstra, S.F. In: *Proceedings of a Conference on Forage Conservation Towards 2000*. Ed. By Pahlow, G and Honig, H., Braunschweig, Germany, 1991, pp. 48-70.
14. Choi, C.W.; Kim, S.C.; Hwan, S.S.; Choi, B.K.; Ahn, H.J.; Lee, M.Y.; Park, S.H. *Plant Sci.* **2002**, *163*, 1161-1168.
15. Rice-Evans, C.; Miller, N.J.; Paganga, G. *Free Radic. Biol. Med.* **1996**, *20*, 933-956.
16. Rhodes, M.J.C.; Price, K.R. *Food Chem.* **1996**, *57*, 113-117.
17. Coulon, S.; Chemardin, P.; Gueguen, Y.; Arnaud, A; Galzy, P. *Biochem. Biotech.* **1998**, *74*, 105-114.
18. Sue, M.; Ishibara, A.; Iwamura, H. *Planta* **2000**, *210*, 432-438.
19. Amoa-Awua, W.K.A.; Appoh, F.E.; Jakobsen, M. *Int. J. Food Microbiol.* **1996**, *31*, 87-98.
20. Gueguen, Y.; Chemardin, P.; Labrot, P.; Arnaud, A.; Galzy, P. *J. Appl. Microbiol.* **1997**, *82*, 469-476.
21. Lei, V.; Amoa-Awua, K.W.A.; Brimer, L. *Int. J. Food Microbiol.* **1999**, *53*, 169-184.
22. Andlauer, W.; Stumpf, C.; Fürst, P. *Biochem. Pharm.* **2001**, *62*, 369-374.
23. Morand, C.; Manach, C.; Crespy, V.; Remesy, C. *BioFactors* **2000**, *12*, 169-174.
24. Yamamoto, N.; Moon, J.H.; Tsushida, T.; Nagao, A.; Terao, J. *Arch. Biochem. Biophy.* **1999**, *372*, 347-354.
25. Morand, C.; Crespy, V.; Manach, C.; Besson, C.; Demigne, C.; Remesy, C. *Am. J. Physiol.* **1998**, *275*, R212-219.
26. Gee, J.M.; DuPont, M.S.; Rhode, M.J.C.; Johnson, I.T. *Free Radic. Biol. Med.* **1998**, *25*, 12-25.
27. Moon, J.H.; Tsushida, T.; Nakahara, K.; Terao, J. *Free Radic. Biol. Med.* **2001**, *30*, 1274-1285.
28. Spencer, J.P.E.; Chowrimootoo, G.; Choudhury, P.; Debnam, E.S.; Srai, S.K.; Rice-Evans, C. *FEBS Letters* **1998**, *458*, 224-230.

Chapter 6

Antioxidant Capacity of Phenolics from Canola Hulls as Affected by Different Solvents

M. Naczk[1], R. Amarowicz[2], R. Zadernowski[3], and F. Shahidi[4]

[1]Department of Human Nutrition, St. Francis Xavier University, Antigonish, P.O. Box 5000, Nova Scotia B2G 2W5, Canada
[2]Division of Food Science, Institute of Animal Reproduction and Food Research of Polish Academy of Sciences, Olsztyn, Poland
[3]Faculty of Food Science, Warminsko-Mazurski University, Olsztyn, Poland
[4]Division of Biochemistry, Memorial University of Newfoundland, St. John's, Newfoundland A1B 3X9, Canada

Potential use of canola hulls as a source of natural antioxidants was explored. The antioxidant activity of 80-100% (v/v) methanol and 70-80% (v/v) acetone extracts of canola hulls in a β-carotene-linoleate model system was comparable to that displayed by butylated hydroxyanisole (BHA). The scavenging effect of the above extracts, at 40 μg/assay on (DPPH) radical, was over 95%.

Rapeseed species include *Brassica napus, Brassica campestris,* and *Brassica junce,* commonly known as rape, turnip rape, and leaf mustard, respectively. In Canada and Europe the seeds of rape and turnip rape are known as rapeseed. Traditional rapeseed varieties contained from 22 to 66% erucic acid in their oil. In 1979, the name canola was adopted in Canada for genetically modified rapeseed varieties containing less than 2% erucic acid in their oil and no more than 30 μmoles/g of one or any combination of the four aliphatic glucosinolates (gluconapin, progoitrin, glucobrassicanapin, and napoleiferin) in its defatted meal. In 1997, this definition was revised to less than 1% of erucic acid in the oil and 18 μmoles/g of one or any combination of the above four aliphatic glucosinolates. The US Food and Drug Administration recognized rapeseed and canola as different species in 1985 (*1*).

© 2005 American Chemical Society

Advances in dehulling of canola/rapeseed (*2-4*) may, in the near future, bring about the introduction of dehulling to the canola/rapeseed industry. The seeds of canola/rapeseed contain 14-18% hulls. Hulls may contain up 20% oil, 19.1% crude proteins [Nx6.25], 4.4% minerals and 48% dietary fiber (Table I). Other constituents include simple sugars and oligosaccharides, polyphenolics, phytates and residual polar lipids (*5*). Phenolic acids and their derivatives as well as soluble and insoluble condensed tannins are the predominant phenolic compounds found in canola and rapeseed. Canola and rapeseed hulls have been reported to contain up to 1000 mg phenolic acids, from 89 to 1847 mg soluble condensed tannins and between 1913 and 6213 mg insoluble condensed tannins per 100g sample (*6-8*). Therefore, the use of hulls, after dehulling, as a potential source of natural antioxidants may be of interest.

Table I. Chemical Composition of Seeds, Cotyledons and Hulls from Excel Canola Variety

Seed Fraction	Yield	Oil	Protein [Nx6.25]	Ash	Dietary Fiber	Total Analyzed
Whole seed	100	42.1	26.8	3.8	17.3	90.0
Cotyledon	74	46.3	28.3	3.7	11.3	89.6
Mixed fines	12	41.7	26.7	3.8	18.0	90.2
Hulls	14	14.0	19.1	4.4	48.0	91.6

NOTE: Units are percent dry weight.
SOURCE: Adapted from ref. 8.

Alcoholic extracts of rapeseed meal (*9*) and rapeseed cakes (*10*) exhibited strong antioxidant activity in a β-carotene-linoleate model system. Wanasundara and Shahidi (*11*) reported that the antioxidant activity of ethanolic extracts of canola meal in canola oil was equivalent to that of TBHQ and stronger than that of BHA, BHT and BHA/BHT/ monoacylglycerol citrate (MGC). The compound 1-*O*-β-D-glucopyranosyl sinapate was the most active phenolic component of these extracts (*12*). Recently, Amarowicz et al. (*13*) reported that crude tannin extracts isolated from high-tannin canola hulls exerted significantly (P≤0.025) greater scavenging activity against DPPH radical than those from low-tannin rapeseed hulls. Crude tannins extracts of canola hulls contained 10-40 times more condensed tannins than those of rapeseed hulls. Therefore, synergism of phenolics with one another and/or other components present in the extract may be responsible for a stronger free radical scavenging activity of crude tannins extracts isolated from high-tannin canola hulls. Later, Amarowicz *et al.* (*14*) isolated five major phenolic fractions from non-tannin canola hull phenolics using Sephadex LH-20 and 95% (v/v) ethanol as a mobile phase. Of these, only fractions I, III, and V exhibited marked scavenging effects for DPPH radical.

The objective of this study was to investigate the antioxidative potential of phenolics extracted from Cyclone canola hulls. Effects of different solvent systems on the extraction of phenolics were further investigated.

Materials and Methods

Cyclone canola hulls were prepared according to the procedure described by Sosulski and Zadernowski (2). Hulls were extracted with hexane for 12 h using a Soxhlet apparatus and then dried at room temperature. The defatted hulls were extracted twice at room temperature with 0, 80, 70, 50 and 30% (v/v) aqueous acetone or methanol using a Waring Blender (Waring Products Division, Dynamics Corporation of America, New Hartford, CT) for 2 min at maximum speed. The extracts were combined, evaporated to near dryness under vacuum at 40°C and then lyophilized.

The total content of phenolic compounds in the extracts was estimated using the Folin-Denis reagent (15) and expressed as sinapic acid equivalents. The content of tannins in the extracts was also determined using the modified vanillin assay (16) and expressed as mg condensed tannins per 1 g of extract. The protein precipitating capacity of phenolic extracts was determined as described by Naczk et al. (17). The protein precipitating potential of phenolics in the extracts was estimated using the protein precipitation assay of Hagerman and Butler (18) and the dye-labeled bovine serum albumine (BSA) assay of Asquith and Butler (19) as modified by Naczk et al. (20).

The antioxidant activity of phenolic extracts was determined using a β-carotene-linoleate model system (21). Methanolic solutions (0.2 mL) containing 1 mg of phenolic extract or 0.5 mg of butylated hydroxyanisole (BHA) were added to a series of tubes containing 5 mL of a prepared emulsion of linoelate and β-carotene stabilized by Tween 40. Immediately after the addition of emulsion to tubes, the zero-time absorbance at 470 nm was recorded. Samples were kept in a water bath at 50 °C and their absorbances recorded over a 120 min period at 15 min intervals.

Scavenging effect of phenolic extracts for α,α−diphenyl-β-picrylhydrazyl (DPPH) radical was monitored according to the method of Hatano et al. (22) as detailed by Chen et al. (23). A 0.1 mL of methanolic solution containing from 20 to 100 μg of phenolic extract was mixed with 2 mL of distilled water and then added to a methanolic solution of DPPH (1mM, 0.25 mL). The mixture was vortexed for 1 min, then left to stand for 30 min at room temperature and its absorbance read at 517 nm.

Statistical analysis of data was carried out using the SigmaStat v.3.0 (SSPS Science Inc., Chicago, IL). Each extract, for the purpose of statistical analysis, was referred to as treatment. The statistical analysis of all treatments was perfomed using the ANOVA test. In addition, the t-test was employed among treatments when a statistically significant difference (P≤0.05) was found using the ANOVA test.

The results presented in tables and figures are mean values of at least three experiments. Results for treatments followed by the same superscript letter in a graph legend are not significantly different (P>0.05; t-test).

Results and Discussion

The crude extracts of canola hull phenolics comprised from 5.6 to 13.4 % of the weight of the hulls. The yield of the extract was affected by the content of water in the solvent system used for extraction of phenolics from canola hulls (Table II).

Table II. The Yield (%) of Phenolic Extracts Isolated from Canola Hulls

Water Content [%, v/v]	Methanol	Acetone
70	13.4	13.2
50	10.4	16.1
30	7.7	7.8
20	6.5	5.6
0	7.5	-

The total content of phenolics in crude extracts was determined by the Folin-Denis assay and expressed as sinapic acid equivalent per gram of extract, as this reagent is sensitive to all classes of phenolics. On the other hand, the content of condensed tannins was estimated by the vanillin and proanthocyadin methods that are commonly used for their determination. The proportions of total phenolics and condensed tannins in crude extracts are shown in Tables III and IV, respectively. The acetone-water solvent systems were more efficient solvents for extraction of phenolics than corresponding methanol-water systems. Of these, 70 and 80% (v/v) acetone was the most effective solvent for the extraction of canola hull phenolics.

A number of methods have been proposed for estimation of protein precipitating potential of plant phenolic extracts. Of these assays, the dye-labeled BSA assay developed by Asquith and Butler (*19*) and the protein precipitation assay developed by Hagerman and Butler (*18*) were chosen for determination of protein precipitation potential of crude phenolic extracts isolated from canola hulls. The protein precipitation assay allows for the estimation of the amount of phenolics complexing with proteins, while the dye-labeled BSA assay measures the amount of protein precipitated by phenolics.

The protein precipitation potential of phenolic extracts from canola hulls was expressed as the slopes of lines (titration curves) depicting the amount of phenolic-protein complex precipitated as a function of the amount of extract added to the reaction mixture. According to Naczk et al. (20) the slope of the titration curve is a more meaningful measure of protein-precipitating potential of plant phenolics extracts than the measurement carried out at one nonstandardized phenolic/protein ratio (18, 19). The numerical values of the slope of titration curves are shown in Table V. Only 70 and 80% (v/v) acetone extracts of canola hulls phenolics exhibited strong affinities for protein. According to Porter and Woodruffe (24) the ability of phenolics to precipitate proteins is influenced by the degree of their polymerization. The chemical structures of canola hull phenolics are still unknown and more detailed analyses of these compounds are needed.

The crude extracts from plant materials are complex mixtures of phenolics with different molecular structures. It is well established that the antioxidant potential phenolics is strongly influenced by their molecular structures (25, 26). Phenols with a second hydroxyl group located in the *ortho* and *para* positions display stronger antioxidant activity than those with a second hydroxyl group in the *meta* position (27, 28).

The DPPH and β-carotene-linoleate assays are commonly used for determination of antioxidant potentials of crude extracts of plant phenolics. The DPPH assay used in this study measures the change in the absorbance at 517 nm after 30 min of reaction at room temperature. The absorption at this characteristic wavelength disappears as the reaction between the antioxidant molecules and radicals progresses. The ratio of the decrease in the absorbance of DPPH radical solution at this wavelength to that in the absence of phenolics at 517 nm is a measure of the radical scavenging activity of the antioxidant employed (29). This procedure was modified by Brand-Williams et al. (25) to take into account the different kinetic behavior of antioxidants. However, the modified methodology is not suitable for evaluation of scavenging activity of crude phenolics extracts, because knowledge of molecular structures of compounds involved is essential.

Figure 1 shows the scavenging activity of canola hulls extracts on DPPH radicals. The acetone extracts of canola hulls displayed significantly ($P<0.05$) greater scavenging effect of DPPH than methanol extracts. The scavenging effect of acetone extracts from canola hulls, at a level 0.3 mg, was 20 and 94% for 30 and 80% (v/v) acetone extracts, respectively. On the other hand the scavenging effect of 30 and 80% (v/v) methanolic extracts from canola hulls, at similar levels, was only 7 and 53%, respectively. It should also be noted that 100% methanolic extract of canola hulls displayed scavenging effect of 72 and 94%, at levels of 0.3 and 0.5 mg, respectively. The scavenging effect of 70 and 80% canola hulls extracts on DPPH was similar to that reported for epigallocatechin (EGC) and theaflavin monogallate (23), but greater than that found for epigallocatechin gallate (EGCG), at a level of 10 μM, (23) and pouching tea extract, at a dose of 2 mg (30).

Figure 2 shows the effect of phenolic extracts of canola hulls on the coupled

Table III. Total Phenolic Content in Canola Hull Extracts[1]

Water Content [%, v/v]	Methanol	Acetone
70	15.1±0.1	21.2±0.2
50	23.9±0.2	35.7±0.3
30	36.5±0.6	94.3±3.0
20	40.4±0.2	136.2±2.7
0	61.9±0.8	-

[1] Units are mg sinapic acid equivalents per gram of extract.

Table IV. Total Content of Condensed Tannins in Canola Hull Extracts[1]

Water Content [%, v/v]	Methanol		Acetone	
	VAN	PROANTH	VAN	PROANTH
70	3.5±0.6	16.8±1.0	10.8±0.1	32.9±2.0
50	4.4±0.2	25.0±0.6	40.1±1.4	95.3±2.0
30	6.9±0.2	41.4±0.9	127.0±2.5	224.4±9.2
20	6.0±0.2	45.6±1.6	238.0±5.3	296.9±8.6
0	79.8±0.3	169.0±8.6	-	-

1 VAN, Vanillin assay; units are in mg canola tannins per gram of extracts; and PROANTH, proanthocyanidin assay; units are in absorbance units per gram of Extract.

Table V. Protein Precipitating Capacity of Canola Hulls Extracts[1]

Water Content [%, v/v]	Methanol		Acetone	
	PP	DLP	PP	DLP
70	ND	0.70	ND	ND
50	ND	2.40	0.10	3.44
30	0.03	3.04	0.37	9.54
20	0.06	4.32	0.82	9.66

[1] PP, protein precipitation assay; and DLP, dye-labeled assay; ND, not detected.

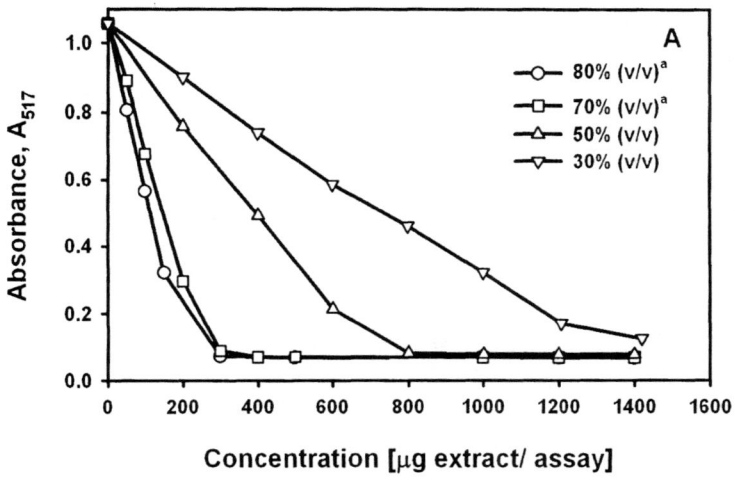
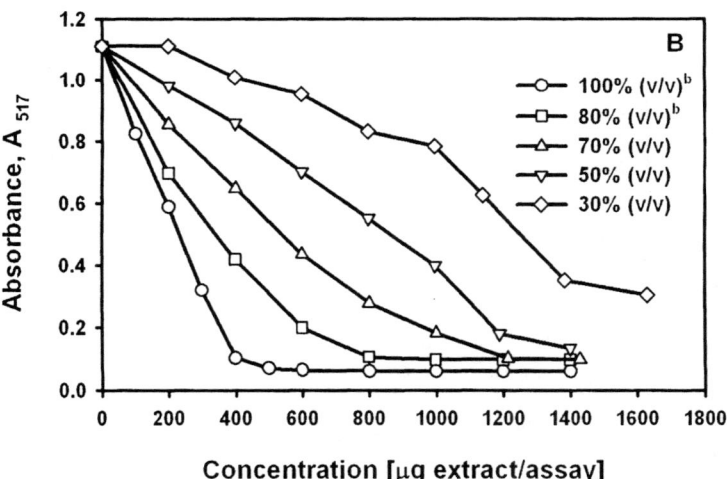

Figure 1. DPPH radical scavenging of canola hull phenolics. (A) acetone-water extracts; (B) methanol-water extracts. Treatments followed by the same superscripts are not statistically different (P>0.05).

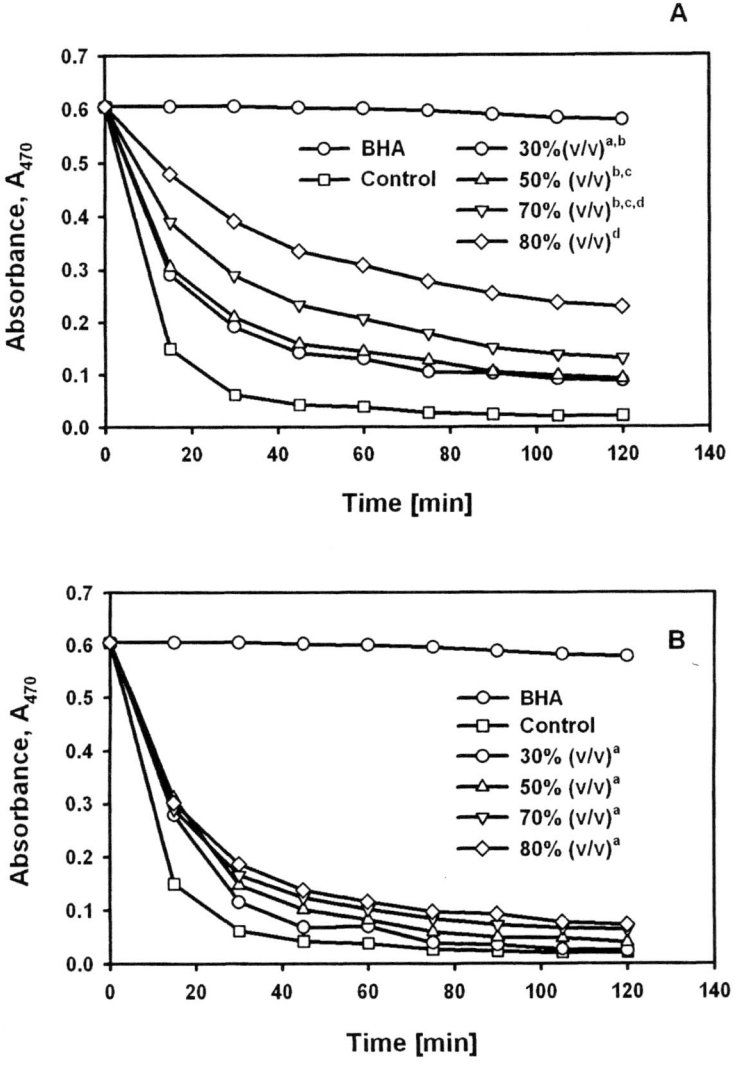

Figure 2. Antioxidant activity of canola hull phenolics in a β-carotene-linoleate model system, as measured by changes in the absorbance values at 470 nm. (A) acetone-water extracts; (B) methanol-water extracts. Treatments followed by the same superscripts are not statistically different (P>0.05).

oxidation of linoleic acid and β-carotene as compared to that of BHA. Acetone-water extracts of canola hull phenolics exhibited good antioxidant activity against the bleaching β-carotene after 120 min of reaction, similar to that reported by Wanasundara *et al.* (*11*) for fraction IV of canola meal phenolics after 30 min of reaction. On the other hand, methanolic extracts displayed a very poor antioxidant activity against the bleaching of β-carotene. However, the antioxidant activities of all canola hull extracts were lower than that of BHA in a β-carotene-linoleate model system.

The acetone extracts of canola hull phenolics exhibited stronger antioxidant properties than their corresponding methanolic extracts. This may be due to (i) the difference in the total phenolics and tannins contents (Table III), (ii) the diversity in molecular structures of potential phenolic antioxidants present in the extracts, and (iii) different kinetic behavior of potential phenolic antioxidants. Acetone extracts contained 3-40 times more condensed tannins than methanolic extracts. Therefore, tannins may contribute to a greater antioxidant activity exhibited by acetone extracts. More research is needed to isolate and identify the active phenolic components of canola hulls.

Acknowledgment

Marian Naczk thanks the Natural Sciences and Engineering Research Council (NSERC) of Canada for financial support in the form of a research grant.

References

1. *Antinutrients and Phytochemicals*; Shahidi, F., Ed.; American Chemical Society: Washington, DC, 1997, pp. 157-170.
2. Sosulski, F.W.; Zadernowski, R. *J. Am. Oil Chem. Soc.* **1981**, *58*, 96.
3. Greilsamer, B. *Proceedings of Sixth International Rapeseed Congress*, Paris, France, 1983, pp. 1496-1501.
4. Diosady, L.L.; Tar, C.G.; Rubin, L.J.; Naczk, M. *Acta Alimen.* **1986**, *16*, 167.
5. Naczk, M.; Nichols, T., Pink, D., Sosulski, F. *J. Agric. Food Chem.* **1994**, *42*, 2196.
6. Krygier, K.; Sosulski, F.; Hogge, L. *J. Agric. Food Chem.* **1982**, *30*, 334.
7. Naczk, M.; Nichols, T.; Pink, D.; Sosulski, F. *J. Agric. Food Chem.* **1994**, *42*, 2196.
8. Naczk, M.; Amarowicz, R.; Pink, D.; Shahidi, F. *J. Agric. Food Chem.* **2000**, 48, 1758.

9. Nowak, H.; Kujawa, K.; Zadernowski, R.; Roczniak, K. B.; Kozlowska, H. *Fett. Wiss. Technol.* **1994**, *94*, 149.
10. Amarowicz, R.; Fornal, J.; Karamac, M. *Grasas y Aceites* **1995**, *46*, 354.
11. Wanasundara, U.N.; Shahidi, F. *J. Am. Oil Chem. Soc.* **1994**, *71*, 817.
12. Wanasundara, U.N.; Amarowicz, R.; Shahidi, F. *J. Agric. Food Chem.* **1994**, *42*, 1285.
13. Amarowicz, R.; Naczk, M.; Shahidi, F. *J. Am. Oil Chem. Soc.* **2000**, *77*, 957.
14. Amarowicz, R.; Naczk, M.; Shahidi, F. *J. Agric. Food Chem.* **2000**, *48*, 2755.
15. Swain, T.; Hillis, W.E. *J. Sci. Food Agric.* **1959**, 10, 63.
16. Price, M.L.; Van Scoyoc, S.; Butler, L.G. *J. Agric. Food Chem.* **1978**, *26*, 1214.
17. Naczk, M.; Amarowicz, R.; Zadernowski, R.; Shahidi, F. *J. Am. Oil Chem. Soc.* **2001**, *78*, 1173.
18. Hagerman, A.E.; Butler, L.G. *J. Agric. Food Chem.* **1978**, *12*, 1243.
19. Asquith, T.N.; Butler, L.G. *J. Chem. Ecol.* **1985**, *11*, 1535.
20. Naczk, M.; Amarowicz, R.; Zadernowski, R.; Shahidi, F. *J. Am. Oil Chem. Soc.* **2001**, 1173.
21. Miler, H.E. *J. Am. Oil Chem. Soc. 1971*, *48,* 91.
22. Hatano, T.; Kagawa, H.; Yashura, T.; Okuda, T. *Chem. Pharm. Bull.* **1988**, *36*, 2090.
23. Chen, C-W.; Ho, C-T. *J. Food Lipids* **1995**, *2*, 46.
24. Porter, L.J.; Woodruffe, J. *Phytochemistry* **1984**, *25*, 1255.
25. Brand-Williams, W.; Cuvelier, M.E.; Berset, C. *Food Sci. Technol.* **1995**, *28*, 25.
26. Bondet, V.; Brand-Williams, W.; Berset, C. *Food Sci. Technol.* **1997**, *30*, 609.
27. Rice-Evans, C.A.; Miller, N.J.; Paganga, G. *Free Rad. Biol. Med.* **1996**, *20*, 933.
28. Cuvelier, M.E.; Richard, H.; Besret, C. *Biosci. Biotechnol. Biochem.* **1992**, *56*, 324.
29. Yoshida, T.; Mori, K.; Hatano, T.; Okumara, T.; Vehara, T.; Komagoe, K.; Fujita, Y.; Okuda, T. *Chem. Pharm. Bull.* **1989**, *37*, 1919.
30. Yen, C.-G.; Chen, H.-Y. *J. Agric. Food Chem.* **1995**, *43*, 27.

Chapter 7

Antioxidant Activity of Polyphenolics from a Bearberry-Leaf (*Arctostaphylos uva-ursi* L. Sprengel) Extract in Meat Systems

Ronald B. Pegg[1], Ryszard Amarowicz[2], and Marian Naczk[3]

[1]Department of Applied Microbiology and Food Science, University of Saskatchewan, College of Agriculture, 51 Campus Drive, Saskatoon, Saskatchewan S7N 5A8, Canada
[2]Division of Food Science, Institute of Animal Reproduction and Food Research, Polish Academy of Sciences, 10–747 Olsztyn, ul. Tuwima 10, Poland
[3]Department of Human Nutrition, St. Francis Xavier University, Antigonish, Nova Scotia B2G 2W5, Canada

Bearberry-leaf (*Arctostaphylos uva-ursi* L. Sprengel) extract possesses marked antioxidant activity in model and meat systems. A crude ethanolic extract of bearberry leaves was dechlorophyllized using a silicic acid column, and then fractionated by Sephadex LH-20 column chromatography using ethanol (95%, v/v) and acetone (50%, v/v) as the mobile phases. According to a mass balance, the ethanol fraction comprised 79.2% of the starting material, while the acetone fraction consisted of 9.7%. The content of total phenolics for the fractions and subfractions ranged from 2135 to 9110 Abs_{725} units/g extract. Even though the acetone fraction was only *ca.* 10% of the crude extract, its vanillin response was five times greater than that of the ethanol fraction. According to a mass balance of this fraction, *ca.* 50% of the polyphenolics remained in the aqueous phase when partitioned between water and ethyl acetate.

Results showed that the crude bearberry-leaf extract, and its fractions inhibited TBARS formation in cooked meat systems after seven days of refrigerated storage by 97.0, 49.1 and 100%, respectively, when added at a 200-ppm concentration. The ethanol fraction exhibited a classical dose response: when incorporated in meat systems at levels of 200- and 500-ppm before thermal processing, TBARS development was inhibited by 49.7 and 93.9%, respectively, after seven days. This [ethanol] fraction was further subdivided into vanillin-positive constituents; the vanillin-positive fraction possessed weak antioxidant activity in meat model systems, and in some cases demonstrated a slight pro-oxidant effect. The acetone fraction also showed a classical dose response. When added to meat systems at levels of 25-, 50- and 100-ppm, TBARS formation was inhibited by 36.7, 91.4 and 100%, respectively, after seven days of refrigerated storage. It was interesting to note that a subfraction from the acetone product, which was soluble in ethyl acetate but did not have a positive reaction with vanillin, imparted strong antioxidant activity in meat systems. Therefore, vanillin-positive reaction constituents (*i.e.*, condensed tannins) are not solely responsible for the antioxidant activity observed from the bearberry-leaf extract.

Spices and herbs are not only used to flavor foods, but they also offer preservation against microbiological and oxidative degradation. Constituents in spices and herbs have antioxidant activity and can retard rancidity resulting from lipid oxidation and thereby help to preserve the wholesomeness of food products. Butylated hydroxyanisole (BHA), butylated hydroxytoluene (BHT), propyl gallate (PG) and *tert*-butylhydroquinone (TBHQ) are examples of some common synthetic antioxidants used by food processors in their formulations. Yet, concerns have been raised over the safety of some of these food additives (*1*). Moreover, the concept of synthetic and chemical does not fit in with the consumers' growing interest toward using constituents that are either natural or organic in food products. There are a number of natural antioxidants added to foods, some of which are depicted in Figure 1. Ascorbic acid and α-tocopherol (*i.e.*, vitamin C and E) are the best-known water- and lipid-soluble antioxidants, respectively; they are added to food products as well as being found naturally in them. In addition, carotenoids, simple phenolic acids and phenolics from the flavonoid family, such as rutin (*i.e.*, a flavone), naringin (*i.e.*, a flavonone), kaempferol (*i.e.*, a flavonol), genistein (*i.e.*, an isoflavone) and (+)-catechin (*i.e.*, a flavanol) are routinely found in the foods we eat.

Researchers continue to search for novel sources of natural antioxidants from plants for use in food systems. Amarowicz *et al.* (*2,3*) screened selected plant

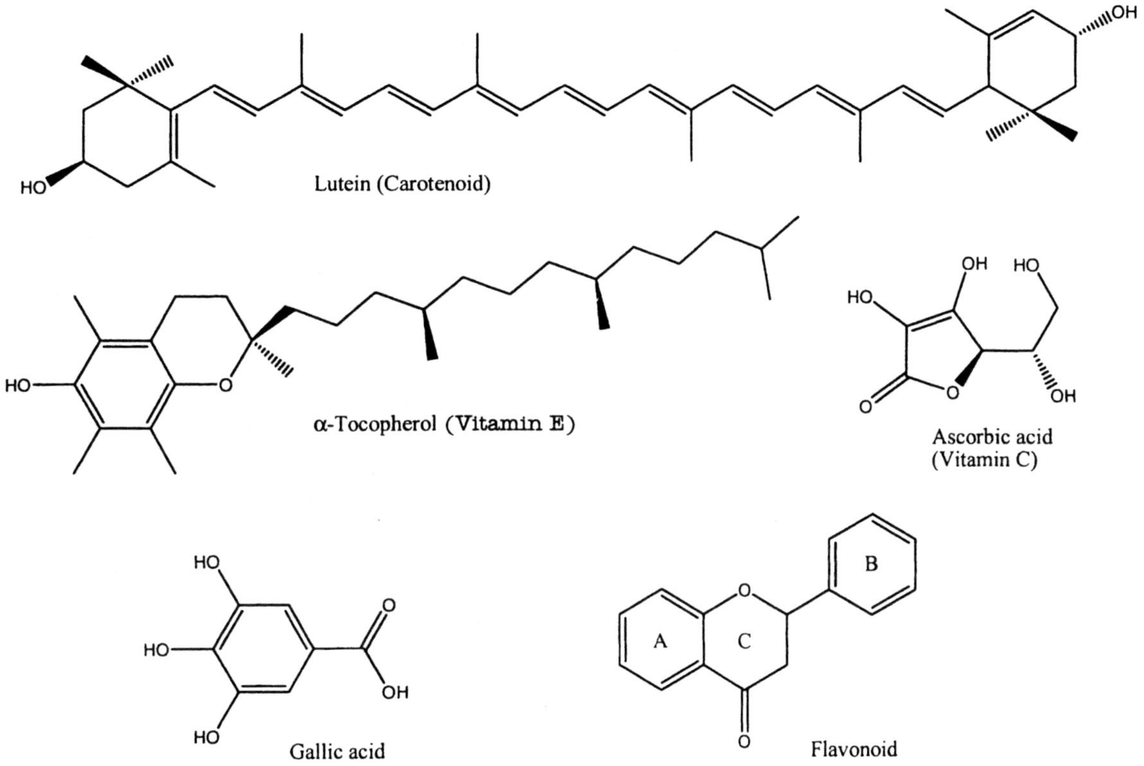

Figure 1. Some typical natural antioxidants present in and added to food products.

species for their phenolic content, as it was realized that polyphenolics can function as strong antioxidants. Emphasis was placed, but not limited to, species thriving on the Canadian Prairies. From this research, it was determined that a crude ethanol extract from bearberry (*Arctostaphylos uva-ursi* L. Sprengel) leaves possessed marked antioxidant activity in both model and food systems. Bearberry is one of the most ubiquitous herbs on the Canadian Prairies. It comprises arbutin (5-15%), variable amounts of methyl arbutin (up to 4%) and small quantities of the free aglycones. Other constituents include gallic acid, *p*-coumaric acid, syringic acid, galloylarbutin and up to 20% gallotannins and catechol types, as well as some flavonoids, notably glycosides of quercetin, kaempferol and myricetin.

The antioxidant activity of the crude bearberry-leaf extract was assessed by a number of chemical assays including a β-carotene-linoleic acid (linoleate) model system, reducing power, scavenging effect of the DPPH• free radical, liposome model system, scavenging capacity of hydroxyl free radicals (HO•) by electron paramagnetic resonance spectroscopy and ability to curb lipid oxidation in meat model systems. Some results from these tests are reported by Amarowicz *et al.* (*3*). Amarowicz and Pegg (*4*) fractionated the crude bearberry-leaf extract and found that the tannin constituents play a significant role toward the antioxidant activity observed in both model and food systems.

Tannins are complex secondary metabolites of plants, which are widely distributed in foods and feeds. They comprise gallic acid esters or flavan-3-ol polymers and have molecular weights typically in the range of 500 to 3,000 Da (*5*), although the existence of larger tannins has been reported. Tannins can form soluble and insoluble complexes with proteins; the formation of these depends not only on the size, conformation and charge of the protein molecules, but also the molecular weight, length and flexibility of the tannins involved (*6*). The precipitation of a protein–tannin complex results from the formation of a sufficiently hydrophobic surface on the complex. At low protein concentrations, precipitation is due to the formation of a hydrophobic monolayer of polyphenols on the protein surface. At higher concentrations, however, a hydrophobic surface results from the combination of complexing of polyphenols on the protein surface and crosslinking of different protein molecules with polyphenols (*7*). Tannins bound to proteins have been shown to retain their antioxidant activity and may provide persistent antioxidant activity in the gastrointestinal tract when consumed (*8*).

The purpose of this study is to investigate further the tannin constituents fractionated from the bearberry-leaf extract as well as the antioxidant activity they can afford to a meat model system.

Materials and Methods

Preparation of Bearberry-Leaf Extract

Dried bearberry leaves were ground in a coffee mill (Moulinex Corporation, Toronto, ON). Prepared material was transferred to dark-colored flasks, mixed with 95% (v/v) ethanol at a material-to-solvent ratio of 15:100 (m/v) and placed in a shaking Magni Whirl constant-temperature bath (Blue M Electric Company, Model MSG-1122A-1, Blue Island, IL) at 50°C for 30 min. Afterwards, the slurry was filtered through Whatman No. 1 filter paper and the residue was re-extracted twice more. Combined supernatants were evaporated to dryness under vacuum at < 40°C using a Büchi Rotavapor/Water bath (Models EL 131 and 461, respectively, Brinkmann Instruments [Canada] Ltd., Mississauga, ON). The crude preparation was then dechlorophyllized according to Pegg *et al.* (*9*) on a silicic acid column using hexanes and 95% (v/v) ethanol as mobile phases. Dried extracts were stored at 4°C in air until further analyzed.

Fractionation of the Crude Bearberry-Leaf Extract

Approximately 4 g of the crude dechlorophyllized bearberry-leaf extract was suspended in 20 mL of 95% (v/v) ethanol and then applied to a chromatographic column (4.5 × 18 cm) packed with Sephadex LH-20 which had been equilibrated with ethanol. The phenolics on the column were exhaustively "washed" with ethanol at a flow rate of 400 mL/h and then eluted with 50% (v/v) acetone at a flow rate of 300 mL/h. Eluates from each solvent system were pooled and the organic solvent removed under vacuum at < 40°C using the Rotavapor. The collected tannins from the acetone fraction were lyophilized to remove residual water.

One hundred milligrams of each fraction was dissolved in 25-mL water and transferred to a 100-mL separatory funnel. Twenty-five milliliters of ethyl acetate were added, the contents shaken and phases left to separate for *ca.* 20 min. Three liquid phases were evident: a clear, upper ethyl acetate layer, a turbid middle layer, and a bottom water layer. The ethyl acetate layer was collected by pipette and labeled as either EtOH Fr. II or Acetone Fr. II, while the turbid middle layer was carefully removed from the separatory funnel and labeled as either EtOH Fr. III or Acetone Fr. III. Another 25-mL portion of ethyl acetate was added to the funnel and the extraction/collection procedure was repeated nine more times. Solvents from each fraction were removed using the Rotavapor.

Total Phenolics in the Bearberry-Leaf Extract and its Fractions

The total phenolics' content in the bearberry-leaf extract and its fractions was estimated by a colorimetric assay based on procedures described by Swain and Hillis (*10*) and Naczk and Shahidi (*11*). Briefly, a 0.5-mL aliquot of plant material dissolved in methanol was pipetted into a test tube containing 8 mL of distilled water. After mixing the contents, 0.5 mL of the Folin & Ciocalteu's phenol reagent and 1 mL of a saturated solution of sodium carbonate were added. The contents were vortexed for 15 s and then left to stand at room temperature for 30 min. Absorbance measurements were recorded at 725 nm using a Milton Roy Spectronic Genesis 5 spectrophotometer (Fisher Scientific Co., Nepean, ON). Estimation of the phenolic compounds was carried out in triplicate; the results were averaged and expressed as Abs_{725} units/g extract or fraction.

Tannin Contents in the Bearberry-Leaf Extract and its Fractions

The content of condensed tannins in extracts was determined colorimetrically using the modified vanillin assay (*12*) and the proanthocyanidin assay (*13*). For the vanillin assay, either 5 mL of 0.5% (w/v) vanillin reagent prepared with 4% concentrated HCl in methanol (sample) or 5 mL of 4% concentrated HCl in methanol (blank) was added to 1 mL of methanolic solution of the bearberry-leaf extract or its fractions and then mixed well. After standing for 20 min in the dark at room temperature, absorbance measurements of the blank and samples were recorded at 500 nm. The absorbance of the blank was subtracted from that of the sample to account for potential interference from pre-existing chromophores. (+)-Catechin was used as the standard; from the curve constructed the results were expressed as the content of tannins in mg of (+)-catechin equivalents per g extract or fraction. For the proanthocyanidin assay, 1 mL of methanolic solution of the bearberry-leaf extract or its fractions was added to 10 mL of *n*-butanol-HCl reagent (*i.e.*, 0.7 g of $FeSO_4 \cdot 7H_2O$ was dissolved in 25 mL of concentrated HCl containing a small volume of *n*-butanol; the solution was then made up to 1 L with *n*-butanol) and mixed well. The mixture was heated in sealed ampules for 2 h in a boiling water bath and then allowed to cool to room temperature. The absorbance of the solution was measured at 550 nm against a reagent-only blank. Results were expressed as absorbance units per g extract or fraction (Abs_{550}/g).

Protein–Tannin Complex Formation Determinations

The effect of tannins at various concentrations resulting in the formation of an insoluble protein–tannin complex was assayed by the protein precipitation method of Hagerman and Butler (*14*) with a pH modification as described by Naczk *et al.*

(*15*). A series of methanolic solutions of the bearberry-leaf extract or its fractions were prepared at concentrations ranging from 0.5 to 3 mg/mL. From the series, 1 mL of each was taken and mixed with 2 mL of a standard bovine serum albumin (BSA, Fraction V, Sigma-Aldrich Canada, Ltd., Oakville, ON) solution (*i.e.*, 1 mg of protein/mL in 0.2 M acetate buffer, pH 4.0, and containing 0.17 M sodium chloride). After 15 min of standing at room temperature, the solution was centrifuged for 15 min at maximum speed using an IEC clinical centrifuge. The supernatant was discarded and the surface of the pellet and tube walls were carefully washed with the acetate buffer (pH 4.0), without disturbing the pellet. The pellet was then dissolved in 4 mL of a sodium dodecyl sulfate (SDS) – triethanolamine solution (*i.e.*, 1% SDS and 5% (v/v) triethanolamine in distilled water); one milliliter of ferric chloride reagent (*i.e.*, 0.01 M $FeCl_3 \cdot 6H_2O$ in 0.01 M HCl) was added to it and mixed. Fifteen minutes after addition of the ferric chloride reagent, the absorbance of the resultant solution was read at 510 nm against a reagent blank. Absorbance readings at 510 nm multiplied by 10 were plotted against the quantity of extract employed in the assay. The protein precipitating capacity of the tannins was expressed as the slope of the resultant linear "titration curve" and reported as Abs_{510} units × 10/mg extract.

The biological activity of tannins in the bearberry-leaf extract and its fractions was assayed by the dye-labeled protein assay of Asquith and Butler (*16*). The BSA employed was covalently labeled with Remazol brilliant blue R (*i.e.*, a blue dye) according to Rinderknecht *et al.* (*17*) and as modified by Asquith and Butler (*16*), to provide a substrate which could be detected spectrophotometrically. Briefly, 1 mL of methanolic solution of the bearberry-leaf extract or its fractions (0.1-2.0 mg/mL) was added to 4 mL of blue BSA solution containing 2 mg of protein/mL in 0.2 M phosphate buffer, pH 3.8, as modified by Naczk *et al.* (*18*). The solution was mixed vigorously at 1000 rpm for 5 min at room temperature using a shaker Type 50000 Maxi-Mix III (Barnstead/Thermolyne, Dubuque, IA). The protein–tannin complex so formed was separated by centrifugation for 15 min at 5000 rpm. The supernatant was discarded and the surface of the pellet and tube walls were washed with the phosphate buffer, without disturbing the pellet. The pellet was then dissolved in 3.5 mL of 1% (w/v) SDS solution containing 5% (v/v) triethanolamine and 20% (v/v) 2-propanol. The absorbance was measured at 590 nm against an appropriate blank. As there is a fixed quantity of protein in the assay, the percent dye-labeled BSA precipitated due to protein–tannin complex formation was plotted against the amount of crude extract or fraction added to the reaction mixture. The protein precipitating capacity of the tannins was then expressed as the slope of the resultant linear "titration curve" and reported as the percent dye-labeled protein preciptated/mg extract or fraction.

2-Thiobarbituric Acid Reactive Substances (TBARS) Determination

To monitor the antioxidant efficacy of the bearberry-leaf extract and its fractions, the development of 2-thiobarbituric reactive substances (TBARS) over time was monitored using a pork model system. TBARS of stored cooked meat were analyzed using a trichloroacetic acid/phosphoric acid extraction assay. Briefly, a 5.0-g portion from a ground pork system, which had been treated with the crude bearberry-leaf extract or one of its fractions and then thermally processed, was weighed into a sample bag. Fifty milliliters of a 20% (w/v) solution of trichloroacetic acid (TCA) containing 1.65% of 85% (w/w) phosphoric acid were added to the bag *via* a Brinkmann dispensette, and the contents were blended for 2 min. A 50-mL aliquot of ice cold distilled water was added to the slurry and blended for an additional minute. The mixture was filtered through a Whatman No. 1 filter paper into a 100-mL volumetric flask and made to mark with 1:1 (v/v) TCA:water. Five milliliters of the filtrate were pipetted into a polypropylene tube followed by 5 mL of 0.02 M 2-thiobarbituric acid (TBA) reagent. The tube was capped and contents vortexed. A blank was prepared using 5 mL each of 1:1 (v/v) TCA:water and TBA reagent. All tubes were heated in a boiling water bath for 35 min. After color development, the tubes were removed from the bath, cooled in ice water for *ca.* 5 min and then absorbance readings taken at 532 nm using the Milton Roy spectrophotometer. A standard curve was prepared using dilutions from a 0.2 mM 1,1,3,3-tetramethoxypropane working standard as described by Pegg (*19*).

The antioxidant capacities of the extracts were expressed as the percent inhibition of lipid oxidation on day seven using the following general equation:

$$\% Inhibition = (1 - \frac{Sample\ absorbance_t - TBHQ\ absorbance_{t=1}}{Nonadditive\ absorbance_t - TBHQ\ absorbance_{t=1}}) \times 100$$

where, sample absorbance$_t$ is the absorbance reading for the TBA-TBARS complex at 532 nm of the test sample at day t (in this case, t=7); TBHQ absorbance$_{t=1}$ is the absorbance reading for the synthetic antioxidant treated control at day 1; and non-additive absorbance$_t$ is the absorbance reading for the control sample containing no additives at day t.

Results and Discussion

Preparation of the crude extract from bearberry leaves, its dechlorophyllization and fractionation on a lipophilic Sephadex LH-20 column, followed by liquid-liquid fractionation using ethyl acetate:water (1:1, v/v) are depicted in Figure 2. Ethanol fraction I (*i.e.*, EtOH Fr. I) comprises the phenolics which eluted from the column when 95% (v/v) was used as the mobile phase. Thin layer chromatography plates (*i.e.*, silica gel with a fluorescent indicator and a mobile phase of chloroform:methanol:water, 65:35:10, v/v/v) with a ferric chloride spray and UV lamp detection were employed to indicate when no more relevant phenolics had eluted from the column. About 1 L of ethanol had passed through the Sephadex LH-20 column, at which point the mobile phase was changed over to 50% (v/v) acetone. A similar approach as described above was employed when collecting Acetone Fr. I, except that vanillin spray was used instead of ferric chloride to indicate when all of the tannins had eluted from the column. According to a mass balance, EtOH Fr. I comprised 79.2% of the crude dechlorophyllzied bearberry-leaf extract and Acetone Fr. I consisted of 9.7%. EtOH Fr. I was fractionated further using ethyl acetate:water (1:1, v/v): EtOH Fr. II represents the phenolics extracted into ethyl acetate while EtOH Fr. III is the product of the middle cloudy layer between the upper ethyl acetate and bottom water phases of the separatory funnel. An analogous fractionation of Acetone Fr. I was performed.

The content of total phenolics and condensed tannins in the bearberry-leaf extract and fractions so obtained are presented in Table I. Owing to the fact that the dominant phenolics in each fraction differ based upon the chromatography employed to obtain them, the results of total phenolics are expressed as Abs_{725} units per gram extract or fraction instead of as an equivalent to a monomeric unit from some polyphenolic. For example, the crude bearberry-leaf extract contains a substantial amount of arbutin, while Acetone Fr. I possesses more tannins. Consequently, construction of a calibration curve using an arbutin standard would be more appropriate in the first situation while using a flavanol such as (+)-catechin might be preferable in the latter. The content of total phenolics ranged from a low of 2135 Abs_{725} units/g for EtOH Fr. II to a high of 9110 Abs_{725} units/g for Acetone Fr. I. Use of 50% (v/v) acetone as a mobile phase assisted in fractionating and concentrating the tannin constituents from the bearberry-leaf extract. Previous research has shown that a crude extract from bearberry leaf contains a significant amount of polyphenolics, notably tannins (*4*).

The results of the vanillin and proanthocyanidin assays are quite interesting. The vanillin assay is widely used for quantitative measurement of condensed tannins,

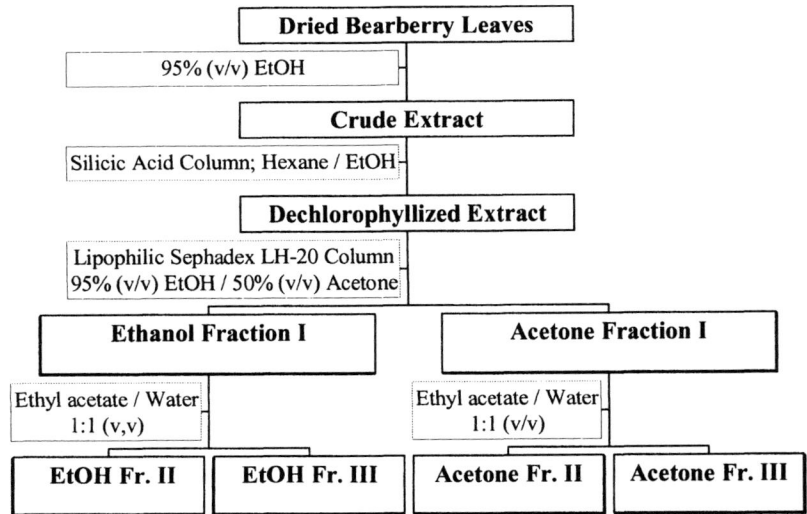

Figure 2. Flow diagram for the preparation of the crude bearberry-leaf extract and its fractions.

Table I. Bearberry Fractions

Bearberry Fractions	Total Phenolics[a]	Condensed Tannins[b]	
		Vanillin	Proanthocyanidin
Crude Extract	4700±47	232±2	130±1
EtOH Fr. I	3360±161	124±4	160±4
EtOH Fr. II	2135±57	ND	ND
EtOH Fr. III	3480±61	847±46	607±37
Acetone Fr. I	9110±129	619±9	392±15
Acetone Fr. II	8380±132	ND	ND
Acetone Fr. III	3110±182	101±7	trace

NOTE: Results are mean values of at least three determinations ± standard deviation.
[a]expressed as Abs_{725} units/g extract or fraction.
[b]Vanillin assay is expressed as mg (+)-catechin equivalents/g extract or fraction; Proanthocyanidin assay is expressed as Abs_{550} units/g extract or fraction; ND – not detected.

or its monomeric components. The assay appears to be highly specific for a narrow range of flavanols and dihydrochalcones which have a single bond at the 2,3 position and free *meta*-orientated hydroxyl groups on the B ring (*20*). A positive reaction is indicated by the appearance of a light pink to deep cherry red coloration. Vanillin-positive compounds were detected in the crude bearberry-leaf extract as well as EtOH Fr. I and Acetone Fr. I. Vanillin-positive compounds from EtOH Fr. I and Acetone Fr. I were not extracted by ethyl acetate (*i.e.*, they were absent from EtOH Fr. II), but were concentrated in the cloudy ethyl acetate/water and water layers. A marked concentration effect was noted for EtOH Fr. III. A parallel story is evident from the results of the proanthocyanidin assay: data are reported as Abs_{550} units/g extract or fraction as opposed to a standard equivalent. The proanthocyanidin assay is carried out in *n*-butanol/HCl and depends on an acid-assisted hydrolysis of the interflavan bonds of condensed tannins to produce anthocyanidins. The presence of 3-deoxyproanthocyanidins, flavan-3,4-diols and flavan-4-ols in extracts will also produce heat stable anthocyanidins (*21*); these will react with the ferric ions employed in the colorimetric assay and therefore overestimate absorbance readings. The yield of the reaction depends on the HCl concentration, temperature and length of the reaction time, proportion of water in the reaction mixture, and presence of transition metal ions as well as the length of the proanthocyanidin chain (*13*).

Selection of an appropriate standard for quantification is difficult as plant tannins are mixtures of polymeric compounds which differ in their sensitivity toward reagents used in their determination. The longer the polymer chain, the more anthocyanidin pigment is formed (*22*); thus, less of the red condensation product from the vanillin reaction is produced due to an increased steric hinderance toward the vanillin reagent (*23*). Assays of purified tannins have shown that the use of catechin equivalents overestimates the true tannin content (*12*). For example, authors reported that a tannin extract two times as concentrated as another will have twice the ΔA_{500}, but will also appear to have more than twice as much tannin based on the catechin standard curve. Nevertheless, the vanillin assay is routinely standardized against catechin, due to its structural relationship with condensed tannins. Table 1 reports the contents of condensed tannins from the vanillin assay expressed as mg catechin eq./g extract or fraction and from the proanthocyanidin assay as Abs_{550} units/g extract or fraction. The content of condensed tannins in the crude bearberry-leaf extract and its fractions ranged from 101 to 847 mg catechin eq./g extract and from 130 to 607 Abs_{550} units/g extract or fraction for the vanillin and proanthocyanidin assays, respectively. These levels are substantially greater than those found in the hulls of canola (*13*).

Although a number of protocols are available to estimate the protein precipitating capacity of tannins, the protein precipitation method described by Hagerman and

Butler (*14*) and the dye-labeled BSA assay of Asquith and Butler (*16*) were selected for quantification of the protein precipitating potential of tannins from the bearberry-leaf extract and its fractions. The protein precipitation method estimates the amount of precipitated protein-bound tannins (*i.e.*, quantity of the insoluble protein–tannin complex formed), whereas the dye-labeled BSA assay permits the direct measurement of the amount of protein precipitated by tannins present in an extract or fraction. In the first method a "titration curve," showing the amount of protein–tannin complex formed *vs* increasing quantities of extract or fraction added to a solution containing a fixed amount of protein, is plotted. Quantitative incorporation of tannins in the precipitate is assured by using at least twice as much protein as tannin, by weight (*14*). Due to the linear nature of the relationship, the slope is calculated. A larger slope denotes that the tannin constituents in the extract or fraction had a greater affinity to bind with the model protein (*i.e.*, BSA), resulting in the formation of an insoluble protein–tannin complex. Results of the protein precipitating capacity of various fractions of bearberry-leaf phenolics are presented in Table II and indicated that bearberry polyphenolics are very effective protein precipitants. Acetone Fr. I was almost three times more effective than the crude extract at precipitating protein resulting in the formation of protein–tannin complexes. Knowing that Acetone Fr. I and EtOH Fr. I represent 9.7 and 79.2% of the crude extract, respectively, a calculated slope of 1.4 instead of 1.8 would be expected for the crude extract. The degree of polymerization of the polyphenolics and the presence of low-molecular-weight phenolics in crude preparations probably affect the extent of precipitation, thereby resulting in slight discrepancies from predicted and actual slope values. Nonetheless, the data generated from the protein precipitation method is in line with the results of the mass balance. Even though EtOH Fr. III was richer in condensed tannins than Acetone Fr. I, as evident by the vanillin and proanthocyanidin assays (see Table 1), it was not as effective as Acetone Fr. I at precipitating BSA and forming a protein–tannin complex. This suggests that other constituents besides condensed tannins – presumably hydrolyzable tannins – are present in Acetone Fr. I and facilitate protein precipitation. Hagerman and Butler (*14*) noted that their protocol was rapid, reproducible and could be used with either condensed or hydrolyzable tannins. When comparing the slope from the titration curve of Acetone Fr. I (*i.e.*, 5.10) with those of acetone fractions for beach pea, canola hulls, evening primrose and faba bean (*i.e.*, in the range of 2.5 to 11.3, from reference *6*), it can be seen that the fraction from the bearberry-leaf extract was stronger at precipitating BSA than all except for evening primrose. Caution must be exercised, however, when comparing data in the literature. Besides modifications in methodology, a marked difference exists when results are expressed in Abs_{510} units × 10/mg extract or fraction compared to when they are expressed as Abs_{510} units/mg catechin equivalents.

A similar story is seen from the results of the dye-labeled BSA assay. The numerical values of the slopes from the linear plots of % blue BSA precipitated *vs* increasing quantities of extract or fraction added to a solution containing a fixed amount of the dye-labeled protein ranged from 12.2 to 98.4. Introduction of the hydrophobic dye reduces the solubility of the BSA and makes it precipitate more readily than unlabeled BSA (*16*). Consequently, formation of a precipitate should not be interpreted as indicative of the presence of tannin unless corroborated by the protein precipitation assay. As with the data of the protein precipitation method, Acetone Fr. I was strongest at precipitating the blue BSA. The quantity of tannin present in an extract or fraction will affect its affinity for BSA resulting in precipitation, but so will the degree of polymerization. According to Porter and Woodruffe (*24*), the ability of condensed tannins to precipitate proteins depends on their molecular weight. It is particularly interesting to note that EtOH Fr. II, which was devoid of condensed tannins, had no affinity to complex with and precipitate the dye-labeled BSA. Furthermore, hydrolyzable tannins which were present in Acetone Fr. I (results not shown) are most likely absent from EtOH Fr. II. Clearly, further research is needed to characterize the tannin constituents from the bearberry-leaf extract.

Results of the antioxidant activity of the dechlorophyllized bearberry-leaf extract and its fractions in cooked pork systems after 7 days of refrigerated storage are summarized in Figure 3. The active constituents present in the crude extract and its fractions showed excellent thermal stability during the cooking of the meat systems. This is significant, as a good number of natural antioxidants lose their capacity to quench free radicals or to chelate metal ions when subjected to food processing operations such as cooking. TBARS values of cooked pork model systems treated with a synthetic antioxidant (*i.e.*, TBHQ) as well as the bearberry-leaf extract and its fractions, at various concentrations, were compared. TBARS values of pork systems treated with the bearberry-leaf preparations were much lower than those of the control with no additive, thereby indicating protection to the meat lipids from constituents of the extract against autoxidation. Such data is better presented by reporting the % inhibition of lipid oxidation after a specified storage period in relation to two control systems: one without any additive in which autoxidation runs its course, and the second with the synthetic antioxidant at a suitable level to completely curb oxidation. Although an antioxidative efficacy was noticeable at 100- and 200-ppm levels, the crude bearberry-leaf extract at 200-ppm addition inhibited TBARS formation by 97% after 7 days of storage at 4°C storage. When the addition level of the extract was increased to 500 ppm, there was 100% inhibition of lipid oxidation over the same period.

The ethanol and acetone fractions recovered from the Sephadex LH-20 column gave some interesting data. EtOH Fr. I most likely contained simple phenolic

Table II. Protein precipitating capacity of various fractions of bearberry phenolics expressed as a slope value from titration curves[a]

Bearberry Fractions	Protein Precipitation Assay Abs × 10/mg extract or fr.	Dye-labeled BSA Assay %PP/mg extract or fr.
Crude Extract	1.82±0.03	16.7±0.5
EtOH Fr. I	1.12±0.02	12.2±0.6
EtOH Fr. II	ND	ND
EtOH Fr. III	3.55±0.06	57.3±4.9
Acetone Fr. I	5.10±0.15	98.4±4.9
Acetone Fr. II	2.20±0.47	54.9±1.8
Acetone Fr. III	2.45±0.10	31.3±3.2

[a]titration curves are lines that reflect the amount of proteins or tannins precipitated as a protein–tannin complex with increasing quantities of extract or fraction added to a solution containing a known amount of protein. Abs × 10 is the absorbance at 510 nm multiplied by 10; %PP is % dye-labeled BSA precipitated; ND – not detected.

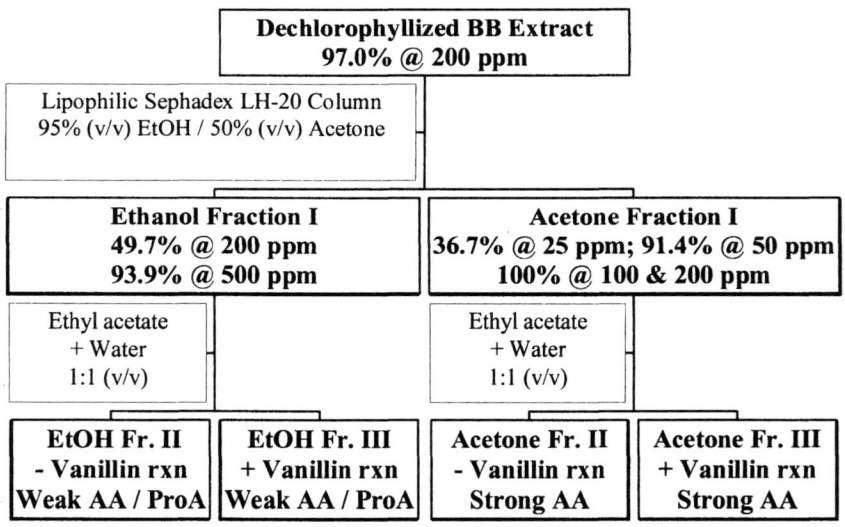

Figure 3. Block diagram reporting the % inhibition of lipid oxidation of cooked pork systems treated with the crude bearberry-leaf extract and its fractions after 7 days of refrigerated storage. AA – antioxidant activity; ProA – pro-oxidant activity; - Vanillin rxn – negative reaction with the vanillin reagent; + Vanillin rxn – positive reaction with the vanillin reagent.

acids, arbutin, methyl arbutin, gallic acid as well as condensed tannins, as indicated by the vanillin assay. Its antioxidant activity was weaker than that of the crude extract. Higher addition levels ~500 ppm were necessary to achieve the protection against autoxidation afforded by the crude bearberry-leaf extract at a 200-ppm concentration. On the other hand, Acetone Fr. I possessed a much stronger antioxidant activity than its precursor. As shown from the protein precipitating capacity data in Table III, this fraction was an effective protein precipitant and rich in hydrolyzable and condensed tannins. At a 50-ppm addition level, 91.4% protection in the cooked pork system was achieved after 7 days of refrigerated storage; when addition levels of 100 and 200 ppm were employed, complete protection against lipid oxidation, as assessed by the TBARS assay, was achieved. Such results demonstrate the concentration dependence of additive inclusion in relation to observed antioxidant activity.

The liquid-liquid fractionation of EtOH Fr. I and Acetone Fr. I gave some interesting results in terms of observed antioxidant activity. Both EtOH Fr. II and III imparted weak antioxidant activity to the pork systems when employed at 200- and 500-ppm levels. In fact, when lower levels than these were incorporated in the meat matrix, a slight pro-oxidant activity was observed. What is most interesting, however, is that EtOH Fr. III possessed a positive reaction with the vanillin test, thereby indicating the presence of flavan-3-ol end groups of condensed tannins. One would anticipate that such structural units should impart antioxidant activity, as observed for Acetone Fr. I which is rich in condensed tannins. This, however, was not the case. Concerning the data for Acetone Fr. II and III: both possessed strong antioxidant activity when applied to meat systems, but Fr. II gave a negative response with the vanillin reaction, again indicating the absence of condensed tannins. Hence, there are other constituents in the acetone fraction – presumably hydolyzable tannins – which imparted protection to the cooked pork systems against lipid oxidation. Further research on characterizing the tannins responsible for the observed antioxidant activity in food systems, such as the meat system described in this study, is in progress.

References

1. Valentão, P.; Fernandes, E.; Carvalho, F.; Andrade, P. B.; Seabra, R. M.; Bastos, M. L. *J. Agric. Food Chem.* **2002**, *50*, 4989-4993.
2. Amarowicz, R.; Barl, B.; Pegg, R. B. *J. Food Lipids* **1999**, *6*, 317-329.
3. Amarowicz, R.; Pegg, R. B.; Rahimi-Moghaddam, P.; Barl, B.; Weil, J. A. *Food Chem.* **2004**, *84*, 551-562.
4. Amarowicz, R.; Pegg, R. B. *J. Agric. Food Chem.* **2004**, *submitted*.

5. White, T. *J. Sci. Food Agric.* **1957**, *8*, 377-385.
6. Naczk, M.; Amarowicz, R.; Zadernowski, R.; Shahidi, F. *Food Chem.* **2001**, *73*, 467-471.
7. Shahidi, F.; Naczk, M. *Phenolics in Food and Nutraceuticals*; CRC Press: Boca Raton, FL, 2004.
8. Riedl, K. M.; Carando, S.; Alessio, H. M.; McCarthy, M.; Hagerman, A. E. In *Free Radicals in Food. Chemistry, Nutrition, and Health Effects*; Morello, M. J.; Shahidi, F.; Ho, C-T., Eds.; ACS Symposium Series No. 807; American Chemical Society: Washington, DC, 2002; pp 188-200.
9. Pegg, R. B.; Barl, B.; Amarowicz, R. U.S. Patent 6,660,320, 2003.
10. Swain, T.; Hillis, W. E. *J. Sci. Food Agric.* **1959**, *10*, 63-68.
11. Naczk, M.; Shahidi, F. *Food Chem.* **1989**, *31*, 159-164.
12. Price, M. L.; Van Scoyoc, S; Butler, L. G. *J. Agric. Food Chem.* **1978**, *26*, 1214-1218.
13. Naczk, M.; Nichols, T.; Pink, D.; Sosulski, F. *J. Agric. Food Chem.* **1994**, *42*, 2196-2200.
14. Hagerman, A. E.; Butler, L. G. *J. Agric. Food Chem.* **1978**, *26*, 809-812.
15. Naczk, M.; Oickle, D.; Pink, D.; Shahidi, F. *J. Agric. Food Chem.* **1996**, *44*, 2144-2156.
16. Asquith, T. N.; Butler, L. G. *J. Chem. Ecol.* **1985**, *11*, 1535-1544.
17. Rinderknecht, H.; Geokas, M. C.; Silverman, P.; Haverback, B. *J. Clin. Chim. Acta* **1968**, *21*, 197-203.
18. Naczk, M.; Pegg, R. B.; Amarowicz, R.; Pink, J. *Proceedings of the 47th International Congress of Meat Science and Technology, Volume II*; 47th International Congress of Meat Science and Technology: Kraków, Poland, August 26-31, 2001, pp 194-195.
19. Pegg, R. B. In *Current Protocols in Food Analytical Chemistry;* Wrolstad, R. E.; Acree, T. E.; An, H.; Decker, E. A.; Penner, M. H.; Reid, D. S.; Schwartz, S. J.; Shoemaker, C. F.; Sporns, P., Eds.; John Wiley & Sons: New York, NY, 2001; pp D2.4.1-D2.4.18.
20. Sarkar, S. K.; Howarth, R. E. *J. Agric. Food Chem.* **1976**, *24*, 317-320.
21. Porter, L. J.; Hrstich, L. N.; Chan, B. G. *Phytochem.* **1985**, *25*, 223-230.
22. Scalbert, A. In *Plant Polyphenols: Synthesis, Properties, Significance*; Hemingway, R. W.; Laks, P., Eds.; Plenum Press: New York, NY, 1992; pp 259-281.
23. Mole, S.; Waterman, P. G. *Oecologia* **1987**, *72*, 137-147.
24. Porter, L. J.; Woodruffe, J. *Phytochem.* **1984**, *23*, 1255-1256.

Chapter 8

Beans: A Source of Natural Antioxidants

Terrence Madhujith and Fereidoon Shahidi

Department of Biochemistry, Memorial University of Newfoundland,
St. John's, Newfoundland A1B 3X9, Canada

Antioxidant efficacy of beans with different colors were studied. Beans are well recognized for their macronutrients, but little is known about their bioactive components. Beans supply many bioactives, once classified as antinutrients, in minor amounts, but these may contribute to beneficial metabolic and physiological effects. Pulses, including beans, are known to possess hypoglycemic, hypocholesterolemic, antimutagenic and anticarcinogenic as well as other therapeutic effects. Antioxidants in beans might also contribute to their cardiovascular and anticarcinogenic effects. Antioxidant potential, including inhibition of human LDL oxidation, as well as prevention of DNA double strand breakage of different beans is described in this contribution.

Introduction

Foods that contain physiologically active ingredients with health benefits above their basic nutrition are known as functional foods (*1*). Beans (*Phaseolus vulgaris* L.), staple foods in Mexico, Central and South America as well as some African countries, are a good source of protein, vitamins, minerals (*2*) and polyphenolic compounds (*3*) which have been shown to be responsible for a myriad of health benefits. In view of the presence of a number of phytochemicals, beans are receiving increasing attention as a functional food.

Once known as the poor man's meat, beans are now presented as staple foods for vegetarians and most health organizations encourage their frequent consumption (*4*). Intake of beans has been linked to reduce risk of diabetes and obesity (*5*). Beans, among other pulses, are known as one of the foods with low glycemic index (glycemic index [GI] refers to the blood glucose raising potential of carbohydrate foods after consumption) (*6*). The protein, fiber, starch, vitamins, minerals and other components of beans contribute to the cardiovascular disease (*7*). Pulses, including beans, contain a wide range of nutrients and non-nutritive bioactives such as protease inhibitors, saponins, lectins, galactosides and phytates that may be protective against cancer when consumed in sufficient quantities (*8*). Table I lists the bioactive compounds present in pulses.

Table I. Bioactive Compounds in Pulses and Their Health Effects

Compound	*Health Effect*
Protease inhibitors	
Phytic acid	
Phytoestrogen	Anticarcinogenic
Lignans	
Saponins	
Phytic acid	
Lectins	Glucose lowering effect
Amylase inhibitors	
Tannins	
Saponins	
Phytosterols	Hypolipaemic effect
Phytic acid	
Isoflavones	

From Ref. *37* and *38*.

Yokota *et al.* (*9*) reported that oral administration of fermented bean crude residues repressed experimentally induced inflammation in rats. de Meja *et al.* (*10*) indicated that methanolic extracts of beans inhibited mutagenicity in *Salmonella typhimurium*. Beans contain significant quantities of polyphenolic compounds such as flavonoids, phenolic acids and lignans, but are typically low in ascorbic acid, β-carotene, and α-tocopherol (*11*). Bean seed coat contains a number of anthocyanin pigments that exhibit antioxidative activity. Tsuda *et al.* (*12*) indicated that anthocyanin pigments (cyanidin-3-*O*-β-D-glucoside, pelargonidin 3-*O*-β-D-glucoside and delphinidin 3-*O*-β-D-glucoside) isolated from seed coat s of *Phaseolus vulgaris* had marked antioxidant activity in liposomal and rat liver microsomal systems, an inhibitory effect on malonaldehyde levels by UV radiation and radical scavenging effect against hydroxyl and superoxide anion radicals. Anthocyanins and anthocyanidins, which are metabolic products of flavanones (*13*), constitute a major portion of antioxidants in beans (*14*). Most of the naturally occurring anthocyanins and anthocyanidins carry hydroxyl groups at the C-4`position. Anthocyanins such as cyanidin and delphinidin, found in beans, contain a hydroxyl group at C-3` position while pelargonidin does not (*15*). Figure 1 depicts the structures of anthocyanin pigments identified in seed coats of beans. Tsuda *et al.* (*14*) investigated the antioxidant activity of cyanidin 3-glucoside in linoleic acid, liposome, rabbit erythrocyte membrane ghost and rat liver microsomal systems. Their results in an *in-vitro* system suggested that these pigments play a role in the prevention of lipid oxidation. The pigments pelargonidin 3-*O*-β-glucoside and delphinidin-3-*O*-β-glucoside isolated from red and black beans do not possess any antioxidative activity at pH 7.0 while cyanidin-3-*O*-β-glucoside exhibits antioxidative activity, but all three pigments have strong activity under acidic conditions (*16*). This suggests that antioxidative activity is related to the stability of the flavylium cation (*16*), which is generally quite stable under acidic conditions (17). Cyanidin may be produced from cyanidin glucosides by hydrolysis with β-glucosidase of intestinal bacteria after ingestion, indicating that cyanidin rather than cyanidin-3-glucoside may act as an antioxidant in living systems (*16*). In addition to anthocyanins, beans contain procyanidins and phenolic acids. Drumm *et al.* (*18*) identified ferulic, *p*-coumaric, sinapic and cinnamic acids (Figure 1) in four varieties of beans (navy, dark red kidney, pimanto and black turtle soup).

Depending on the cultivar, beans contain a variety of flavonoids/anthocyanins and may have different coat colors. Four colors are most common; red, brown, black and white (*15*). The antioxidative activity of extracts of pea beans has been evaluated by different authors (*10,12,14,16,19-21*). Tsuda *et al.* (*12*) evaluated the antioxidative activity of a number of common beans using linoleic acid autoxidation system (Table II). The effect of crude extracts of hulls of navy bean on the oxidative stability of edible oils was reported by Onyeneho and Hettiarachchy (*22*) who suggested that extracts of navy bean hulls can be used commercially to inhibit oxidation of vegetable oils. Although beans are cultivated throughout the world and consumed in both eastern and western dishes, little attention has been paid to their antioxidant and antigenotoxic potentials. Thus, in this study we examined the antioxidant potential of several bean extracts.

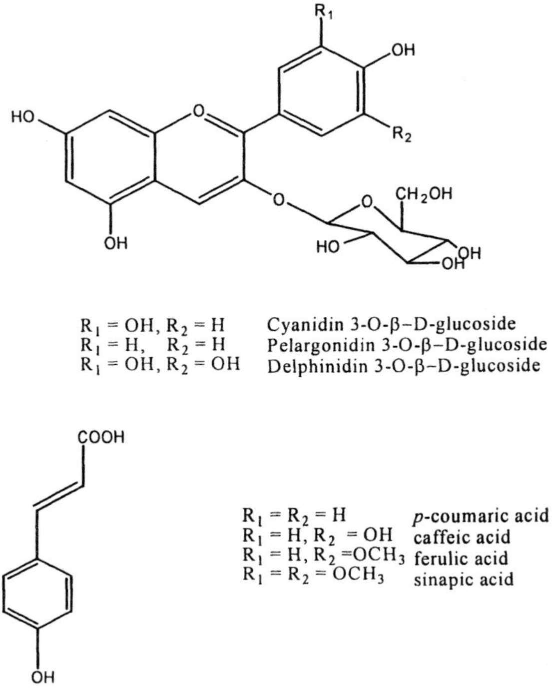

R₁ = OH, R₂ = H   Cyanidin 3-O-β-D-glucoside
R₁ = H, R₂ = H   Pelargonidin 3-O-β-D-glucoside
R₁ = OH, R₂ = OH   Delphinidin 3-O-β-D-glucoside

R₁ = R₂ = H   *p*-coumaric acid
R₁ = H, R₂ = OH   caffeic acid
R₁ = H, R₂ = OCH₃   ferulic acid
R₁ = R₂ = OCH₃   sinapic acid

Figure 1. Structures of anthocyanins and phenolic acids in bean seed coats.

Table II. Antioxidative Activity of Edible Beans

Common Name	Species	Antioxidative Activity[1]
Kidney bean	*P. vulgaris* L. cv. Honkintoki	++
Kidney bean	*P. vulgaris* L. cv. Ohtebo	+
Scarlet runner bean	*P. coccineus* L.	+
Hyacinth bean	*Lablab purpureus* L. Sweet	+
Winged bean	*Psophocarpus tetragonolobus* L.	+
Horse gram	*Macrotyloma uniflorus* (Lam.)	+

[1] Antioxidative activity in a linoleic acid autoxidation system using the thiocyanate method. Adapted from Ref. *12*.

Evaluation of Antioxidant Activity of Bean Extracts

Beans with different coat colors (red, brown, black and white) were obtained from a local grocery store in Singapore and evaluated for their total phenolic content and Trolox equivalent antioxidant capacity. Phenolic compounds present in defatted bean samples were extracted with 80% acetone under reflux conditions. The resulting slurries were centrifuged and the supernatants collected. The residue was re-extracted with 80% acetone and the supernatants combined and desolventized *in vacuo*. The resulting concentrated solutions were lyophilized and used in the experiments.

The total phenolic content was determined, essentially according to an improved version of the procedure explained by Singleton and Rossi (*23*), and expressed as catechin equivalents. The method explained by van den Berg (*24*) was used to determine the Trolox equivalent antioxidant capacity (TEAC) of the extracts.

Among primary catalysts that initiate *in-vivo* and *in-vitro* oxidation, transition metal ions and complexes containing metals have been identified. Transition metal ions, such as iron and copper participate in direct and indirect initiation of lipid oxidation (*25*). Therefore, metal chelation can be regarded as one of the important characteristics of certain antioxidants. In our study, the metal chelation efficacy of the bean extracts was determined using the method explained by Terasawa *et al.* (*26*).

Total Phenolic Content

Preliminary studies in this work and literature data indicated the presence of catechin and catechin-related compounds in beans. Therefore, the total phenolic contents were expressed as catechin equivalents. In all four bean types, hulls contained a higher amount of total phenolics when compared to those of their whole seed counterparts. Thus, the phenolic compounds are mainly concentrated in the seed coats. Total phenolic content of extracts of black bean hulls (LHE) was highest while that of white hull extract (WHE) was the lowest with the following trend: black > brown > red > white (See Table III). This trend is different from that observed for the whole bean extracts, possibly due to the proportional difference in the hull content in different beans. As white bean extracts did not contain a considerable amount of total phenolics, they were not studied any further.

Table III. Total Phenolic Content (TPC) of Bean Extracts[1]

Sample	TPC, mg Catechin Equivalents Per Gram Extract
Red whole bean extract (RWE)	93.6 ± 2.1^c
Brown whole bean extract (BWE)	91.4 ± 1.6^c
Black whole bean extract (LWE)	44.0 ± 2.5^b
Whole white bean extract (WWE)	4.9 ± 0.8^a
Red hull extract (RHE)	223.5 ± 1.9^d
Brown hull extract (BHE)	253.2 ± 2.3^e
Black hull extract (LHE)	270.0 ± 1.6^f
White hull extract (WHE)	6.7 ± 0.9^a

[1] Results reported are mean values of three determinations ± standard deviation. Means in each column sharing the same superscript are not significantly ($p>0.05$) different from one another.

Trolox Equivalent Antioxidant Capacity

TEAC value of a compound represents the concentration of a Trolox solution that has the same antioxidant capacity of a known or unknown compound or a mixture of compounds. Therefore, TEAC value can be used to rank the efficacy of an unknown compound or known antioxidants. Whole seed extracts were used at the same concentration while the hull extracts were diluted to a final concentration of 0.05 mg/mL. Figure 2 shows the TEAC values of whole seed and hull extracts of beans. Whole bean extracts showed low TEAC values ranging from 4.64 to 8.84 (whole bean extracts are 4.64 to 8.84 times as effective as Trolox) whereas TEAC values of hull extracts varied from 40.74 to 46.68. In both cases, the red bean extract showed the highest TEAC value followed by extracts of brown and black beans. In both whole bean and hull extracts, at the same concentration, TEAC values were in the order of red > brown > black. TEAC values of red and brown bean hulls were about 5.5 times higher than those of their corresponding whole seed extracts while black hulls were about 9 times more effective than black whole seed extracts.

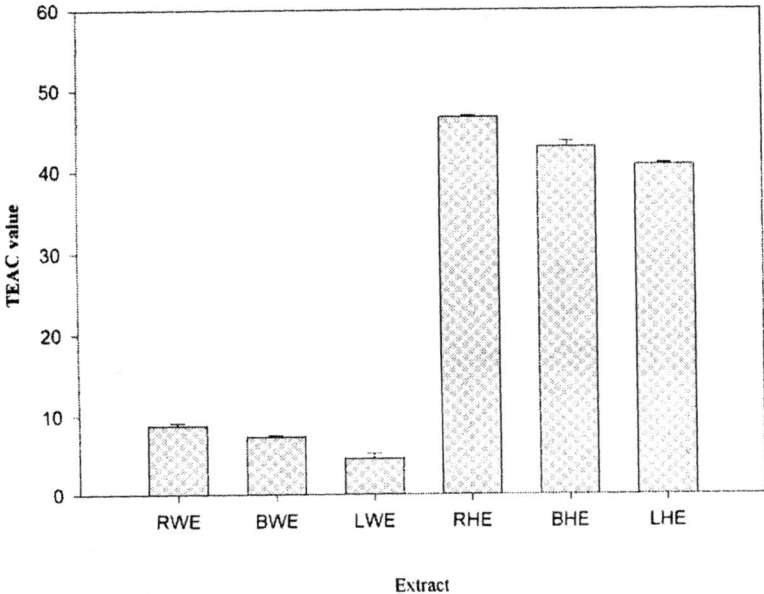

Figure 2. Trolox equivalent antioxidant activity of whole seed and hull extracts of beans. Abbreviations as given in Table III.

Metal Chelation Capacity

The chelation of Fe^{2+} by 50 ppm catechin equivalents of phenolics in RWE and BWE was moderate and at 58 and 56%, respectively. This is slightly lower than that for catechin (60%); meanwhile chelation capacity of LWE was significantly ($p<0.05$) lower than that of RWE and BWE at 50 ppm. When used at 100 ppm, catechin exhibited nearly 100% chelation of Fe^{2+} while RWE and BWE showed 72 and 69% chelation, respectively. LWE showed 60% chelation at 100 ppm level, but this was significantly ($p<0.05$) lower than that of other extracts used in this study at the same concentration (Table IV). It is evident that bean crude extracts show better Fe^{2+} chelation efficacy compared to borage and evening primrose extracts (23). The phenolic compounds serve as main chelators of metal ions although non-phenolic constituents in the crude extracts may also participate in sequestering of metal ions (27).

Table IV. Fe^{2+} Concentration (μM) and Percentages of Chelated Ferrous Ion by Bean Extracts at 50 and 100 ppm Concentrations as Catechin Equivalents[1]

Additive	50 ppm		100 ppm	
	μM	%	μM	%
Catechin	240±4^d	60	395±8^d	99
RWE	232±6^c	58	289±4^c	72
BWE	224±2^b	56	276±7^b	69
LWE	156±5^a	39	240±5^a	60

[1] The initial concentration of ferrous is 400μM. Results reported are mean values of three determinations ± standard deviation. Means in each column sharing the same superscript are not significantly (p>0.05) different from one another. Abbreviations are given as in Table III.

Inhibition of Oxidation of Human Low Density Lipoprotein (LDL) by Bean Extracts

Oxidation of LDL *in-vivo* contributes to the pathology of atherosclerosis (28). Thus, there has been an increased interest in investigating the role of natural antioxidants in preventing oxidation of LDL and membrane lipids. It has been reported that anthocyanins derived from grape juices (29), wines (30), and berries (31) are major compounds contributing to antioxidant activity, thus preventing oxidation of LDL. Measurement of hLDL oxidation was achieved using EDTA-free hLDL and by mixing it with different levels of extracts (32). The oxidation was initiated by the addition of 10 μM $CuSO_4$. After incubating the mixture at 37°C for 20 h, the mixtures were tested for their content of conjugated dienes.

Conjugated dienes can be used as an indicator of LDL oxidation as affected by antioxidants. Recent studies have shown that Cu^{2+}-mediated oxidation of LDL can exhibit different kinetics depending on the concentration of Cu^{2+}. Propagation can proceed when antioxidants are depleted, at high Cu^{2+} concentrations or when they are present at low concentrations (33).

All extracts tested and catechin showed almost 100% inhibition of oxidation of LDL cholesterol at 50 ppm (as catechin equivalents). At 2, 5 and 10 ppm, the protection rendered by RWE and BWE against hLDL oxidation was not significantly (p<0.05) different, but the protection by LWE was significantly (p< 0.05) lower than that of RWE and BWE. All three bean extracts showed a significantly (p<0.05) higher activity against LDL oxidation than catechin at both 2 and 5 ppm. This behavior may be attributed to the synergistic effect of the mixture of natural compounds present in bean extracts.

Prevention of DNA Strand Breakage by Bean Extracts

Oxidants, produced as by-products of mitochondrial electron transport and products from lipid peroxidation that escape the numerous antioxidant defense systems, can cause damage to cellular macromolecules, including DNA. Damage to DNA can lead to mutation and cancer (34). In our study, plasmid supercoiled DNA (pBR 322) was mixed with bean extracts at different concentrations and the resultant peroxyl radical, generated using AAPH, was scavenged by the extracts. The reactants were incubated at 37°C for 1h and subsequently electrophoresed in order to estimate DNA band intensities (35). The protective effect of extracts and catechin was calculated based on relative band intensities.

The DNA retention capacities of RWE, BWE and catechin were not significantly ($p>0.05$) different from one another against AAPH-derived peroxyl radical damage at a 5 ppm level while that of LWE was low. At all other concentrations tested (10, 50 and 100 ppm as catechin equivalents), the three extracts and catechin showed very high and similar ($p>0.05$) DNA retention. Radicals cleave supercoiled plasmid DNA (form I) to nicked circular DNA (form II) or at higher concentrations of radicals to linear DNA (form III). The presence of peroxyl radical resulted in a dramatic scission of supercoiled DNA. The radical concentration used in the present study was not sufficient enough to destroy the nicked circular DNA which is more difficult to be destroyed as compared to DNA of form I. As the concentration of antioxidative extract was increased the protective effect against nicking of supercoiled DNA also increased. In the absence of any antioxidant, it may be expected that a peroxyl radical abstracts a hydrogen atom from a nearby DNA to generate new radicals, which in turn evokes a free radical chain reaction resulting in the destruction of the DNA molecule. However, in the presence of antioxidants, this chain reaction is terminated by abstracting a hydrogen atom from the antioxidant molecule (32).

Alternatively, antigenotoxic/genotoxic properties can be evaluated by comet assay which involves determination of single strand breaks (SSBs) in the DNA (36). Raab et al. (21) evaluated a number of pigments isolated from beans using comet assay where hydrogen peroxide, *tert*-butyl hydroperoxide, and linoleic acid hydroperoxide were used to induce DNA damage. The expected correlation between the antioxidative properties of the flavanols investigated and their influence on DNA damage caused by the above reactive oxygen species was not found. Possible prooxidative effects of flavanols might have been responsible for the observed pattern.

Conclusions

This present study clearly demonstrated that beans, especially those with colored coats, possess a strong antioxidant activity. The antioxidants from beans effectively suppress the oxidation of human LDL cholesterol, which could act as a culprit in the development of coronary heart diseases. They also protect DNA against radical damage. The experiments performed in this study were *in-vitro* and hence it is necessary to extend these to include investigation of their behavior in *in-vivo* systems.

References

1. Shahidi, F.; Weerasinghe, D. In *Nutraceutical Beverages: Chemistry, Nutrition, and Health Effects*, Shahidi, F., Weerasinghe, D.K., Eds.; AOCS : Washington, DC, **2003**; pp. 1-5.
2. Berrios, J.D.J.; Swanson, B.G.; Cheong, W.A. *Food Res. Intl.* **1999**, *32*, 669-676.
3. Larson, R.A. *Phytochemistry.* **1988**, *27*, 969-978.
4. Leterme, P. *Brit. J. Nutr.* **2002**, *88*, S239-S242.
5. Geil, P.B.; Anderson, J.W. *J. Am. Coll. Nutr.* **1994**, *13*, 549-558.
6. Bornet, F.R.; Billaux, M.S.; Messing, B. *Intl. J. Biol. Macromol.* **1997**, *21*, 207-219.
7. Anderson, J.W.; Hanna, T.J. *Journal of Nutrition* **1999**, *129*, 1457-1466.
8. Mathers, J.C. *Brit. J. Nutr.* **2002**, 88, S273-S279.
9. Yokota, T.; Hattori, T.; Ohishi, H.; Ohami, H.; Watanabe, K. *Lebens.-Wiss. U. Technol.* **1996**, *29*, 304-309
10. de Meja, E.G.; Castano-Tostado, E.; Loarca-Pina, G. *Mutat. Res.* **1999**, 441, 1-9.
11. Ganthovorn, C.; Hughes, J.S. *J. Am. Oil. Chem. Soc.* **1997**, *74*, 1025-1030.
12. Tsuda, T.; Oshima, O.; Kawakishi, S.; Osawa, T. In *Food factors for Cancer Prevention*; Ohigashi, S; Osawa, T.; Terao, J.; Watanabe, S.; Yoshikawa, T. Eds.; Springer Verlag: Tokyo, Japan. **1997**; pp. 318-322.
13. Hall, C. In *Antioxidants in Food: Practical Applications*, Pokorny, J.; Yanishlieva, N.; Gordon, M.; Eds.; CRC Press: Washington, DC. **2001**; pp. 160-169.
14. Tsuda, T.; Watanabe, M.; Ohshima, K.; Norinobu, S.; Choi, S-W.; Kawakishi, S.; Osawa, T. *J. Agric. Food Chem.* **1994**, *42*, 2407-2410.
15. Mazza, G.; Miniati, E. *Anthocyanin in Fruits, Vegetables, and Grains*, CRC Press, London. **2000**; pp. 2-10.

16. Tsuda, T.; Oshima, K.; Kawakishi, S.; Osawa, T. *J. Agric. Food Chem.* **1994**, *42*, 248-251.
17. Brouillard, R. In *The Flavonoids*, Harborne, J.B.; Ed.; Chapman and Hall: London, **1988**; pp. 525-538.
18. Drumm, T.D.; Grag, J.I.; Hasfield, G.C. *J. Sci. Food Agric.* **1990**, *51*, 285-297.
19. Tsuda, T.; Osawa, T.; Nakayama, T.; Kawakishi, S.; Ohshima, K. *J. Am. Oil Chem. Soc.* **1993**, *70*, 909-913.
20. Tuda, T.; Makino, Y.; Kato, H.; Kawakishi, S. *Biosci. Biotech. Biochem.* **1993**, *57*, 1606-1608.
21. Raab, B.; Hempl, J.; Bohm, H. In *Food Factors for Cancer Prevention*. Ohigashi, H.; Osawa, T.; Terao, J.; Watanabe, S.; Yoshikawa, T., Springer-Verlag: Tokyo, Japan, **1997**; pp. 309-312.
22. Onyeneho, S.N.; Hettiarachchy, N.S. *J. Agric. Food Chem.* **1991**, *39*, 1701-1704.
23. Singleton, V.L.; Rossi, J.A. *Am. J. Enol. Vitic.* **1965**, *16*, 144-158.
24. van den Berg, R.; Haenen, G.R.M.M.; van den Berg, H.; Bast, A. *Food Chem.* **1999**, *66*, 511-517.
25. Schaich, K.M. *Critic. Rev. Food Sci. Nutr.* **1980**, *13*, 89-159.
26. Terasawa, N.; Murata, M.; Homma, S. *Agric. Biol. Chem.* **1991**, *55*, 1507-1514.
27. Wettasinghe, M.; Shahidi, F. *Food Res. Intl.* **2002**, *35*, 65-71.
28. Steinberg, D; Parthasarathy, S.; Carew, T.E.; Khow, J.C.; Witztem, J.L. *New Engl. J. Med.* **1989**, *320*, 915-924.
29. Frankel, E.N.; Bosanek, C.A.; Meyer, A.S.; Sillimank, Kirk, L.L, *J. Agric. Food Chem.* **1998**, *6*, 834-838.
30. Tesissers, R.L.; Frankel, E.N.; Waterhouse, A.L.; Peleg, H.; German, J.B. *J. Sci. Food Agric.* **1996**, *70*, 55-61.
31. Abuja, P.M.; Murkovic, M.; Werner, P. *J. Agric. Food Chem.* **1998**, *46*, 4091-4096.
32. Hu, C.; Kitts, D.D. *J. Agric. Food Chem.* **2000**, *48*, 1466-1472.
33. Ziouzenkova, O.; Sevaian, A.; Abuja, M.P.; Ramos, P.; Esterbauer, H. *Free Rad. Biol. Med.* **1998**, *24*, 607-623.
34. Ames, B.N.; Shegenaga, M.K. In *DNA and Free Radicals*. Halliwell, B.; Aruoma, O.I.; Eds.; Ellis Horwood Ltd.: West Sussex, England, **1993**; pp.1-15.
35. Hu, C.; Zhang, Y.; Kitts, D. D. *J. Agric. Food Chem.* **2000**, *48*, 3170-3176.
36. McKelvey-Martin, V.J.; Green, M.H.L.; Schmezer, P.; Pool-Zobel, B.L.; De Meo, M.P.; Collins, A. *Mutat. Res.* **1993**, *288*, 47-63.
37. Champ, J.; Martine, M. *Brit. J. Nutr.* **2002**, *88*, S307-319.
38. Alonso, R.; Grant, G.; Marzo, F. *Nutr. Res.* **2001**, *21*, 1067-1077.

Chapter 9

Antioxidant and Antibacterial Properties of Extracts of Green Tea Polyphenols

Ryszard Amarowicz[1], Ronald B. Pegg[2], Gary A. Dykes[3], Agnieszka Troszynska[1], and Fereidoon Shahidi[4]

[1]Division of Food Science, Institute of Animal Reproduction and Food Research, Polish Academy of Sciences, 10–747 Olsztyn, ul. Tuwima 10, Poland
[2]Department of Applied Microbiology and Food Science, University of Saskatchewan, College of Agriculture, 51 Campus Drive, Saskatoon, Saskatchewan S7N 5A8, Canada
[3]Food Science Australia, P.O. Box 3312, Tingalpa DC, Queensland 4173, Australia
[4]Department of Biochemistry, Memorial University of Newfoundland, St. John's, Newfoundland A1B 3X9, Canada

Crude extracts of phenolic compounds were prepared from seven Chinese green tea preparations. The contents of individual catechins were chromatographed and quantified in all extracts by HPLC, while the proanthocyanidins fraction was isolated by column chromatography. The content of total catechins ranged from 38.5 to 82.2%, with EGCG being dominant in six of the extracts; EGC was found to be the main catechin in the remaining preparation. The antioxidant activity of crude extracts, the proanthocyanidins fraction and a reconstituted mixture from catechin standards were evaluated in a meat model system. The % inhibition of TBARS development varied from 27.9 to 55.7% (*i.e.*, based on extract addition of 25 ppm), 62.6 to 88.7% (*i.e.*, 50 ppm), 78.7 to 100% (*i.e.*, 100 ppm) and 94.3 to 100% (*i.e.*, 200 ppm). The proanthocyanidins fraction was less active relative to the crude extract from which it was derived. Results from experiments with the reconstituted mixture of catechins show that there are other constituents present in green tea extract (GTE) which impart antioxidant activity or, at least, exert a synergistic effect with the endogenous catechins. Antimicrobial activity was observed for all extract preparations using a TYSG broth model. Minimum inhibitory concentration data indicated that all extracts possessed marked antimicrobial activity against the *Salmonella Typhimurium* strain used in this study, whereas intermediate activity was observed against the *Pseudomonas fragi*, *Listeria monocytogenes*, *Staphylococcus aureus* and *Lactobacillus plantarum* strains used in this study.

. Green tea (*Camellia sinensis* L.) has been used as a daily beverage and crude medicine in China for thousands of years. Its processing is important not only economically as a bulk commodity but also from a dietary perspective. Tea leaves contain caffeine, theanine (*N*–ethyl L-glutamine), vitamins, saponins as well as a group of low-molecular-weight polyphenolic compounds known as tea catechins; these latter compounds are present in green and black tea beverages at 30-42% and 3-10% levels based on the total dry matter, respectively (*1*). Green tea differs from black tea in that the enzymes responsible for the fermentation during black tea processing (*i.e.*, polyphenol oxidases) are deactivated by steaming prior to maceration of harvested leaves (*2*). The steaming treatment protects against the degradation of vitamins and thus, their contents in green tea are much higher than those in semi- or fully-fermented counterparts.

Tea consumption is responsible for antipyretic and diuretic actions in our bodies, but a number of reports exist where a reduced risk of several forms of cancer among populations was observed when significant quantities of green tea were consumed (*3,4*). Within the late twentieth century, pharmacological and epidemiological studies revealed that various constituents of tea possess antioxidative efficacy (*5*), a hypocholesterolemic effect (*6*), as well as antimutagenic (*7,8*), anticarcinogenic (*9-12*) and antitumorigenic activities (*13-15*). Antimutagenic activity has been linked to the ability of catechins from green tea to scavenge free radicals such as HO• and ROO• (*16*).

Green tea contains relatively large amounts of flavanols including (+)-catechin, (+)-gallocatechin (GC), (-)-epicatechin (EC), (-)-epicatechin gallate (ECG), (-)-epigallocatechin (EGC) and (-)-epigallocatechin gallate (EGCG). These secondary metabolites are synthesized in tea leaves through malonic acid- and shikimic acid-metabolic pathways. The catechins are recognized as being important contributors to the antioxidant activity of GTE (*17*). The suppressive effect on oxidation from green tea catechins is concentration-dependent and has been demonstrated in the following: lipid model systems involving the acceleration of oxidation by varying the heat and oxygen supply (*5*); enzyme-catalyzed oxidation studies (*18*); a β-carotene-linoleic acid emulsion (*19*); a sunflower oil emulsion (*20*); heated oils (*21*); assays involving radical-scavenging capacity with free radicals (*22-29*); phospholipid bilayers (*30*); and assays for superoxide- and peroxynitrite-scavenging activity (*31,32*).

Tea possesses antibacterial properties; yet, the reasons for this are not fully understood. In literature, few reports on the antibacterial activity of the polyphenolic constituents of tea exist, and those which do are generally without reference to isolated flavonol components. Some data on the efficacy of individually purified tea polyphenols against foodborne bacteria, such as *Clostridium botulinum*, have been reported by Japanese researchers (*33-35*). It has also been noted that GTE could selectively regulate the growth of clostridial bacteria among various intestinal microorganisms (*36*), and had an *in vivo* effect in establishing and maintaining human intestinal microflora in an ideal balanced state (*37*). Green tea catechins, especially GC, EGC and EGCG, have also been reported to inhibit bacteria such as *Streptococcus mutans*, which are responsible for dental caries (*38-40*).

The aim of this study is to evaluate the antioxidant activity of several different GTE preparations in meat model systems, as a potential source of nutraceutical in a processed food system, and to examine the antibacterial properties of these materials against some specific food-related bacteria.

Experimental

Sample Preparation and Chromatography

Crude extracts of phenolic compounds were prepared from seven Chinese green teas using 95% (v/v) ethanol according to Amarowicz et al. (*41*). One sample (No. 1 in Table I) of leaves was obtained from Anhui Province in the People's Republic of China, three samples (No. 2, 3, 4) from commercial green tea purchased in Poland, and two (No. 5, 6) procured in Canada. In addition, a commercial GTE (No. 7) sold in the United States under the name "Herbal Plus" was investigated. The individual catechins, namely EC, EGC, ECG and EGCG, were chromatographed and determined in all extracts by HPLC using a C_{18} LUNA column (Phenomenex, USA) and a mobile phase consisting of water-acetonitrile-methanol-water (79.5:18:2:0.5, v/v/v/v) (*42*).

Table I. Variation in the Concentration of Individual Catechins from Seven Green Tea Extract Preparations[a]

Sample	EGC	EC	EGCG	ECG	Total
1	15.1	6.6	13.9	2.9	38.5
2	13.4	1.8	33.6	4.0	52.8
3	16.1	2.2	55.2	4.8	78.3
4	20.6	2.8	36.9	4.1	64.4
5	18.9	3.3	52.1	7.8	82.0
6	18.6	3.6	52.2	7.8	82.2
7	7.3	3.9	52.7	7.9	71.8

[a] All data reported as %, w/w

The proanthocyanidins fraction was separated from one of the extracts using low-pressure column chromatography. The extract was applied on a column of silica gel (Sigma-Aldrich Canada Ltd., 40-63 μm, average pore diameter: 60 Å; Chromaflex column, 30 × 250 mm [I.D. × length], Kontes, USA). The catechins were eluted using a mobile phase of chloroform-methanol-water (65:35:10, v/v/v; lower phase) (43), after which the proanthocyanidins were eluted using methanol.

Based on quantitative HPLC data from one sample, a reconstituted extract was formulated from catechin standards at the appropriate levels. This preparation as well as the seven crude tea extracts and the proanthocyanidins fraction were examined for their antioxidant and antimicrobial activities using assays described below.

Antioxidant Assay

To monitor the antioxidant efficacy of tea extracts, the development of 2-thiobarbituric reactive species (TBARS) over time was monitored using a pork model system. TBARS from stored cooked pork systems were analyzed using a trichloroacetic acid/phosphoric acid extraction assay. Briefly, a 5.0-g sample of ground pork, which had been treated with GTE and then thermally processed, was weighed into a sample bag. Fifty milliliters of a 20% (w/v) solution of trichloroacetic acid containing 1.65% of 85% (w/w) phosphoric acid was added to the bag via a Brinkmann dispensette, and the contents were blended for 2 min. A 50-mL aliquot of ice cold distilled water was added to the slurry and blended for an additional minute. The mixture was filtered through a Whatman No. 1 filter paper into a 100-mL volumetric flask and made to mark with a 1:1 (v/v) TCA:water. Five milliliters of the filtrate were pipetted into a polypropylene tube followed by 5 mL of 0.02 M 2-thiobarbituric acid (TBA) reagent. The tube was capped and contents vortexed. A blank was prepared using 5 mL each of 1:1 (v/v) TCA:water and TBA reagent. All tubes were heated in a boiling water bath for 35 min. After color development, the tubes were removed from the water bath, cooled in ice water for *ca.* 5 min and then absorbance readings taken at 532 nm. A standard curve was prepared using dilutions from a 0.2 mM 1,1,3,3-tetramethoxypropane working standard as described by Pegg (*44*).

The antioxidant capacity of the seven GTE preparations was expressed as the percent inhibition of lipid oxidation on day seven using the following general equation:

$$\% \text{ Inhibition} = (1 - \frac{\text{Sample absorbance}_t - \text{TBHQ absorbance}_{t=1}}{\text{Control absorbance}_t - \text{TBHQ absorbance}_{t=1}}) \times 100$$

where, sample absorbance$_t$ is the absorbance reading for the TBA-TBARS complex at 532 nm of the test sample on day t; TBHQ absorbance$_{t=1}$ is the absorbance reading for the synthetic antioxidant treated control on day 1; and control absorbance$_t$ is the absorbance reading for the sample containing no additives on day t.

Antimicrobial Assays

Seven different common food-related bacterial strains were used to assess the antibacterial efficacy of the GTE preparations. *Brochothrix thermosphacta* (ATCC 11509) and *Staphylococcus aureus* (ATCC 25923) were obtained from the American Type Culture Collection, USA. *Listeria monocytogenes* Scott A was acquired from the Centers for Disease Control and Prevention, USA. *Salmonella* Typhimurium, *Escherichia coli* O157:H7, *Pseudomonas fragi*, and *Lactobacillus plantarum* were obtained from Food Science Australia, Australia. Bacteria were recovered from their stock cultures by inoculation into Tryptone Soya Broth (TSB, Oxoid Ltd., UK) followed by incubation for 24 h at 25°C. All antimicrobial tests were performed using TSB supplemented with 2% D-glucose and 2% yeast extract (TYSG). For all experiments, bacteria for inoculation were initially grown for 18 h at 25°C in 50 mL of TYSG with shaking. The minimum inhibitory concentration (MIC) of the compounds was determined using the microdilution method as described by Amsterdam (*45*). Briefly, the compounds were subjected to a two-fold dilution series, starting at a concentration of 1000 µg/mL, with TYSG in microtiter plate wells. The wells containing the diluted compounds were inoculated with a 5-µL bacterial suspension and incubated for 48 h at 25°C. The lowest concentration of the compound that reduced microbial growth by more than 50%, as determined by optical density readings and visual estimation, was taken to be the MIC.

Results and Discussion

The catechins - epigallocatechin (EGC), epicatechin (EC), epigallocatechin gallate (EGCG), and epicatechin gallate (ECG) - were easily chromatographed and quantified from the seven green tea preparations by reversed-phase HPLC as depicted in Figure 1. Results indicated that the content of total catechins in the prepared extracts ranged from 38.5 to 82.2% (w/w) (Table I). EGCG was the main catechin present in six of the seven preparations; only in one was EGC dominant. This is in contrast to the results of Price and Spitzer (2), in which they found that EGC, and not EGCG, was the most prevalent flavanol of nine green teas from China and Japan. In their study, however, an infusion was prepared using 4 g of tea and 250 mL of distilled water. In the present work, 95% (v/v) ethanol was employed to extract polyphenols from tea leaves; hence, the differences in the polarities may account for the variation in EGC and EGCG levels recovered. It is also interesting to note the wide variation in catechin levels based on the source material used. For example, sample No. 1 (*i.e.*, the extract prepared from green tea leaves of Anhui Province) had a markedly low content of EGCG and ECG at 13.9 and 6.6%, respectively, but the highest content of EC at 6.6% from all of the extracts examined. The Herbal Plus extract had the lowest content of EGC (7.3%), but a very high proportion of EGCG and ECG at 52.7 and 7.9%, respectively, in relation to the other prepared extracts. The acquired data are, nevertheless, in accordance with ranges reported in the literature (2). Thus, one can expect ranges in the content of individual catechins as well as total, depending upon the source and processing conditions to which the tea leaves have been subjected.

The TBARS values of cooked pork samples that were treated with either the synthetic antioxidant, TBHQ, or GTE preparations at several concentrations, are reported as mg malonaldehyde equivalents/kg meat and are presented in Figure 2. Antioxidant activity from GTE was observed in the meat systems after thermal processing. This is an important consideration since many bioactives, when added to foods, lose their biological activities during processing operations (*e.g.*, cooking). Using data from Figures 2 and 3 and the equation listed in the experimental section, the percent inhibition of TBARS development was calculated (Tables II and III): the results varied from 27.9 to 55.7% (*i.e.*, based on extract addition of 25 ppm), 62.6 to 88.7% (*i.e.*, 50 ppm), 78.7 to 100% (*i.e.*, 100 ppm) and 94.3 to 100% (*i.e.* 200 ppm).

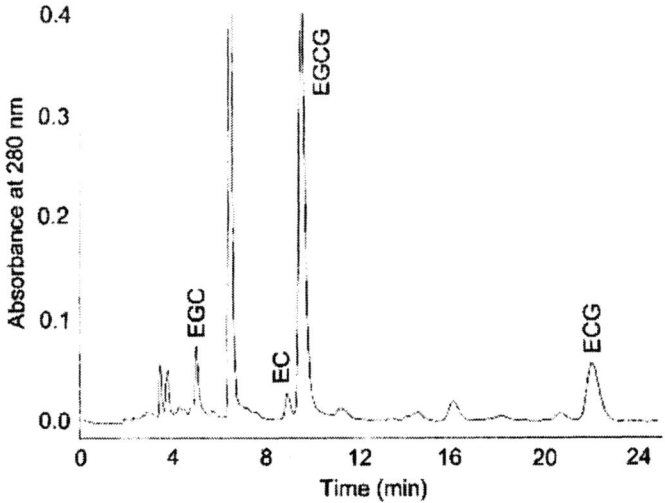

Figure 1. HPLC chromatogram of the separation of catechins from a green tea extract.

At a 25-ppm addition level, the sample of Herbal Plus and the crude extract containing the smallest amount of catechins showed the weakest antioxidant efficacy; however, the percent inhibition of meat lipid oxidation was not strictly dependent upon the content of catechins in each extract. The proanthocyanidins fraction was less active than the crude extract from which it was derived. When employed at a 200-ppm concentration, the reconstituted mixture inhibited TBARS formation to the same degree as that of its counterpart (*i.e.*, its crude extract), but at a 100-ppm addition level its efficacy was markedly less than the crude mixture. Evidently, there are other constituents present in GTE that impart antioxidant activity or, at minimum, exert a synergistic effect with the endogenous catechins. Use of GTE at the 200-ppm addition level or at reduced levels in combination with a chelator system such as sodium tripolyphosphate and sodium ascorbate can protect meats against oxidation similar to that observed from employment of a synthetic antioxidant such as TBHQ (results not shown). Shahidi and Alexander (*46*) also reported similar findings: the inhibitory effect of catechin standards against TBARS development in a meat model system was concentration dependent, being most inhibitory at a 200-ppm meat concentration. The antioxidant activity of catechins was also manifested in the retardation of hexanal formation, a major secondary lipid oxidation product of linoleic acid. Research on the antioxidant capability of catechins has been reported in other muscle food systems, including fish (*47*).

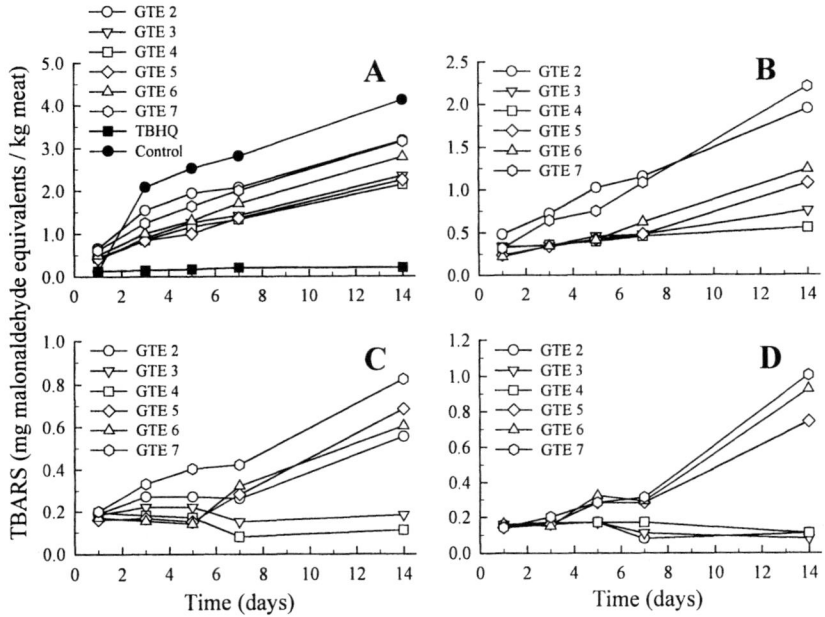

Figure 2. Inhibition of TBARS formation by green tea extract (GTE) preparations (A: 25 ppm; B: 50 ppm; C: 100 ppm; D: 200 ppm).

Table II. Percent Inhibition of Meat Lipid Oxidation by Green Tea Extract (No. 1), Reconstituted Extract, and Proanthocyanidins Fraction

Sample	50 ppm	100 ppm	200 ppm	500 ppm
Crude extract	–	78.7	87.2	–
Reconstituted extract	–	33.8	33.8	–
Proanthocyanidins fraction	18.0	44.0	44.0	86.2

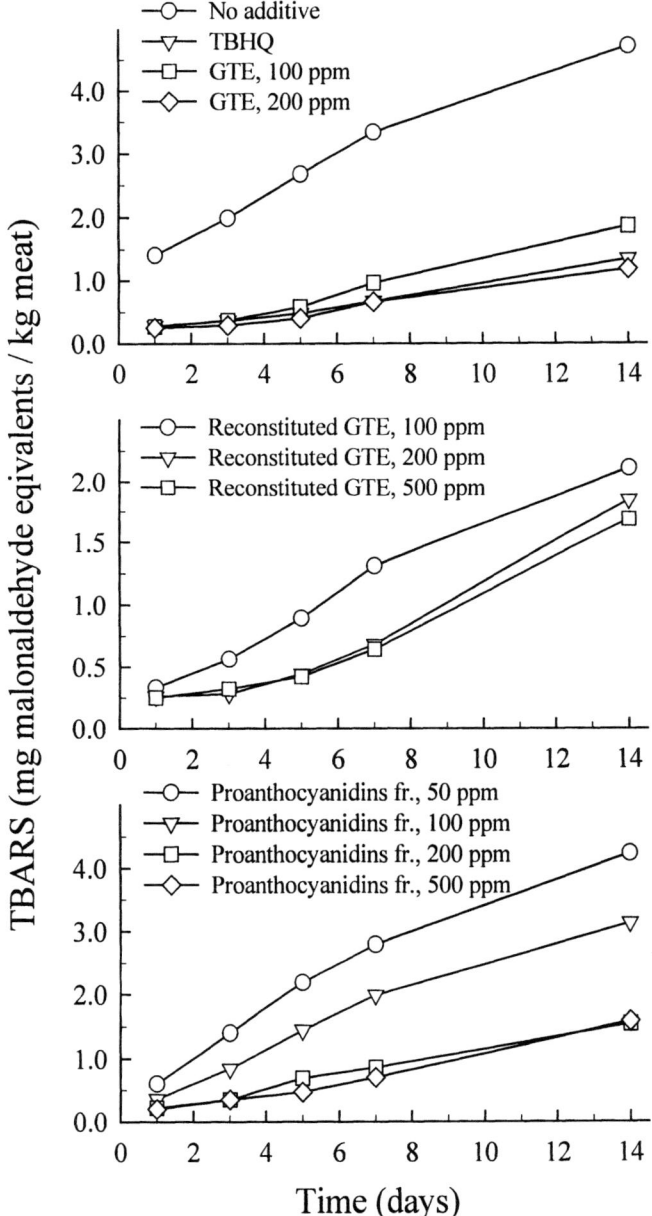

Figure 3. Inhibition of TBARS formation by green tea extract (GTE), reconstituted GTE and proanthocyanidins fraction.

Table III. Concentration Dependence of the % Inhibition of Meat Lipid Oxidation by Green Tea Extract Preparations after Seven Days of Storage

Sample	25 ppm	50 ppm	100 ppm	200 ppm
2	27.9	62.6	96.2	100
3	52.7	87.8	100	100
4	55.7	88.7	100	100
5	54.7	87.8	95.5	95.6
6	41.8	82.6	94.0	95.1
7	30.5	65.3	90.1	94.1

The effects of GTE on some food-related bacterial strains, expressed as the lowest concentration of the extract which reduces microbial growth by more than 50% (*i.e.*, the minimum inhibitory concentration), are reported in Table IV. All extracts were strongly active against the *Salmonella Typhimurium* strain, whereas only intermediate activity was observed for GTE against the *Pseudomonas fragi* strain. Identical MICs were noted in the case of GTE preparations No. 2, 3, 4, 6 and 7 against the *Listeria monocytogenes*, *Staphylococcus aureus* and *Lactobacillus plantarum* strains. Extract No. 5 exhibited an intermediate antibacterial activity against the *Brochothrix thermosphacta* strain. The results of Ikigai et al. (*48*) showed that catechins can act on and damage bacterial membranes: Gram-negative bacteria were more resistance than Gram-positive ones. Aqueous extracts of green teas inhibited a wide range of pathogenic bacteria, including methicillin-resistant *Staphylococcus aureus*. Tea extracts were bactericidal to *Staphylococcus* spp. and *Yersinia enterocolitica* at well below the concentrations found in a "cup of tea" (*49*). Tea catechins exhibited activity against *Helicobacter pylori*, of which EGCG was found to be the strongest (*50*). The lack of antibacterial activity against *Escherichia coli* O157:H7 in this work mirrors our results from a previous study: the GTE preparation, as well as isolated EC and EGC used in our 2000 study had no antimicrobial effect on the bacterial population of an *E. coli* K 12 strain at the concentrations employed (*51*). The proanthocyanidins fraction, on the other hand, did have a pronounced effect on the *E. coli* K 12 strain, as noted by ca. a two-log reduction in the plate count. Whereas EC and EGC had no efficacy against the *E. coli* K 12 strain, EGCG did have a slight one, as seen after 48 h of incubation. Fukai et al. (*35*) reported antibacterial activity of EGCG against phytopathogenic bacteria such as *Erwinia, Clavibacter, Xanthominas* and *Agrobacterium* spp., and this was 10-20 times stronger than that of EC.

Table IV. Antibacterial Activity of Green Tea Extracts Expressed as
Minimum Inhibitory Concentration (µg/mL)

Sample	2	3	4	5	6	7
Listeria monocytogenes	125	125	125	500	125	125
Staphylococcus aureus	125	125	125	250	125	125
Escherichia coli O157:H7	250	250	250	250	250	250
Brochothrix thermosphacta	250	250	250	125	250	250
Pseudomonas fragi	125	125	125	125	125	125
Salmonella Typhimurium	62.6	62.5	62.5	62.5	62.5	62.5
Lactobacillus plantarum	125	125	125	250	125	125

References

1. Graham, H. N. *Prev. Med.* **1992**, *21*, 334-350.
2. Price, W. E.; Spitzer, J. C. *Food Chem.* **1993**, *47*, 271-276.
3. Dreosti, I. E.; Wargovich, M. J.; Yang, C. S. *Crit. Rev. Food Sci. Nutr.* **1997**, *37*, 761-770.
4. Bushman, J. L. *Nutr. Cancer* **1998**, *31*, 151-159.
5. Matsuzaki, T.; Hara, Y. *Nippon Nogeikagaku Kaishi* **1985**, *59*, 129-134.
6. Muramatsu, K.; Fukuyo, M.; Hara, Y. *J. Nutr. Sci. Vitaminol.* **1986**, *32*, 613-622.
7. Weisburger, J. H.; Nagao, M.; Wakabayashi, K.; Oguri, A. *Cancer Lett.* **1994**, *83*, 143-147.
8. Kada, T.; Kaneko, K.; Matsuzaki, S.; Matsuzaki, T.; Hara, Y. *Mutat. Res.* **1985**, *150*, 127-132.
9. Katiyar, S. K.; Agarwal, R.; Wood, G. S.; Mukhtar, H. *Cancer Res.* **1992**, *52*, 6890-6897.
10. Hirose, M.; Hoshiya, T.; Akagi, K.; Futakuchi, M.; Ito, N. *Cancer Lett.* **1994**, *83*, 149-156.
11. Hirose, M.; Akagi, K.; Hasegawa, R.; Yaono, M.; Satoh, T.; Hara, Y.; Wakabayashi, K.; Ito, N. *Carcinogenesis* **1995**, *16*, 217-221.
12. Wang, Z. Y.; Huang, M-T.; Lou, Y-R.; Xie, J-G.; Reuhl, K. R.; Newmark, H. L.; Ho, C-T.; Yang, C. S.; Conney, A. H. *Cancer Res.* **1994**, *54*, 3428-3435.
13. Hara, Y.; Nakamura, K.; Fujino, R.; Hosaka, H.; Kohisae, S.; Asai, H.; Sugiura, M. In *Proceedings of Annual Meeting Japanese Society of Cancer Research*; Nagoya, 1984; p 993.
14. Yamane, T.; Hagiwara, N.; Tateishi, M.; Akachi, S.; Kim, M.; Okuzumi, J.; Kitao, Y.; Inagake, M.; Kuwata, K.; Takahashi, T. *Jpn. J. Cancer Res.*

1991, *82*, 1336-1339.
15. Yamane, T.; Takahashi, T.; Kuwata, K.; Oya, K.; Inagake, M.; Kitao, Y.; Suganuma, M.; Fujiki, H. *Cancer Res.* **1995**, *55*, 2081-2084.
16. Ruch, R. J.; Cheng, S. J.; Klaunig, J. E. *Carcinogenesis* **1989**, *10*, 1003-1008.
17. Huang, S-W.; Frankel, E. N. *J. Agric. Food Chem.* **1997**, *45*, 3033-3038.
18. Yeo, S-G.; Ahn, C-W.; Lee, T-G.; Park, Y-H.; Kim, S-B. *J. Korean Soc. Food Nutr.* **1995**, *24*, 299-304.
19. Amarowicz, R.; Shahidi F. *J. Food Lipids* **1995**, *2*, 47-56.
20. Roedig-Penman, A.; Gordon, M. H. *J. Agric. Food Chem.* **1997**, *45*, 4267-4270.
21. Wanasundara, U. N.; Shahidi, F. *J. Am. Oil Chem. Soc.* **1996**, *73*, 1183-1190.
22. Chen, C-W.; Ho, C-T. *J. Food Lipids* **1995**, *2*, 35-46.
23. Salah, N.; Miller, N. J.; Paganga, G.; Tijburg, L.; Bolwell, G. P.; Rice-Evans, C. *Arch. Biochem. Biophys.* **1995**, *322*, 339-346.
24. Cao, G.; Sofic, E.; Prior, R. L. *J. Agric. Food Chem.* **1996**, *44*, 3426-3431.
25. Nanjo, F.; Goto, K.; Seto, R.; Suzuki, M.; Sakai, M.; Hara, Y. *Free Radic. Biol. Med.* **1996**, *21*, 895-902.
26. Rice-Evans, C. A.; Miller, N. J.; Paganga, G. *Free Radic. Biol. Med.* **1996**, *20*, 933-956.
27. Sawai, Y.; Sakata, K. *J. Agric. Food Chem.* **1998**, *46*, 111-114.
28. Guo, Q.; Zhao, B.; Shen, S., Hou, J.; Hu, J.; Xin, W. *Biochim. Biophys. Acta* **1999**, *1427*, 13-23.
29. Nicoli, M. C.; Calligaris, S.; Manzocco, L. *J. Agric. Food Chem.* **2000**, *48*, 4576-4580.
30. Terao, J.; Piskula, M.; Yao, Q. *Arch. Biochem. Biophys.* **1994**, *308*, 278-284.
31. Yen, G-C.; Chen, H-Y. *J. Agric. Food Chem.* **1995**, *43*, 27-32.
32. Chung, H. Y.; Yokozawa T.; Soung, D. Y., Kye, I. S., No, J. K., Baek, B. S. *J. Agric. Food Chem.* **1998**, *46*, 4484-4486.
33. Hara, Y.; Watanabe, M. *Nippon Shokuhin Kogyo Gakkaishi* **1989**, *36*, 951-955.
34. Hara, Y.; Ishigami, T. *Nippon Shokuhin Kogyo Gakkaishi* **1989**, *36*, 951-955.
35. Fukai, K.; Ishigami, T.; Hara, Y. *Agric. Biol. Chem.* **1991**, *55*, 1895-1897.
36. Ahn, Y-J.; Sakanaka, S.; Kim, M.; Kawamura, T.; Fujisawa, T.; Mitsuoka, T. *Microb. Ecol. Health Dis.* **1990**, *3*, 335-338.
37. Okubo, T.; Ishihara, N.; Oura, A.; Serit, M., Kim, M.; Yamamoto, T.; Mitsuoka, T. *Biosci. Biotech. Biochem.* **1992**, *56*, 588-591.
38. Sakanaka, S.; Kim, M.; Taniguchi, M.; Yamamoto, T. *Agric. Biol. Chem.* **1989**, *53*, 2307-2311.

39. Sakanaka, S.; Sato, T.; Kim, M.; Yamamoto, T. *Agric. Biol. Chem.* **1990**, *54*, 2925-2929.
40. Sakanaka, S.; Shimura, N.; Aizawa, M.; Kim, M.; Yamamoto, T. *Biosci. Biotech. Biochem.* **1992**, *56*, 592-594.
41. Amarowicz, R.; Pegg, R. B.; Rahimi-Moghaddam, P.; Barl, B.; Weil, J. A. *Food Chem.* **2004**, *84*, 551-562.
42. Amarowicz, R.; Shahidi F. *Food Res. Int.* **1995**, *29*, 71-76.
43. Amarowicz, R.; Shahidi, F.; Wiczkowski, W. *J. Food Lipids* **2003**, *10*, 165-177.
44. Pegg, R. B. In *Current Protocols in Food Analytical Chemistry;* Wrolstad, R. E.; Acree, T. E.; An, H.; Decker, E. A.; Penner, M. H.; Reid, D. S.; Schwartz, S. J.; Shoemaker, C. F.; Sporns, P., Eds.; Spectrophotometric measurement of secondary lipid oxidation products (Unit D2.4 - Supplement 1); John Wiley & Sons: New York, NY, 2001; pp D2.4.1-D2.4.18.
45. Amsterdam, D. In *Antibiotics in Laboratory Medicine, Fourth Edition;* Lorian, V., Ed.; Susceptibility testing of antimicrobials in liquid media; Williams & Wilkins: Baltimore, MD, 1996; pp 52-111.
46. Shahidi, F.; Alexander, D. M. *J. Food Lipids* **1998**, *5*, 125-133.
47. He, Y.; Shahidi, F. *J. Agric. Food Chem.* **1997**, *45*, 4262-4266.
48. Ikigai, H.; Nakae, T.; Hara, Y.; Shimamura, T. *Biochim. Biophys. Acta* **1993**, *1147*, 132-136.
49. Yam, T. S.; Shah, S.; Hamilton-Miller, J. M. T. *FEMS Microbiol. Lett.* **1997**, *152*, 169-174.
50. Mabe, K.; Yamada, M.; Oguni, I.; Takahashi, T. *Antimicrob. Agents Chemother.* **1999**, *43*, 1788-1791.
51. Amarowicz, R.; Pegg, R. B.; Bautista, D. A. *Nahrung* **2000**, *44*, 60-62.

Chapter 10

Radical Scavenging Properties of Cold-Pressed Edible Seed Oils

John W. Parry, Kequan Zhou, and Liangli (Lucy) Yu

Department of Nutrition and Food Science, 0112 Skinner Building, University of Maryland, College Park, MD 20742

Cold-pressed cranberry, black raspberry, black caraway, carrot, and hemp seed oils were evaluated for their free radical scavenging capacities against $ABTS^{•+}$, $DPPH^{•}$ and oxygen radical (ORAC), as well as total phenolic contents (TPC). All tested cold-pressed seed oils directly reacted with and quenched $DPPH^{•}$ in the reaction mixture. The cold-pressed black caraway seed and cranberry seed oils at concentrations of 5.3 and 22.6 mg oil equivalent/mL exhibited stronger DPPH radical scavenging activities than that of 50 mM α-tocopherol. The $DPPH^{•}$ scavenging capacity was both time and dose dependent for all tested seed oils. Significant $ABTS^{•+}$ scavenging activities were detected in the cold-pressed seed oils in a range of 9-31 μmole TE/g oil. The greatest ORAC value of 220 μmole TE/g oil was detected in black caraway seed oil, and was followed by that of 160 and 28 μmole TE/g oil for carrot and hemp seed oils, respectively. In addition, the total phenolic contents were 0.1, 0.43, 1.6 and 3.5 mg gallic acid equivalent per gram of oil for the cold-pressed black raspberry, hemp, cranberry and black caraway seed oils, respectively. These results suggest that the cold-pressed edible seed oils may serve as dietary sources for natural antioxidants.

It is well recognized that dietary factors may influence human health and life quality. More consumers are interested in disease prevention and health promotion by dietary improvements. Recently, cold-pressed seed oils, including black caraway, carrot, hemp, cranberry and black raspberry seed oils, have become commercially available. The cold-pressing procedure involves neither heat nor chemicals, and is becoming a more interesting substitute for conventional practices because of consumers' desire for natural and safe food products. Cold-pressed seed oils may retain more natural beneficial components of the seeds such as natural antioxidants. Antioxidants are well recognized for their potential in health promotion and prevention of aging-associated diseases, such as cancer and heart disease (*1,2*). Several chemical mechanisms have been proposed to explain the beneficial effects of antioxidants. These mechanisms included directly reacting with and quenching free radicals, forming chelating complexes with transition metals, reducing peroxides, and stimulating antioxidative defense enzymes. Novel dietary sources of natural antioxidants are desired for health benefits. Several cold-pressed seed oils have been investigated for their fatty acid composition and oxidative stability, as well as other physical properties (*3*). Cold-pressed cranberry and hemp seed oils contain about 20% α-linolenic acid, and may serve as dietary sources for ω-3 fatty acids. Cold-pressed carrot seed oil was high in oleic acid content (80%) and low in total saturated fatty acids. These results showed the potential health benefits from consuming cold-pressed edible seed oils. In addition, the cold-pressed seed oils exhibited excellent oxidative stability compared to commercial soybean and corn oils, suggesting the possible presence of natural antioxidants in the cold-pressed seed oils (*3*). Therefore, the present study was conducted to evaluate the radical scavenging properties of cold-pressed carrot, hemp, black caraway, cranberry, and black raspberry seed oils.

Materials and Methods

Materials.

Cold-pressed, 'extra virgin,' unrefined black caraway, carrot, hemp, cranberry and black raspberry seeds oils were provided by Badger Oil Company (Spooner, WI). Fluorescein (FL), 2,2'-bipyridyl and 2,2-diphenyl-1-

picryhydrazyl radical (DPPH•), 2,2'-bipyridyl, and 6-hydroxy-2,5,7,8-tetramethylchroman-2-carboxylic acid (Trolox) were purchased from Sigma-Aldrich (St. Louis, MO), while 2,2'-azobis (2-amino-propane) dihydrochloride (AAPH) was obtained from Wako Chemicals USA (Richmond, VA). β-cyclodextrin (RMCD) was purchased from Cyclolab R & D Ltd. (Budapest, Hungary). A total antioxidant status kit was purchased from Randox Laboratories Ltd. (San Francisco, CA). All other chemicals and solvents were of the highest commercial grade and used without further purification.

Preparation of antioxidant extract

A measured amount of cold-pressed seed oil was extracted with methanol at ambient temperature. The methanol extract was flushed with nitrogen, and kept in the dark until further analysis. In order to prepare dimethyl sulfoxide DMSO) solution, methanol was removed under vacuum from a known volume of the methanol extract, and the residue was completely dissolved in DMSO. The resulting DMSO solution was also kept in the dark after nitrogen flushing until further analysis.

Radical DPPH scavenging activity

Radical DPPH• scavenging capacity of the antioxidant extract was estimated according to the previously reported procedure (*1*). Freshly made DPPH• solution was added to antioxidant extracts with known concentrations and mixed to start the radical-antioxidant reaction. The initial concentration was 100 μM for DPPH•, and the total volume was 2000 μL for each reaction mixture. The absorbance at 517 nm was determined against a blank of pure methanol at 0, 0.5, 1, 2, 5, 10, 20, 40 and 80 minute of reaction and used to estimate the remaining radical levels according to the standard curve. The dose and time dependencies of carrot oil extract and DPPH• reactions were demonstrated by plotting the percent of DPPH• remaining against time for each level of the carrot seed oil extract tested. Tests were done in triplicate for each antioxidant.

Radical cation $ABTS^{•+}$ scavenging activity

Radical scavenging capacity of each antioxidant extract was evaluated against $ABTS^{•+}$ generated by the enzymatic method using a commercial kit from Randox Laboratories Ltd. (San Francisco, CA) (*2,4*). The absorbance at 734 nm was measured at 1 min of the antioxidant-radical reaction, and used to calculate the trolox equivalent using a standard curve prepared with trolox. The tests were conducted in triplicate for each extract.

ORAC assay

The ORAC assay was performed using fluorescein (FL) as the fluorescent probe following a previously described protocol (5). The DMSO stock solution was diluted with the 7% RMCD in acetone/water (1:1, v/v) to obtain the assay sample. The final assay mixture contained 0.067 µM of FL, 60 mM of AAPH, 300 µL of the assay sample or 7% RMCD containing DMSO for a reagent blank. The total volume was 3000 µL for each reaction mixture. The fluorescence of each assay mixture was determined and recorded once per minute using a Turner Quantech^{TM} Fluorometer (Dubuque, IA). Trolox was used to prepare the standard curve to calculate the trolox equivalent of the cold-pressed seed oil.

Total phenolic contents

The total phenolic content of each oil extract was determined using Folin-Ciocalteu reagent (*1*). The reaction mixture contained 100 µL of the oil extract in DMSO, 500 µL of the Folin-Ciocalteu reagent, and 1.5 mL of 20% sodium carbonate. The final volume was made up to 10 mL with pure water. After two hours of reaction at ambient temperature, absorbance at 765 nm was measured and used to calculate the phenolic contents using a standard curve prepared with gallic acid. Triplicate reactions were performed.

Statistic analysis

Data were reported as mean ± standard deviation (SD). Analysis of variance and least significant difference tests were conducted to identify differences among means. Statistical significance was declared at p<0.05.

Results and Discussion

Antioxidants, including radical scavengers, may protect important molecules from radical attacks and consequently reduce the risk of aging-associated health problems, such as cancer and heart disease. Natural antioxidants are in high demand for preparing functional foods and supplements because of their possible health benefits and consumer preference of chemical free products. Examination of free radical scavenging properties of the selected cold-pressed seed oils will identify the potential utilization of these oils in promoting human health. Cold-pressed carrot, hemp, black caraway, black raspberry and cranberry seed oils were evaluated form their capacities to directly react with and quench stable DPPH•, ABTS+•, and oxygen radicals. Significant radical scavenging capacities were detected in all tested cold-pressed seed oils, along with different total phenolic contents. Total phenolic content was determined for individual seed oil samples because they may contribute to the overall antioxidant properties.

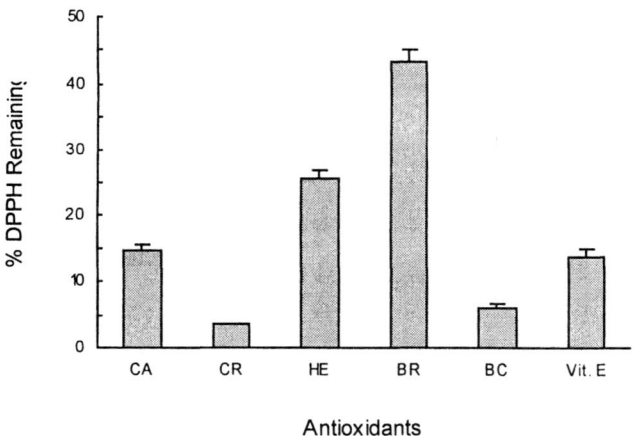

Figure 1. Comparison of radical DPPH scavenging activity. The radical DPPH scavenging capacities of carrot (CA), cranberry (CR), hemp (HE), black raspberry (BR) and black caraway (BC) seed oil extracts are compared with that of 50 mM α-tocopherol (Vit E). The final concentrations were 21.8, 22.6, 21.0, 42.0, and 5.3 mg oil equivalent/mL in the reaction mixtures for carrot, cranberry, hemp, black raspberry and black caraway seed oils, respectively. The final concentration of the DPPH radical was 100 μM in all reaction mixtures, and the total volume was 2000 μL for each reaction mixture. The absorbance at 517 nm of each reaction was measured at minute 40 of the reaction and used to calculate the percent DPPH radical remaining. All tests were conducted in triplicate and the means are used. The vertical bars represent the standard deviation for each data point.

The five cold-pressed seed oils were compared to 50 mM α-tocopherol for their DPPH radical scavenging capacities. All five seed oils showed significant radical scavenging activity against DPPH•, but differed in their potentials to directly react with and quench free DPPH radicals (Figure 1). Black caraway seed oil, at a concentration of 5.3 mg oil equivalent per mL, and cranberry seed oil, at a concentration of 22.6 mg oil equivalent per mL, exhibited greater DPPH• scavenging activity than that observed in 50 mM α-tocopherol, which is 21.5 mg α-tocopherol/mL in the reaction mixture. Among the five seed oils, black raspberry seed oil showed lowest DPPH radical scavenging capacity. Similar dose and time effects on DPPH• scavenging activity were observed for all tested seed oils. The dose and time effects of cold-pressed carrot seed oil against DPPH radicals are reported in Figure 2. Similar kinetics was also detected in antioxidant-DPPH• reaction for all tested cold-pressed seed oil extracts (Figure 3). In addition, the two black raspberry seed oil samples differed in their capacities to quench DPPH radicals, suggesting that the seed quality and processing conditions may significantly alter the radical scavenging properties of cold-pressed seed oils.

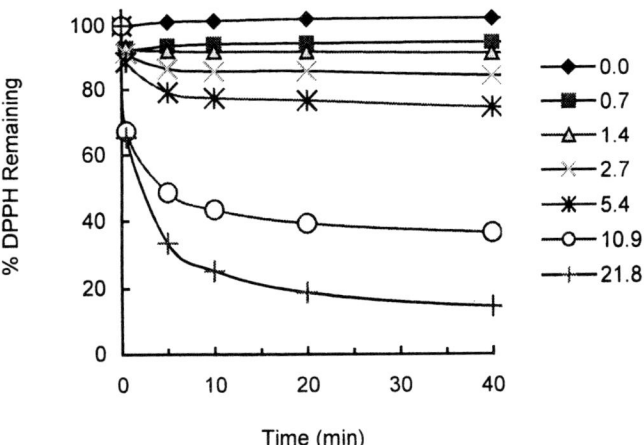

Figure 2. Dose and time effects on DPPH scavenging activity of carrot seed oil extract. 0, 0.7, 1.4, 2.7, 5.4, 10.9 and 21.8 represent the final carrot seed oil extract concentrations of mg oil equivalent/mL in the reaction mixtures. The initial DPPH radical concentration was 100 μM in all reaction mixtures. All tests were conducted in triplicate and the means are used.

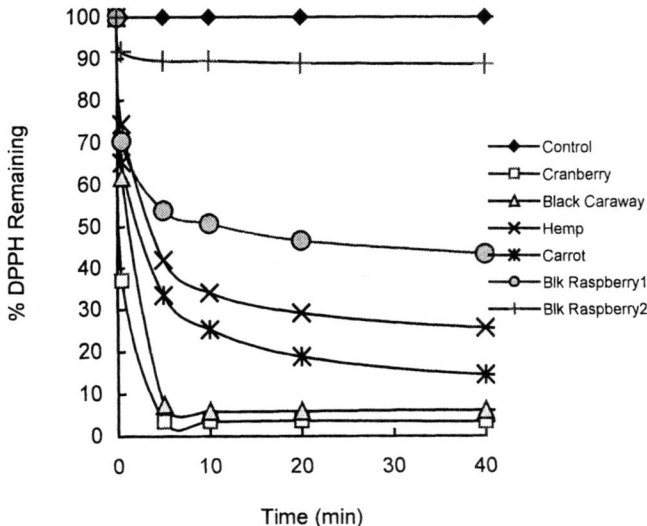

Figure 3. Reaction kinetics of cold-pressed seed oil extracts with DPPH radicals. Cranberry, Black Caraway, Hemp, Carrot, Blk Raspberry 1, Blk Raspberry 2 represent cranberry, black caraway, hemp, carrot, black raspberry sample 1, and black raspberry sample 2, respectively. Control represents the reaction mixture containing no antioxidant. The final concentrations were 22.6, 5.3, 21.0, 21.8, 41.7 and 41.7 for cranberry, black caraway, hemp, carrot, black raspberry sample 1, and black raspberry sample 2, respectively. The initial DPPH radical concentration was 100 µM in all reaction mixtures. All tests were conducted in triplicate and the means are used.

It is noted that the radical system may influence the estimation of free radical scavenging capacity. Two or more radical systems are required to better understand the radical scavenging property of a selected antioxidant. Cation $ABTS^{+\bullet}$ scavenging activities of the cold-pressed cranberry, carrot, hemp and black caraway seed oils were evaluated and reported in Figure 4. Black caraway seed oil had the greatest $ABTS^{+\bullet}$ scavenging capacity followed by cranberry seed oil, on a per weight basis. Interestingly, hemp seed oil showed stronger $ABTS^{+\bullet}$ scavenging capacity than carrot seed oil, but carrot seed oil had stronger DPPH radical scavenging activity than hemp seed oil, under the same experimental conditions. This may be explained by the different chemical mechanisms involved in the two assays.

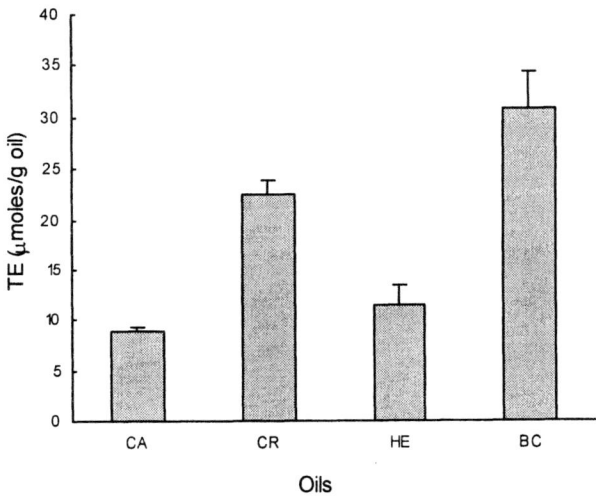

Figure 4. ABTS+• scavenging capacities of the cold-pressed seed oils. ABTS+• scavenging capacities of cold-pressed seed oils were estimated by an enzymatic method. TE stands for trolox equivalent. CA, CR, HE, and BC represent carrot, cranberry, hemp, and black caraway seed oil extracts. All tests were conducted in triplicate and the means are used. The vertical bars represent the standard.

In addition, cold-pressed hemp, carrot, and black caraway seed oils were examined for their oxygen radical absorbing capacities (ORAC). The hemp seed oil had the greatest ORAC value of 220 trolox equivalents (TE) µmoles per gram of oil, and followed by carrot and hemp seed oils, with ORAC values of 160 and 28 TE µmoles/g oil, respectively. These ORAC values are comparable to the average ORAC value of other fruits and vegetables, including that of highbush blueberry (15.9 µmoles TE/g fresh fruit), apple, orange and banana (5.0 – 24.5 µmoles TE/g fresh fruit), and cucumber, cabbage and spinach (0.5 – 179 µmoles TE/g fresh fruit) (6-8). These data suggest that cold-pressed edible seed oils may serve as dietary sources of natural antioxidants and be useful for disease prevention and health promotion. These data also showed the possible value-added utilization of these seeds, especially the fruit seeds, which are the by-products from fruit juice production.

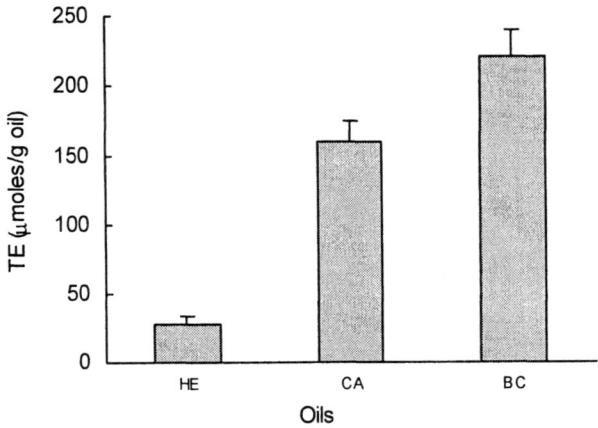

Figure 5. ORAC values of the cold-pressed seed oils. ORAC stands for oxygen radical absorbance capacity, and TE is the trolox equivalent. HE, CA, and BC represent hemp, carrot, and black caraway seed oil extracts, respectively. All tests were conducted in triplicate and the means are used. The vertical bars represent the standard deviation of each data point deviation of each data point.

It is believed that phenolic compounds make important contribution to the overall antioxidant capacity of a selected antioxidant preparation. Total phenolic contents of the cold-pressed cranberry, black raspberry, black caraway, and hemp seed oils were estimated. Phenolics were detected in all tested seed oils (Figure 6). The cold-pressed seed oils differed significantly in their total phenolic contents. Black caraway seed oil contained greatest concentration of total phenolic content among all tested cold-pressed seed oils, while black raspberry seed oil contained lowest phenolics. The phenolic content in cold-pressed carrot seed oil was not available due to the availability of the oil sample.

In conclusion, cold-pressed black caraway, cranberry, carrot, hemp, and black raspberry seed oils may contain significant level of natural radical scavengers and phenolic compounds. Among the five tested cold-pressed edible seed oils, black caraway seed oil had the strongest free radical scavenging activity and the greatest total phenolic contents. These results indicate that cold-pressed edible seed oils may serve as dietary sources of natural antioxidants for disease prevention and health promotion.

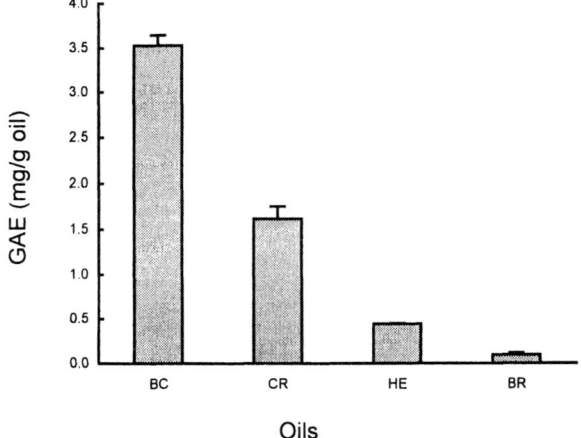

Figure 6. Total phenolic contents of the cold-pressed seed oils. Total phenolic contents of the cold-pressed edible seed oils were expressed as mg of gallic acid equivalent per gram of oil. BC, CR, HE, and BR represent black caraway, cranberry, hemp, and black raspberry seed oil extracts, respectively. All tests were conducted in triplicate and the means are used. The vertical bars represent the standard deviation of each data point.

Acknowledgement

The Badger Oil Company (Spooner, WI) kindly provided donated the cold-pressed black caraway seed oil for this study. The authors would like to thank Mr. Dwayne Adams for his technical assistance.

References

1. Yu, L.; Haley, S.; Perret, J.; Harris, M.; Wilson, J.; Qian, M. J. Agric. Food Chem. 2002, 50, 1619-1624.
2. Yu, L.; Davy, B.; Wilson, J.; Melby, C. L. J. Food Sci. 2002, 67, 2600-2603.
3. Paker, T.D.; Adams, D.A.; Zhou, K.; Harris, M.; Yu, L. J. Food Sci. 2003, 68, 1240-1243.

4. Miller, N. J.; Rice-Evans, C. A. Free Rad. Res. 1997, 26, 195-199.
5. Huang, D.; Ou, B.; Hampsch-Woodill, M.; Flanagan, J. A.; Deemer, E. K. J. Agric. Food Chem. 2002, 50, 1815-1821.
6. Ehlenfeldt, M. K.; Prior, R. L. J. Agric. Food Chem. 2001, 49, 2222-2227.
7. Wang, H.; Cao, G.; Prior, R. L. J. Agric. Food Chem. 1996, 44, 701-705.
8. Cao, G.; Sofic, E.; Prior, R. L. J. Agric. Food Chem. 1996, 44, 3426-3431.

Chapter 11

Honeybush Tea: Chemical and Pharmacological Analyses

Mingfu Wang[1], Rodolfo Juliani[1], James E. Simon[1], Albert Ekanem[1], Chia-Pei Liang[2], and Chi-Tang Ho[2]

[1]New Use Agriculture and Natural Plant Products Program, Department of Plant Biology and Pathology, and [2]Department of Food Science, Rutgers, The State University of New Jersey, 65 Dudley Road, New Brunswick, NJ 08901

The honeybush plant, *Cyclopia spp.* is used to brew a traditional South African tea and is derived from several of the over 20 species of woody legumes of this genus. Honeybush tea has been a popular regional herbal tea and is now becoming available in wider markets in Africa, Europe and the US. Honeybush tea enjoys the reputation as a caffeine-free tea with a pleasant aroma and fruity honey-like flavor. This paper reviews the chemistry, health benefits and the processing of the honeybush tea, and describes analytical methods for the chemical characterization of the plants natural products including the volatile components and polyphenols which could be used for quality control and product profiling. Honeybush is a rich source of phenolic compounds such as mangiferin, isomangiferin, luteolin 7-rutinoside, diosmin, hesperidin, luteolin, and hesperitin. Honeybush contains over 15 aromatic volatiles with α-Terpineol as the major volatile constituent and the monoterpenes largely responsible for the sweet, floral and fruity notes of the tea, while phenylethyl alcohol and 5-methylfurfural imparting sweet and honey notes to the honeybush.

Introduction

The genus *Cyclopia* (Honeybush plant) includes approximately 24 species of woody legumes from South Africa and was first mentioned in the western botanical literature in 1705 (*1-3*). Several *Cyclopia* species are used to brew a traditional African tea-honeybush tea. The *Cyclopia* spp. are woody perennials native to mountainous areas in limited ecological ranges in South Africa (the Western Cape and parts of the Eastern Cape Provinces). The leaves are trifoliate and the flowers bright yellow with a characteristic sweet honey scent from which the name (honeybush) was derived. The seeds are hard shelled. The aerial parts (leaves, flowers and stem) are used for preparation of a beverage drink (honeybush tea). Several species including *C. intermedia, C. sessiliflora, C. genistoides* and *C. maculata* are collected and now cultivated for the production of honeybush tea (*4*). Among these species, it is the *Cyclopia intermedia* which is best known to be rich in phenolic compounds and which serves as a main source of genetic material for honeybush tea production.

The honeybush tea is produced from the aerial parts (leaves, flowers and stem) of the plant. The plant is usually harvested from the wild between May and June. Plants are cut at about 0.3 m above the ground, collected and fermented to produce the tea. The plant material is first gathered from the field and brought into a 'tea court' where the material is then prepared for processing. There are two basic methods of fermentation used in honeybush tea production: (1) the curing and (2) the elevated temperature in a pre-heated oven. Basic curing begins with the placement of the harvested material onto a cement floor within a 'tea court' and placed into mounds or heaps of honeybush plant materials. These mounds or rows of honeybush piles are then covered with canvas or Hessian bags and allowed to ferment for three to five days. The progress of fermentation is observed by changes in the color of the plant materials to dark-brown, and the generation of a floral and sweet honey-like aroma. The mounds are physically turned over from time to time to ensure an even distribution of heat generated by the process so that the entire mound of material regardless of placement within the mound gets fermented as evenly as possible. When fermentation is completed, the materials are spread out onto the same cement floor in the tea court to dry under full sun to both preserve the plant material and to avoid the growth of molds, which is a major threat to the curing-heap method. The second method of fermentation includes forced heat or oven fermentation accomplished by using a pre-heated bakery oven. The oven temperature ranges from 60 to 90 $^{\circ}$C. Prior to fermentation, plant materials are stocked in Hessian bags and scalded with warm water to quicken the process. The oven fermentation method is preferred because it ensures an even distribution of heat, a more rapid and uniform process of fermentation (48 h) and eliminates or reduces contamination by molds. Both methods are used as different communities and processors conduct this under a variety of conditions and the costs involved in both are significantly different. Often both methods are combined in that the fermentation may be completed under the mound method with full sun drying and then the material may be transported to another processor who further dries, cleans, prepares and cuts the honeybush into specific particle sizes for blending, and tea packaging.

Chemical Components of Honeybush Tea

Honeybush tea is an herbal infusion with a reputation to contain no caffeine (7). *Cyclopia* species have been found to be a rich source of phenolic compounds with a xanthone C-glycoside, mangiferin, hesperidin and isosakuranetin as the major phenolics in the unfermented leaves (4). The highest levels of mangiferin (3.61 g/100 g) and hesperidin (1.74 g/100 g) were found in *C. genistoides* and *C. intermedia*, respectively, in four studies species. The fermented aerial part of *C. intermedia* (honeybush tea) was recently reported to contain many phenolic compounds including xanthones, isoflavones, flavones, flavanones, flavonols, coumestans, cinnamic acid derivatives, tyrosol derivatives, benzaldehyde derivatives (2,5-7). (+)-pinitol, a nonphenolic compound was also isolated from *Cyclopia intermedia*.

Xanthones

Two known xanthones, 2-β-D-glucopyranosyl-1,3,6,7-tetrahydroxy- (mangiferin) and 4-β-D-glucopyranosyl-1,3,6,7-tetrahydroxy- (isomangiferin) were identified in honeybush tea (7).

Flavonones

Four flavanones, 5,7,4'-trihydroxyflavanone (naringenin), 5,7,3',4'-tetrahydroxyflavanone (eriodictyol), 5,7,3'-trihydroxy-4'-methoxyflavanone (hesperitin), 5,3'-dihydroxy-4'-methoxy-7-O-rutinosylflavanone (hesperidin), were reported to be isolated and identified from fermented *Cyclopia intermedia* by comparing of the ^{1}H-NMR data of their O-acetyl derivatives with literature in 1998 (5). Four additional flavanone glycosides, prunin (naringenin-7-glucoside), 7-O-β-D-glucopyranosyleriodictyol, 5-O-β-D-glucopyranosyleriodictyol and 5-

O-α-D-rutinosylnaringenin were later reported to be purified from *fermented Cyclopia intermedia* as acetyl derivatives (the extract was acetylated before isolation) (7).

Isoflavones

Six isoflavones, 7-hydroxy-4'-methoxyisoflavone (formononetin), 7-hydroxy-6,4'-dimethoxyisoflavone (afrormosin), 7,3'-dihydroxy-4'-methoxyisoflavone (calycosin), 7-hydroxy-3',4'-methylenedoxyisoflavone (pseudobaptigen), and 7-hydroxy-6-methoxy-3',4'-methylenedioxyisoflavone (fujikinetin), wistin and 7-[O-α-apiofuranosyl-(1″→6″)-β-D-glucopyranosyl]-4-methoxy-isoflavone were identified from a honeybush tea extract and the identification was performed on the respective O-acetyl derivatives (5,7).

Flavonols

Five flavonol glycosides have been identified from honeybush tea (*Cyclopia intermedia*). The isolated compounds included 5-O-α-glucopyranosylkaempferol, 6-C-β-D-glucopyranosylkaempferol, 8-C-β-D-glucopyranosylkaempferol, 3-hydroxy-6-[O-α-apiofuranosyl-(1″→6″)-β-D-glucopyranosyloxy]-3',4'-methylene-dioxyflavonol and 3-O,6-C-di-β-D-glucopyranosylkaempferol (7).

Flavones

Three flavones, 3',4',7-trihydroxyflavone, luteolin and diosmetin were present in a honeybush tea extract as acetyl derivatives (5,7).

C6-C1, C6-C2 and C6-C3 Metabolites

Cinnamic acid, tyrosol, 3-methoxytyrosol, 2-{4-[O-α-apiofuranosyl-(1″→6')-β-D-glucopyranosyloxy]phenyl}-ethanol, 4-[O-α-apiofuranosyl-(1″→2')-β-D-glucopyranosyloxy]benzaldehyde were also present as O-acetyl derivatives from honeybush tea extract (7).

Coumestans

Three coumestans, 3-hydroxy-8,9-methylenedioxy coumestan (medicagol), 3-methoxy-8,9-methylenedioxy-coumestan (flemichapparin) 3-hydroxy-4-methoxy-8,9-methylenedioxy-coumestan (sophoracoumestan B) were identified in a fermented *Cyclopia intermedia* extract (5).

Analytical methods

Two HPLC methods have been reported for the analysis of the phenolic compounds in unfermented *Cyclopia*. The first method was published by Nysschen et al. in 1996 (2). Three compounds, mangiferin and two flavonoids were analyzed in 22 species of *Cyclopia* leaves. The HPLC was run on a Phenomenex IB-sil column (C_{18}, 5 µm, 250*4.6 mm). The flow rate was 1 mL/min, the injection volume was 20 µL and detection wavelength was 280 and 330 nm. The mobile phases were water (1% acetic acid)/methanol gradient (30% to 100 methanol in 20 min).

Later, Joubert et al. (4), reported one reversed-phase HPLC method or the quantitation of some of the phenolic compounds including mangifrin, isomangiferin and hesperidin from four *Cyclopia* species. Mangiferin (2-β-D-glucopyranosyl-1, 3,6,7-tetrahydroxy-9H-xanthen-9-one), hesperitin (5,7,3'-dihydroxy-4'-methoxy-flavanone), hesperidin (5,3'-dihydroxy-4'-methoxy-7-*O*-rutinosylflavone), eriodictyol (5,7,3',4'-tetrahydroxyflavanone), formononetin (7-hydroxy-4'-methoxyisoflavone), luteolin (5,7,3',4'-tetrahydroxyflavone), Isomangiferin (4-β-D-glucopyrosyl-1,3,6,7-tetrahydroxy-9H-xanthen-9-one), and medicagol (3-hydroxy-8,9-methylenedioxycoumestan) were used as standards, but only mangiferin, isomangiferin and hesperidin were found in sufficient amounts to be quantified. Plant material was dried at 40° C and pulverized before analysis. Then, 0.25 g of dry material was extracted by 30 mL of methanol through sonication for half hour. The HPLC was run on a Phenomenex Synergy Max-RP C12 column with TMS end-capping (4 µm, 150*4.6 mm). the detection wavelength was 280 nm, the injection volume was 20 µL and flow rate was 1 mL/min. The solvent was water (2% acetic acid)-acetonitrile gradient system, 0-6 min (12% B), 7 min (18 % B), 14 min (25% B), 19 min (40% B), 24 min (50% B), 29 min (12% B) the total running time is 29 minutes.

Recently, we analyzed the total phenolic contents of honeybush tea and found that the extraction solvent is of critical importance for accurately assessing different phenolic compounds. Using a concentration range of methanol (20% to 100%), we found that 60% methanol was the best solvent concentration to recover the phenolic compounds (Figure 1). The total phenols were then evaluated by the Folin-Ciocalteu method (8). One mL of clear extraction solution prepared from each sample preparation was transferred to 100 mL volumetric flask and swirled with 60-70 mL HPLC grade water. 5 mL of Folin-Ciocalteu's phenol reagent (Sigma, St. Louis, MO) were added and swirled. After 1 min and before 8 min, 15 mL of sodium carbonate solution (20 g in 100 mL) were added and mixed. This was recorded as time zero. Then the volumetric flask was diluted to 100 mL using water. The solution was mixed thoroughly by inverting it several times. After 2 hours (within 1-2 min), the UV absorption, range at 550-850 nm and maximum absorbance about 760 nm, was recorded (HP Model 8453, spectrophotometer) (Agilent Technologies, Inc., Palo Alto, CA) and same solution without the extraction solution served as blank solution. All tests were run in duplicate and averaged. Gallic acid was used as standard for this test.

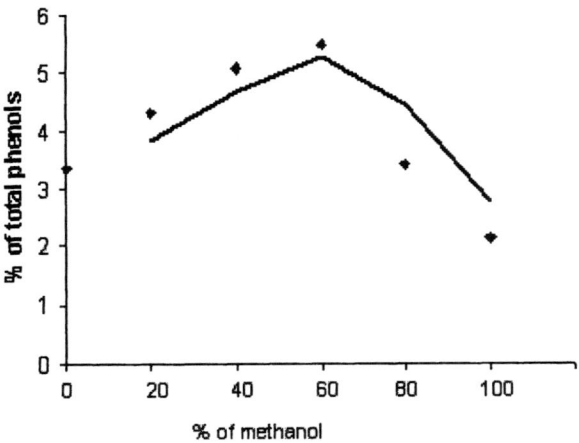

Figure 1. The relationship between extraction solvent and the recovery of the total phenolic content from honeybush extract

A LC/MS method was then developed to analyze phenolic compounds in honeybush tea. Analytical HPLC analysis was performed on a Hewlett-Packard 1100 modular system equipped with an auto-sampler, a quaternary pump system, a photodiode array detector and a HP Chemstation data system with MSD Trap with an electrospray source and software of HP ChemStation, Bruker Daltonics 4.0 and Data Analysis 4.0. but with a 2 to 1 stream splitting for the MSD detector (Agilent Technologies, Inc., Palo Alto, CA). A pre-packed 250×4.6 mm (5 µM particle size) Prodigy ODS3 column (Phenomenex, Torrance, CA) were selected for HPLC analysis. The absorption spectra were recorded from 200 to 400 nm for all peaks. The column temperature was 40 °C and the mobile phase included water (containing 0.1% formic acid, solvent A) and acetonitrile (solvent B) in the following gradient system: initial 4% B, linear gradient to 10% B in 8 min, then linear to 30% B in 22 minutes, then linear gradient to 60% B in 20 min and keep at 60% B for additional 10 min. The total running time was 60 min. The post running time was 10 min. The flow rate was set at 1.1 mL/min. The honeybush tea powdered sample was prepared in 60% methanol. The extract was prepared in aqueous methanol and 300 mg of ground tea powder was transferred into a 50 mL volumetric flask and about 35 mL of 60% methanol were then added and samples were sonicated for 60 min. The flasks were allowed to cool to room temperature and then filled to volume with 60% methanol. Using a disposable syringe and 0.45 µm filter, the samples were filtered into HPLC vials for HPLC analysis. The electrospray mass spectrometer (ESI-MS) was operated under positive ion mode and optimized collision energy level of 60%, scanned from m/z 100 to 700. ESI was conducted using a needle voltage of 3.5 kV. High-purity nitrogen (99.999%) was used as dry gas and flow rate at 10 mL/min, capillary temperature at 350 °C. Helium was used as Nebulizer at 50 psi. The ESI interface and mass spectrometer parameters were optimized to obtain maximum sensitivity. A total of six major phenolic compounds were tentatively identified by comparing the UV and MS spectra with the reference standards and by their $[M+1]^+$ and $[M+Na]^+$ ions. The major phenols are identified as mangiferin (**1**), isomangiferin (**2**), Luteolin 7-rutinoside (**3**), diosmin (**4**), hesperidin (**5**), luteolin (**6**) and hesperitin (**7**). The representative total ion chromatogram and HPLC chromatogram are shown in Figures 2 and 3, respectively.

Analysis of Volatile Components

As the honeybush tea has such a pleasant aroma, we wanted to capture the volatile components by hydrodistillation. Since no visual accumulation of essential oil was observed in the clevenger trap, the volatile components were extracted from the distillation water using hexane. The distillation water was extracted three times with hexane. The hexane was then removed under reduced pressure using a rotary evaporator set a 40 °C. The volatile oil was then re-suspended in two mL of hexane and subjected to gas chromatography/mass spectrometry analysis (GC/MS). The volatile components were identified by GC/MS using a gas chromatograph (Agilent 6890) couple to a mass detector (Agilent 5973) (Agilent Technologies, Inc., Palo Alto, CA) and HP-5 MS (5% Phenyl-95% Methyl Siloxane) column (30.0 m x 250 µm x 0.25 µm) (Agilent Technologies, Inc., Palo Alto, CA) was used. 5 µL sample was injected with the split ratio of 1:50. The injector was set at 220 °C. The temperature program was the following: 60 °C/1 min, and then 4 °C/min, until 200 °C during 15 min. Ionization voltage was 70 eV and ion temperature was 280 °C. An improved method to recovery the aromatic volatiles would be via solvent extraction to minimize any potential hydrolytic alterations in the chemical composition.

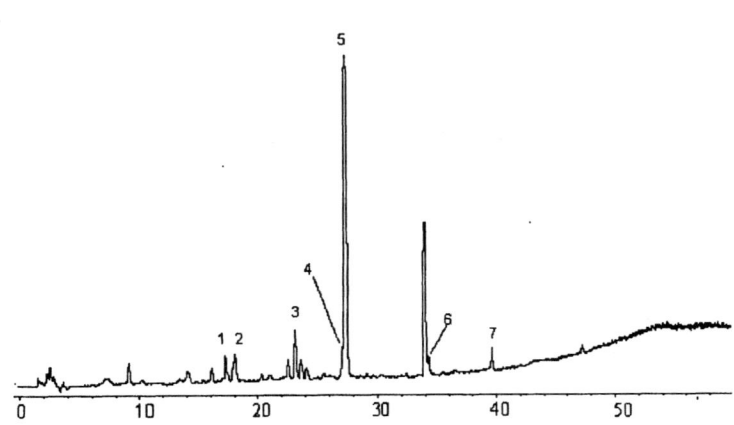

Figure 2. Total ion chromatogram of a honeybush tea sample

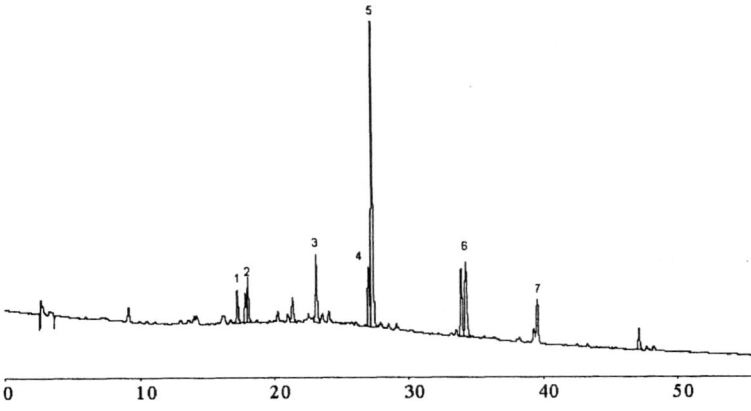

Figure 3. HPLC chromatogram of a honeybush tea sample (330 nm)

Volatile Components

The aroma components were dominated by monoterpene alcohols, of which α-terpineol (28%) was the major component, with minor amounts of linalool (7%), nerol (2%) and geraniol (8%). These monoterpenes are responsible for the sweet, floral and fruity notes of the tea, while other components such as phenylethyl alcohol (3%) and 5-methylfurfural (2.1%) imparted also sweet and honey notes. Other volatiles such as eugenol (6%), linalool oxides (7%), and methyl-heptenol (3%) were also detected.

Table 1 - Chemical Composition of the Volatile Components from Honeybush

Component	*Relative percentage*
Benzaldehyde	0.2
5-Methylfurfural	2.1
6-methyl-5-hepten-2-one	1.0
6-methyl-5-Hepten-2-ol	2.8
2,4 Heptadienal	1.8
Benzene acetaldehyde	3.1
(E)-Linalool oxide	3.8
Trans-Linalool oxide	2.7
Linalool	6.5
Phenylethyl alcohol	2.9
α- Terpineol	27.9
Nerol	2.2
Carvone	0.6
Geraniol	8.0
Eugenol	6.4
Methyleugenol	1.0
Total % Volatiles Identified	72.8

Bio-activities

Honeybush tea was found to have antioxidant and free radical scavenging activities (9,10). The superoxide anion radical scavenging ability of the aqueous extract from honeybush tea was also found to be affected by *Cyclopia* species, the fermentation process and final total phenol contents in the extract.

Unfermented honeybush tea also showed stronger superoxide scavenging ability than the fermented product (*10*). Aqueous extracts of both fermented and unfermented honeybush tea (*Cyclopia intermedia*) were found to possess antimutagenic activity against 2-acetylaminofluorene (2-AAF) and aflatoxin B(1) [AFB(1)]-induced mutagenesis using tester strains TA98 and TA100 in the presence of metabolic activation and the unfermented tea exhibited the higher protective effect than the fermented tea (*11*).

Several pharmacological studies have reported on the activities of mangiferin, the major phenolic in honeybush tea. This compound has been found to be a superopxide scavenger and found to inhibit expression of both iNOS and TNF-α genes, suggesting that it may be of potential value in the treatment of inflammatory and/or neurodegenerative disorders (*12*). Other researchers suggest this compound to have potential to protect streptozocin-induced oxidative damage to cardiac and renal tissues in rats (*13*), an anti-diabetic effect (*14-17*) and cancer chemopreventive activity (18).

Conclusion

Honeybush tea is a rich source of bioactive polyphenolic compounds with the major ones being mangiferin and hesperidin. Honeybush tea is a very promising caffeine-free tea, that imparts a gold brown color to the tea, and pleasant aroma characterized by its floral, fruity and honey notes with potential health benefits. This tea has great potential for wider acceptance into larger regional and global marketplaces once consumers and traders become familiar with this tasty indigenous and under recognized herbal tea.

Acknowledgements

Authors wish to express their thanks and appreciation to Elton Jefthas and Jacky Goliath, ASNAPP-South Africa for providing the honeybush materials and honeybush tea samples used in our chemical analyses and also in sharing their experiences in honeybush with us. We thank the Haarlem community in the Western Cape Province of South Africa who, with assistance from the ASNAPP project, were the first to actually cultivate this plant rather than exclusively collect the material from the wild. We also recognize Jerry Brown, USAID project officer, for his support and encouragement. This work was conducted as part of the Agri-Business in Sustainable African Natural Plant Products Program (ASNAPP) with funding from the USAID (Contract Award No. HFM-O-00-01-00116), the New Use Agriculture and Natural Plant Products Program and the New Jersey Agricultural Experiment Station, Rutgers University.

References

1. Kies, P. *Bothalia* **1951**, *6*, 161-176.
2. De Nysschen, A.M.; Van Wyk, B.E.; Van Heerden, F.R.; Schutte, A.L. *Biochem. Syst. Ecol.* **1996**, *24*, 243-246.
3. Bond, P.; Goldblatt, P. *J. South African Botany* **1994**, *13* (suppl.), 1-455.
4. Joubert, E.; Otto, F.; Grüner, S.; Weinreich B. *Eur. Food Res. Technol.*, **2003**, *216*, 270-273.
5. Ferreira, D.; Kamara, B.I.; Brandt, E.V.; Joubert, E. *J. Agric. Food Chem.* **1998**, *46*, 3406-3410.
6. Kamara, B.I. Ph.D Dissertation, University of the Orange Free State, Bloemfontein, South Africa, 1999.
7. Kamara, B. J.; Brandt, E. V.; Ferreira, D.; Joubert, E. *J. Agric. Food. Chem.* **2003**, *51*, 3874-3879.
8. Wang, M.; Simon, J. E.; Aviles, Irma, F.; He, K.; Zheng, Q.Y.; Tadmor, Y. *J. Agric. Food Chem.* **2003**, *51*, 601-608.
9. Lindsey, K. L.; Motsei, M. L.; Jager, A. K. *J. Food Sci.* **2002**, *67*, 2129-2131.
10. Hubbe, M.E. and Joubert, E. In: Johnson IT, Fenwick GR (eds). Dietary Anticarcinogens and Antimutagens. Chemical and Biochemical aspects. Royal Chemistry Society: Cambridge, 2000, pp. 242-244.
11. Marnewick, J. L.; Gelderblom, W. C. A.; Joubert, E. *Mutation Res.* **2000**, 471, 157-166.
12. Manuel, L. J.; Ezequiel A.; Alberto, A, J.; Gonzalez, S. I.; Francisco, O. *Biochem. Pharm.* **2003**, *65,* 1361-1371.
13. Muruganandan, S.; Gupta, S.; Kataria, M.; Lal, J.; Gupta, P. K. Toxicology **2002**, *176*, 165-173.
14. Toshihiro, M.; Hiroyuki, I.; Itsuko, H.; Naoki, I.; Motoshi, K.; Masayoshi, K.; Eriko, I.; Yasuhiro, K.; Minoru, O.; Torao, I.; Keiichro, T. *Phytomed.* **2001**, *8*, 85-87.
15. Toshihiro, M.; Naoki, I.; Motoshi, K; Hiroyuki, I.; Masayoshi, K.; Yasuhiro, K.; Hiroshi, S.; Torao, I.; Keiichro, T. *Biomed. Res.* **2001**, *22*, 249-252.
16. Toshihiro, M; Naoki, I.; Motoshi, K.; Hiroyuki, I.; Masayoshi, K.; Yasuhiro, K.; Torao, I.; Minoru, O.; Keiichro, T. *Biol. Pharm. Bull.* **2001**, *24*, 1091-1092.
17. Naoki, I; Toshihiro, M.; Motoshi, K.; Tomomi, T.; Hiroyuki, I.; Masayoshi, K.; Yasuhiro, K.; Eriko, I.; Hiroshi, S.; Torao, I.; Yutaka, S.; Keiichro, T. *Biomed. Res.* **2000**, *21*, 221-223.
18. Yoshimi, N.; Matsunaga, K.; Katayama, M.; Yamada, Y.; Kuno, T.; Qiao, Z.; Hara, A.; Yamahara, J.; Mori, H. *Cancer Lett.* **2001**, 163, 163-170.

Chapter 12

Combined Inhibitory Effects of Catechins with Fe^{3+} on the Formation of Potent Off-Odorants from Citral

Toshio Ueno[1], Hideki Masuda[1], and Chi-Tang Ho[2]

[1]Material Research and Development Laboratories, Ogawa and Company, Ltd., 15-7 Chidori, Urayasushi, Chiba 279-0032, Japan
[2]Department of Food Science, Rutgers, The State University of New Jersey, 65 Dudley Road, New Brunswick, NJ 08901-8520

The degradation of citral in an acid buffer (pH 3.0) during storage at 40 °C resulted in the formation of potent off-odorants, *p*-cresol (**1**) and *p*-methylacetophenone (**2**), and other oxidation products including *p*-cymen-8-ol (**3**). The formation behavior of these oxidation products from citral with the addition of catechins was investigated in both the absence and presence of ferric ion (Fe^{3+}). In the absence of Fe^{3+}, all tested catechins, including (-)-epigallocatechin (EGC), (-)-epigallocatechin gallate (EGCg), and (+)-catechin, strongly inhibited the formation of **2**, whereas the formation of **1** was increased with the added catechins except for (+)-catechin. In contrast, the addition of EGC and EGCg in the presence of Fe^{3+} strongly inhibited the formation of **1** as well as **2** and significantly increased the formation of **3**. Thus, the presence of Fe^{3+} was found to be an important factor for inhibiting the formation of both off-odorants **1** and **2** with added catechins. The resulting formation of **3** may be explained by the Fe^{3+}-induced oxidation of catechins to phenoxy radicals and by subsequent their reaction with radical intermediates during the formation of **1** and **2** from citral.

The stability of citral, a key component of lemon aroma, under acidic aqueous conditions is a challenging issue in the field of flavor chemistry. Under such conditions, citral easily degrades by a series of cyclization and oxidation reactions to form a variety of degradation products (*1-8*). Consequently, not only is the fresh lemon-like odor of citral lost, but undesirable off-odors develop. Among degradation products of citral, *p*-cresol and *p*-methylacetophenone were reported to be the most potent off-odorants (*6, 7*). To inhibit the formation of these off-odorants from citral under acidic aqueous conditions, we previously investigated the addition of various antioxidants and found that the catechins from green tea strongly inhibited the formation of *p*-methylacetophenone (*9, 10*). However, the addition of the catechins to the citral solution resulted in an increase in the formation of *p*-cresol.

In this chapter, we report the simultaneous inhibition of formation of both *p*-cresol and *p*-methylacetophenone from citral with added catechins in the presence of ferric ion (Fe^{3+}). Mechanisms for the combined actions of catechins with Fe^{3+} are proposed on the basis of the formation behavior of other oxidation products from citral.

Experimental

Chemicals

Citral was purchased from Polarome International (Jersey City, NJ). (+)-Catechin, (-)-epicatechin (EC), (-)-epigallocatechin (EGC), (-)-epicatechin gallate (ECg), and (-)-epigallocatechin gallate (EGCg) were purchased from Funakoshi (Tokyo, Japan). Commercially available standards for identification of citral degradation products were obtained as follows: *p*-cresol and *p*-methylacetophenone: Nacalai Tesque (Kyoto, Japan); *p*-cymen-8-ol and α,*p*-dimethylstyrene: Sigma-Aldrich Japan (Tokyo, Japan); *p*-cymene: Wako Pure Chemical Industries (Osaka, Japan). 4-(2-Hydroxy-2-propyl)benzaldehyde was synthesized from *p*-bromobenzaldehyde dimethylacetal (Sigma-Aldrich Japan, Tokyo, Japan) according to the method described by Creary and Wang (*11*). 8-Hydroperoxy-*p*-cymene was synthesized from *p*-cymene-8-ol (Sigma-Aldrich Japan, Tokyo, Japan) as described previously (*10*).

Model Reactions

A solution containing 10 mg/L of citral in an acidic buffer (0.1 M citric acid-0.2 M sodium hydrogen phosphate, pH 3.0) with or without 60 mg/L of added (+)-catechin, EC, EGC, ECg, or EGCg was prepared. One hundred milliliters of the prepared solution were transferred to a 100-mL glass bottle, and the sample bottle was sealed with a Teflon liner and a screw cap. The sample was stored in a dark incubator at 40 °C for two weeks. For studies on the combined effects of catechins with Fe^{3+}, 1 mL of a series of ferric chloride hexahydrate aqueous solutions (5 x 10^{-6}-5 x 10^{-3} M) was added to 100 mL of the sample solutions. The sample bottle was sealed and then stored for one week under the same conditions described above.

Preparation of Analytical Samples

The degradation products of citral were extracted with dichloromethane (30 mL x 2). One milliliter of 0.1% (w/v) n-pentadecane in dichloromethane was added to the extract as the internal standard. The extract was dried over sodium sulfate, concentrated *in vacuo* to ~5 mL, and further concentrated under a stream of nitrogen to ~200 μL.

Gas Chromatography-Mass Spectrometry (GC-MS)

A Hewlett-Packard 5890 SERIES II gas chromatograph equipped with an HP-5972 mass selective detector and a DB-1 fused silica capillary column (60 m x 0.25 mm i.d.; film thickness of 0.25 μm; J&W Scientific) was used. The operating conditions were as follows: injector temperature, 250 °C; helium carrier gas flow rate, 1 mL/min; oven temperature program, 60 °C, raised at 3 °C/min to 210 °C (40 min); 1 μL of sample was injected using a split ratio of 1:50; ionization voltage, 70 eV; ion source temperature, 140 °C. The identification of the citral degradation products was based on either of the following criteria: a) Mass spectrum and retention index (RI) in agreement with those of the authentic compounds; b) Mass spectrum in agreement with that of the Wiley mass spectral database (Agilent Technologies, 2000) and the RI in agreement with the literature value (5).

Gas Chromatography (GC)

An Agilent 6890 N gas chromatograph equipped with a flame ionization detector (FID) and a DB-1 fused silica capillary column (60 m x 0.25 mm i.d.; film thickness of 0.25 μm; J&W Scientific) was used. The operating conditions were as follows: injector temperature, 100 °C; detector temperature, 250 °C; nitrogen carrier gas flow rate, 1 mL/min; oven temperature program, 80 °C, raised at 3 °C/min to 210 °C (15 min); 1 μL of sample was injected using a split ratio of 1:50. The amount of citral degradation products was estimated by computing the areas versus that of the internal standard (n-pentadecane). The response factors of all compounds to the FID were assumed to be the same.

Results and Discussion

Formation Pathways of Oxidation Products from Citral under Acidic Aqueous Conditions

During the storage under acidic aqueous conditions, citral (neral and geranial) was almost totally converted to its cyclization products, including the potent off-odorants *p*-cresol (**1**) and *p*-methylacetophenone (**2**), and other oxidation products **3-5** (**Figure 1**). Mechanisms for the acid-catalyzed cyclization of citral have been summarized in an earlier review (*8*). Regarding subsequent formation of oxidation products **1-5**, possible pathways shown in **Figure 2** were proposed in our previous studies (*9, 10*) on the basis of the following considerations:

- Compound **1** and *p*-cymen-8-ol (**3**) were confirmed to be formed by the acid-catalyzed decomposition of 8-hydroperoxy-*p*-cymene (**4**) (*9*), which is most likely derived from the corresponding peroxy radical **7** by abstracting a hydrogen atom from other compounds.

- Compound **2** and 4-(2-hydroxy-2-propyl)benzaldehyde (**5**) were suggested to be formed via the *tert*-alkoxy radical **6**, since the formation of **6** under acidic aqueous conditions by means of the Fe^{2+}-induced decomposition of **4** yielded both **2** and **5** (*10*). The formation of **2** and **5** from the alkoxy radical **6** can be explained, respectively, by β-fragmentation, and by acid-catalyzed radical transformation and subsequent oxidation.

- The alkoxy radical **6** is probably derived from the peroxy radical **7** in the absence of Fe^{2+}. This can be explained by the bimolecular self-coupling of **7** followed by the decomposition of the resulting tetroxide (*12-14*), or by the addition reaction of **7** to olefins, such as *p*-mentha-1,5-trien-8-ol (**8**), followed by elimination of **6** (*15-17*).

- The peroxy radical **7** was postulated to be formed by the oxidation of *p*-mentha-1,4(8),5-triene (**9**), an intermediate in the formation of *p*-cymene (**10**) from **8**. Compound **9** was reported to rapidly oxidize to **4** on brief exposure to air (*18*), and this facile oxidation can be reasonably attributed to the gain of aromaticity.

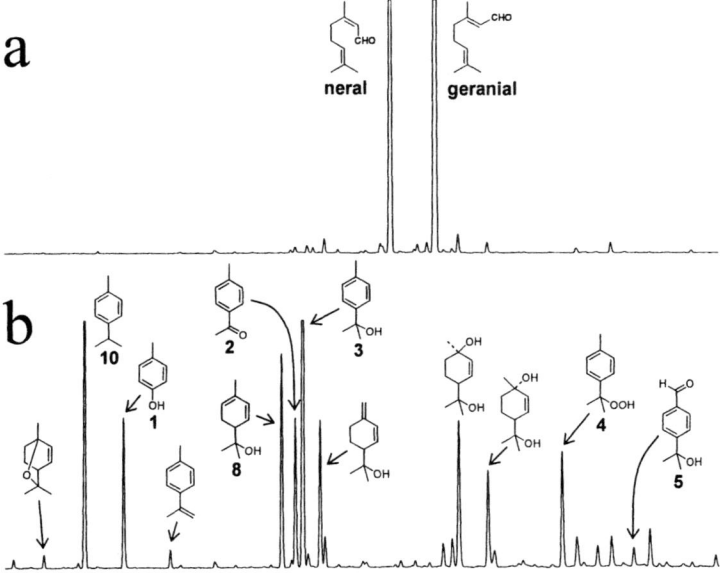

Figure 1. Gas chromatographic analysis of the acidic buffer solution (pH 3.0) of citral (10 mg/L) (a) before and (b) after stored at 40 °C for two weeks.

Effects of Added Catechins on the Formation of Oxidation Products from Citral in the Absence of Fe^{3+}

Figure 3 shows the effects of added (+)-catechin, (-)-epicatechin (EC), (-)-epigallocatechin (EGC), (-)-epicatechin gallate (ECg), and (-)-epigallocatechin gallate (EGCg) on the formation of oxidation products **1-5** from citral under acidic aqueous conditions. All the tested catechins strongly inhibited the formation of both **2** and **5**. Instead, the formation of **1** and **4** increased with the added catechins, except for (+)-catechin. The formation of **3** was increased with the addition of all the tested catechins. The gallated catechins (ECg and EGCg) strongly increased the formation of **1** and **4**, and the non-gallated catechins ((+)-catechin, EC, and EGC) strongly enhanced the formation of **3**. Thus, the

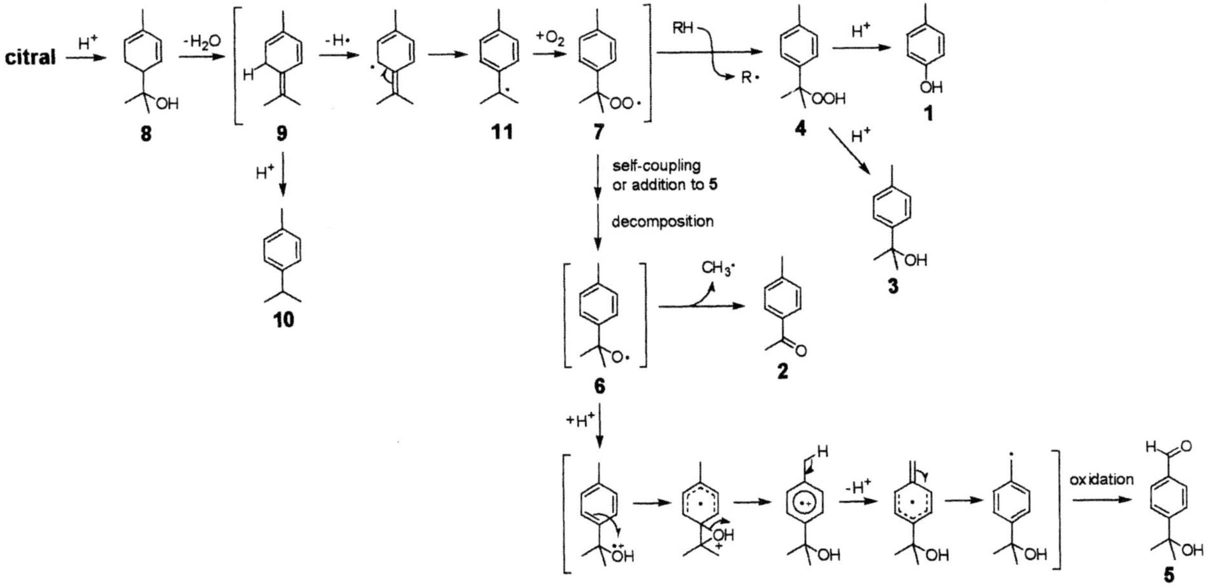

Figure 2. Proposed formation pathways of oxidation products 1-5 from citral under acidic aqueous conditions (9, 10)

formation inhibition of **2** and **5** with added catechins resulted in an increase in the formation of **1** and **4** and/or that of **3**.

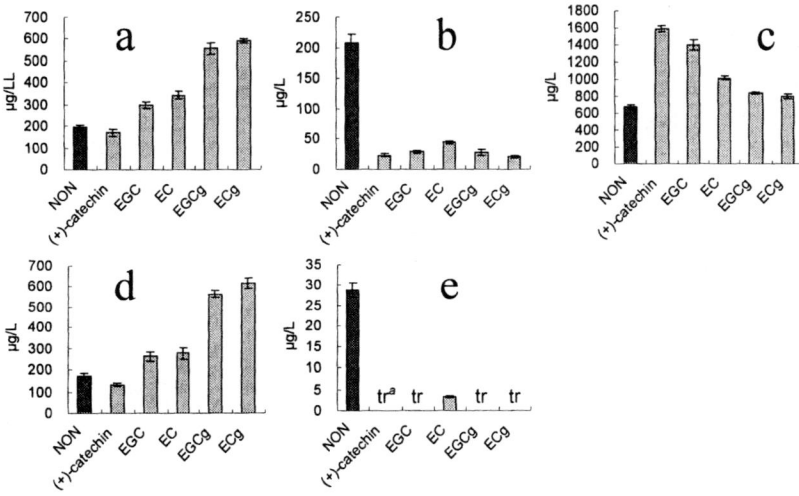

Figure 3. Formation behavior of oxidation products (a) 1, (b) 2, (c) 3, (d) 4, and (e) 5 from citral (10 mg/L) with added catechins (60 mg/L) in the absence of Fe^{3+} stored in an acidic buffer (pH 3.0) at 40 °C for two weeks. Each value is the mean of five experiments ± standard deviation. [a] tr = trace.

The observed formation behavior of oxidation products **1-5** from citral with added catechins can be partly understood on the basis of their formation pathways (**Figure 4**). The significantly increased formation of **1** and **4** with addition of gallated catechins may be explained by an H-atom transfer from the catechins to the peroxy radical **7**. The importance of the 3-galloyl moiety of the catechins on their H-atom donor abilities has been well established in the literature (*19-24*). The formation of **2** and **5** could be competitively inhibited by the increased formation of **4**. However, it remains unclear why the gallated catechins did not increase the formation of **3** as much as they increased the formation of **1** and **4**. Correspondingly, it is also unclear why the non-gallated catechins increased the formation of **3** much more than they increased the formation of **1** and **4**. These questions about the formation behavior of **3** suggest that an unknown mechanism as well as the acid-catalyzed decomposition of **4** could be involved in the formation of **3**. The non-gallated catechins might accelerate this unknown formation mechanism of **3**, so that the formation of **2** and **5** would be competitively inhibited. In the case of the gallated catechins, the same unknown mechanism for the formation of **3** might be partially inhibited by the increased formation of **4**. This can account for the result that the gallated catechins did not increase the formation of **3** as much as they increased the formation of **1** and **4**.

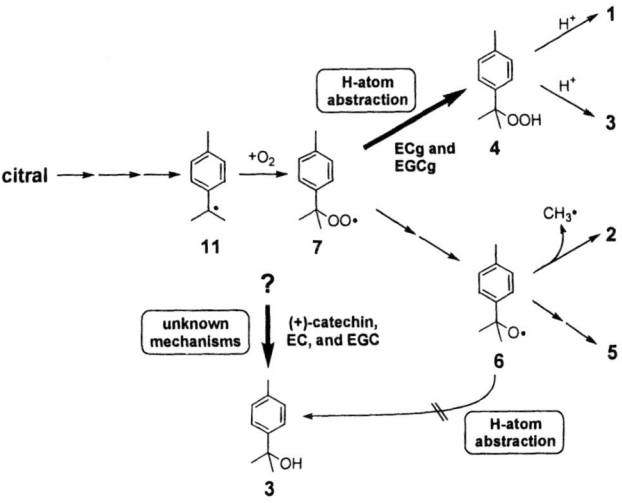

Figure 4. Possible mechanisms for the formation behavior of oxidation products 1-5 from citral with added catechins in the absence of Fe^{3+}.

Insights into the Increased Formation of *p*-Cymen-8-ol (3) with Added Catechins

The mechanism by which non-gallated catechins enhanced the formation of **3** is probably not H-atom transfer from catechins to the alkoxy radical **6** (**Figure 4**). It seems difficult, especially in aqueous solution, for the catechins to trap *tert*-alkoxy radicals such as **6** prior to their β-fragmentation. According to kinetic studies, the rate constants for β-fragmentation of *tert*-alkoxy radicals increase with the polarity of the solvents (25). Meanwhile, the H-atom donor abilities of phenolic compounds are dramatically decreased by H-bonding interactions with polar solvents (26, 27). For example, as for the *tert*-butoxy radical, the rate constants in aqueous solution have values of 1.4×10^6 s^{-1} for β-fragmentation (28) and 1.35×10^8 M^{-1}s^{-1} for H-atom abstraction from (+)-catechin (29). The latter value can be multiplied by the concentration of (+)-catechin in our experiments (60 mg/L, i.e., 2.0×10^{-4} M) to give 2.7×10^4 s^{-1}. This means that β-fragmentation of *tert*-butoxy radical can be ~50-fold faster than the H-atom abstraction from 60 mg/L of (+)-catechin in aqueous solution.

More rational explanations for the increased formation of **3** with added catechins could be the trapping of the *tert*-alkyl radical **11** or the peroxy radical **7** by mechanisms other than H-atom donation. Possible mechanisms for this assumption are shown in **Figure 5**. H-atom abstraction from catechins would generate the corresponding phenoxy radicals, which might react with radical **11** or **7**. A coupling reaction between the alkyl radical **11** and phenoxy radicals derived from catechins followed by subsequent hydrolysis of the resulting ethers would produce compound **3**. With this reaction, the phenoxy radicals would be reduced to the corresponding catechins. On the other hand, a coupling reaction between peroxy radical **7** and the catechin-derived phenoxy radicals followed by decomposition of the coupling products would also yield compound **3**. In this case, the phenoxy radicals would be oxidized, and the seven-membered B-ring anhydrides of the catechins might be expected as the resulting oxidation products. It is reported that these B-ring anhydrides of the catechins are obtained from the reaction of EGC and EGCg with peroxy radicals derived from 2,2'-azobis(2,4-dimethylvaleronitrile) (30).

Effects of EGC and EGCg on the Formation of Oxidation Products from Citral in the Presence of Fe^{3+}

The assumption that the phenoxy radical derived from catechins might react with radical **11** or **7** implies the possible formation inhibition of both off-odorants **1** and **2** by promoting catechin oxidation in the reaction system (**Figure 6**). Promoting the one-electron oxidation of catechins (ArOH) would increase the formation of the corresponding phenoxy radicals (ArO•), which might effectively trap radial **11** or **7** to give compound **3**, so that the formation of both **1** and **2** could be competitively inhibited. The promotion of catechin oxidation may be achieved by incorporation of iron, a common constituent in foods and beverages, into the reaction system. Under acidic aqueous conditions, the ferric state of iron (Fe^{3+}) can easily oxidize catechol derivatives such as (+)-catechin to the corresponding phenoxy radicals (31). Based on these hypotheses, the effect of Fe^{3+} on the formation of the oxidation products from citral with added catechins was investigated.

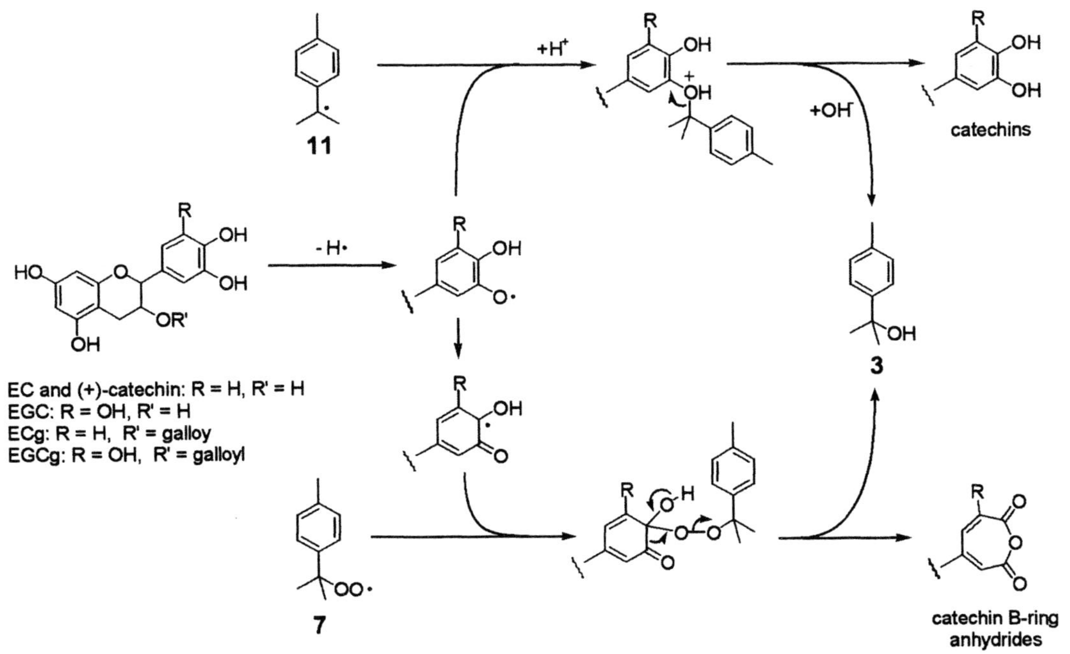

Figure 5. Possible mechanisms for the trapping of citral-derived radical intermediates 7 and 11 with phenoxy radicals derived from the catechins, and the resulting formation of compound 3.

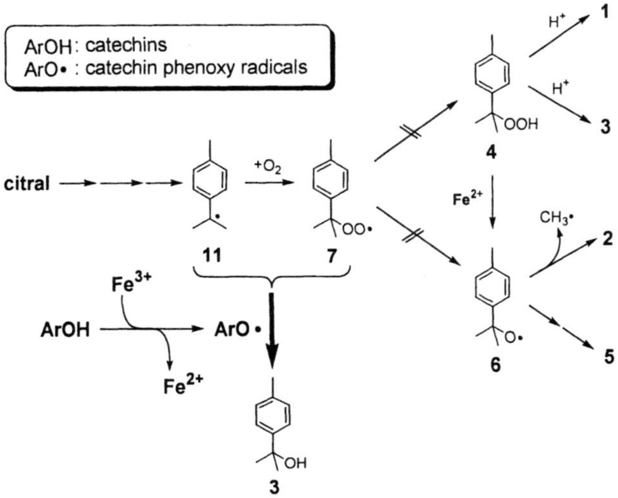

Figure 6. Hypothetical formation inhibition of both off-odorants 1 and 2 from citral with added catechins in the presence of Fe^{3+}.

The formation behavior of oxidation products **1-5** from citral with added EGC and EGCg was significantly affected by the presence of Fe^{3+} (**Figure 7**). The inhibited formation of **2** and **5** with added EGC and EGCg in the absence of Fe^{3+} recovered with the addition of Fe^{3+} with a maximum recovery at 5×10^{-7} M of Fe^{3+}. When the concentration of Fe^{3+} was higher than 5×10^{-7} M, the formation of **2** and **5** again inhibited. The increased formation of **1** and **4** with added EGC and EGCg without Fe^{3+} was dramatically decreased with the addition of Fe^{3+}. The formation of **3** with added EGC and EGCg was then significantly increased with the addition of Fe^{3+}. As a consequence, when a sufficient amount of Fe^{3+} (5×10^{-6} and 5×10^{-5} M) was added, EGC and EGCg inhibited the formation of both off-odorants **1** and **2**, along with **4** and **5**, and instead increased the formation of **3**.

In this reaction system, the catechins would reduce Fe^{3+} to ferrous ion (Fe^{2+}) that can decompose hydroperoxide **4** to compounds **2** and **5** via the alkoxy radical **6** (*10*) (**Figure 6**). This seems to explain the result that the addition of Fe^{3+} around 5×10^{-7} M recovered the inhibited formation of **2** and **5** with added EGC and EGCg. However, with further addition of Fe^{3+}, the formation of catechin-derived phenoxy radicals (ArO•) would be promoted such that radical **7** or **11** could be totally trapped prior to the formation of **4** and **6**. This can account for the formation inhibition of oxidation products **1, 2, 4**, and **5** and for the resulting formation of **3** with the addition of higher amounts of Fe^{3+}. The contribution of compound **3** to the off-odors derived from citral would not be significant, since this compound was not detected by the GC-Olfactometry of the stored citral solution in a previous study (9). Thus, the presence of Fe^{3+} was found to be an important factor inhibiting the formation of off-odors derived from citral with added catechins.

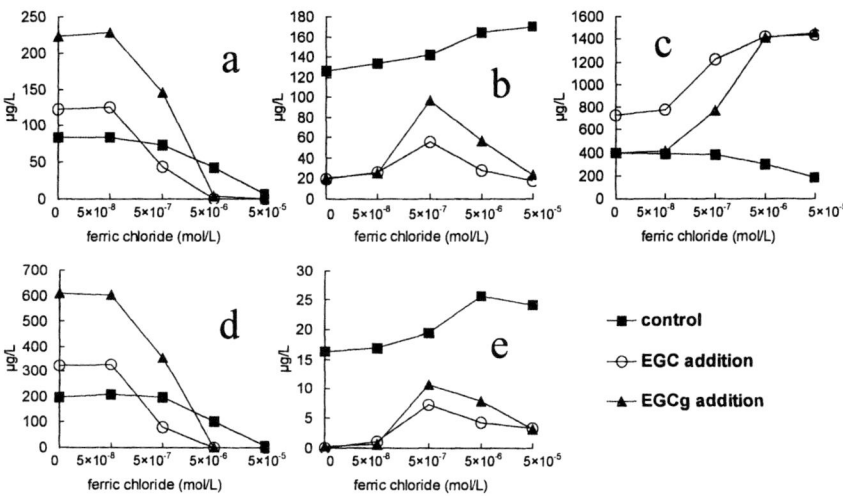

Figure 7. Formation behavior of oxidation products (a) 1, (b) 2, (c) 3, (d) 4, and (e) 5 from citral (10 mg/L) with added EGC and EGCg (60 mg/L) in the presence of Fe^{3+} stored in an acidic buffer (pH 3.0) at 40 °C for one week.

Conclusions

This study provides insight into the action mechanisms of added catechins on the formation of *p*-cresol (**1**) and *p*-methylacetophenone (**2**), potent off-odorants derived from citral under acidic conditions. The presented data indicate that the presence of Fe^{3+} is an important factor inhibiting the formation of both **1** and **2** with added catechins. Phenoxy radicals generated from catechins in the presence of Fe^{3+} might react with radical intermediates during the formation of oxidation products from citral. The proposed mechanisms can explain the resulting formation of *p*-cymene-8-ol (**3**), a non-off-odor oxidation product from citral.

References

1. Clark, B. C.; Powell, C. C.; Radford, T. *Tetrahedron* **1977**, *33*, 2187-2191.
2. McHale, D.; Laurie, W. A.; Baxter, R. L. In *Proceeding of the 7th International Congress of Essential Oils*, Kyoto, 1977; Japan Flavor & Fragrance Manufacturers Association: Tokyo, Japan, 1979; pp 250-253.
3. Kimura, K.; Iwata, I.; Nishimura, H. *Agric. Biol. Chem.* **1982**, *46*, 1387-1389.
4. Kimura, K.; Nishimura, H.; Iwata, I.; Mizutani, J. *J. Agric. Food Chem.* **1983**, *31*, 801-804.
5. Peacock, V. E.; Kuneman, D. W. *J. Agric. Food Chem.* **1985**, *33*, 330-335.
6. Schieberle, P.; Ehrmeier, H.; Grosch, W. Aroma compounds resulting from the acid catalyzed breakdown from citral. *Z. Lebensm. Unters. Forsch.* **1988**, *187*, 35-39.
7. Schieberle, P.; Grosch, W. *J. Agric. Food Chem.* **1988**, *36*, 797-800.
8. Clark, Jr., B. C.; Chamblee, T. S. In *Off-flavors in Foods and Beverages*; Charalambous, G., Ed.; Elsevier: Amsterdam, the Netherlands, 1992; pp 229-285.
9. Ueno, T.; Masuda, H.; Muranishi, S.; Kiyohara, S.; Sekiguchi, Y.; Ho, C.-T. In *Flavour Research at the Dawn of the Twenty-first Century*; Le Quéré, J. L., Etiévant, P. X., Eds.; Proceedings of the 10th Weurman Flavour Research Symposium; Lavoisier: Cachan, France, 2003; pp 128-131.
10. Ueno, T.; Masuda, H.; Ho, C.-T. *J. Agric. Food Chem.*, accepted.
11. Creary, X.; Wang, Y.-X. *J. Org. Chem.* **1992**, *57*, 4761-4765.
12. Blanchard, H. S. *J. Am. Chem. Soc.* **1959**, *81*, 4548-4552.
13. Bartlett, P. D.; Traylor, T. G. *J. Am. Chem. Soc.* **1963**, *85*, 2407-2410.
14. Bartlett, P. D.; Guaraldi, G. *J. Am. Chem. Soc.* **1967**, *89*, 4799-4801.
15. Mayo, F. R. *Acc. Chem. Res.* **1968**, *1*, 193-201.
16. Ingold, K. U. Peroxy radicals. *Acc. Chem. Res.* **1969**, *2*, 1-9.
17. Howard, J. A. In *Peroxyl Radicals*; Alfassi, Z. B., Ed.; Wiley: New York, 1997; Chapter 10.
18. Andemichael, Y. W.; Wang, K. K. *J. Org. Chem.* **1992**, *57*, 796-798.
19. Yoshida, T.; Mori, K.; Hatano, T.; Okumura, T.; Uehara, I.; Komagoe, K.; Fujita, Y.; Okuda, T. *Chem. Pharm. Bull.* **1989**, *37*, 1919-1921.
20. Salah, N.; Miller, N. J.; Paganga, G.; Tijburg, L.; Bolwell, G. P.; Rice-Evans, C. *Arch. Biochem. Biophys.* **1995**, *322*, 339-346.

21. Guo, Q.; Zhao, B.; Li, M.; Shen, S.; Xin, W. *Biochim. Biophys. Acta* **1996**, *1304*, 210-222.
22. Guo, Q.; Zhao, B.; Shen, S.; Hou, J.; Hu, J.; Xin, W. *Biochim. Biophys. Acta* **1999**, *1427*, 13-23.
23. Nanjo, F.; Goto, K.; Seto, R., Suzuki, M.; Sakai, M.; Hara, Y. *Free Radical Biol. Med.* **1996**, *21*, 895-902.
24. Nanjo, F.; Mori, M.; Goto, K.; Hara, Y. *Biosci., Biotechnol., Biochem.* **1999**, *63*, 1621-1623.
25. Avila, D. V.; Brown, C. E.; Ingold, K. U.; Lusztyk, J. *J. Am. Chem. Soc.* **1993**, *115*, 466-470.
26. Avila, D. V.; Ingold, K. U.; Lusztyk, J.; Green, W. H.; Procopio, D. R. *J. Am. Chem. Soc.* **1995**, *117*, 2929-2930.
27. Valgimigli, L.; Banks, J. T.; Ingold, K. U.; Lusztyk, J. *J. Am. Chem. Soc.* **1995**, *117*, 9966-9971.
28. Erben-Russ, M.; Michel, C.; Bors, W.; Saran, M. *J. Phys. Chem.* **1987**, *91*, 2362-2365.
29. Bors, W.; Heller, W.; Michel, C.; Saran, M. *Methods Enzymol.* **1990**, *186*, 343-355.
30. Valcic, S.; Burr, J. A.; Timmermann, B. N.; Liebler, D. C. *Chem. Res. Toxicol.* **2000**, *13*, 801-810.
31. Danilewicz, J. C. *Am. J. Enol. Vitic.* **2003**, *54*, 73-85.

Chapter 13

Influence of Flavonoids on the Thermal Generation of Aroma Compounds

Devin G. Peterson and Vandana M. Totlani

Department of Food Science, The Pennsylvania State University,
215 Borland Laboratory, University Park, PA 16802

The chemical properties of flavonoids not only make them unique with respect to their bioactivity but also as reactants during thermal processing (flavor development). The characteristic aromas of numerous food products are highly dependent on thermally catalyzed reactions (i.e. Maillard and lipid oxidation). Matrix parameters which influence the generation (kinetics) of these aroma-active compounds are central to the quality of such products. This study will focus on the influence of two particular dietary phytonutrient compounds, epicatechin and epigallocatechin gallate on the thermal generation of aroma compounds in model systems and food products. Understanding how phytonutrients (i.e. epicatechin) behave as reactants and alter the thermally catalyzed reactions responsible for the aroma development would assist the food industry in producing flavorful phytonutrient enhanced value added products. Epicatechin, in general, negatively influenced the generation of aroma compounds produced by Maillard-type and lipid oxidation reactions (i.e. aldehydes, pyrazines, furanones, etc).

The modification of foods for optimal health represents one of the newest frontiers in food/nutrition science. The use of plant-based food products with enhanced concentrations of chemopreventive phytonutrients, for example, is a promising new strategy for health promotion (*1,2*). Numerous phytochemicals, such as phenols, flavonoids, terpenes, glucosinolates and others have been reported to have antioxidant, anticarcinogenic, and a wide range of tumor-blocking activities (*3-6*). Of the known phytonutrients, the flavonoids have been the primary focus of most physiological studies (*7*).

In order for phytonutrient enhanced food products (functional foods) to have a positive impact on health, however, they ultimately need to have positive flavor properties for consumption. Flavor is considered to be the greatest single factor in determining our choice of food (*8*) and therefore can be considered a critical parameter of human nutrition. Thus, the manufacture of food products with enhanced concentrations of flavonoids should not impart any negative flavor properties or ideally would even improve them.

Historically the affects of flavonoids on the flavor properties of foods have primarily focused on the taste attributes (i.e. bitter, astringent) and to a large extent have been negative (*7,9-12*). The level of flavonoids in plant material can be present in substantial amounts (0.5-1.5%) (*13*) and, consequently contribute to consumer rejection due to excessive bitterness. However, the manufacture of food products with health-promoting quantities of flavonoids does not automatically have to result in consumer rejection due to negative taste attributies. Previous nutritional studies have reported an efficacious dose of flavanols and procyanidins to be 148 mg (*14*) and 220 mg (*15*), which if delivered in a liquid food matrix (i.e. beverage) could be present at concentrations less than a tenth of a percent (based on a 234 mL serving size). Furthermore, the bitterness attributes of phytochemicals can often be masked or minimized to an acceptable level with fat, sugar or salt (*16*).

In addition to the taste attributies of flavonoids, the chemical properties of these dietary phytonutrients may influence the flavor attributes of flavonoid containing food products and consequently consumption. The basic structure of flavonoids is illustrated in Figure 1, as well as two common phytonutrients, epicatechin and epigallocatechin gallate (flavanols) for reference. Furthermore, the potentially chemically reactive forms of these compounds with respect to Maillard-type reactions are shown in Figure 2. Review of these structures suggests chemical properties about flavonoids that could be potentially reactive with respect to the thermal generation of aroma compound and include:

(1) relatively strong antioxidant property/free radical terminator (aromatic ring with hydroxyl groups);
(2) hydroxyl groups can be oxidized to reactive carbonyl compounds; and
(3) a chemical structure which can form hydrophobic, hydrogen, ionic and covalent bonds.

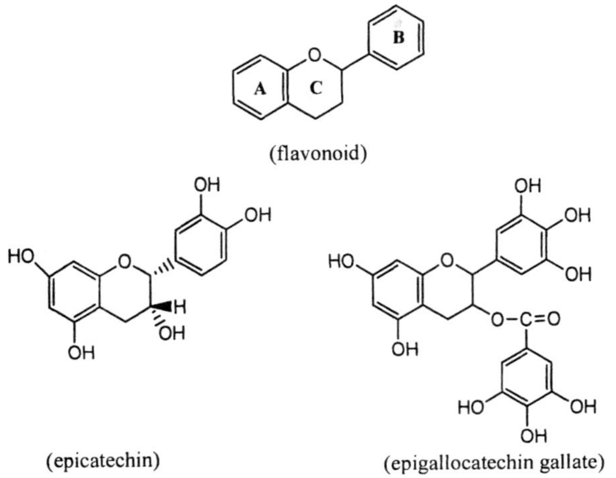

Figure 1. Basic structure of flavonoids and two common flavanols (epicatechin and epigallocatechin gallate)

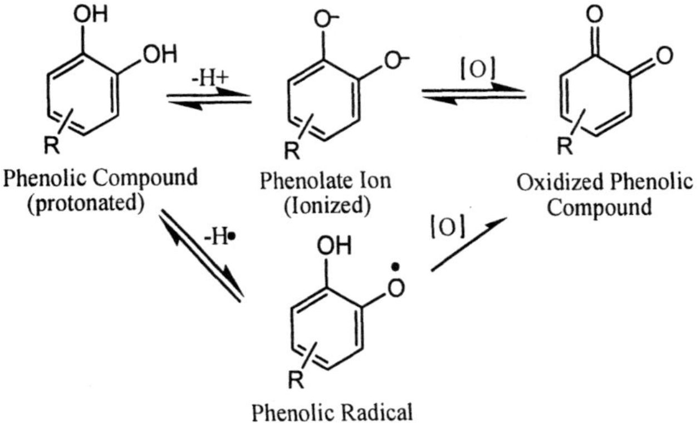

Figure 2. Potentially chemically reactive forms flavonoids - Maillard reaction

The influence of flavonoids (strong antioxidant molecules – free radical terminator) on Maillard chemistry with respect to flavor formation has not been previously defined. This may be attributed to the fact that radical mechanisms are not part of the classical Hodge sequence (*17*) and therefore have not been traditionally considered important with respect to formation of Maillard-type aroma compounds. However, it has been well documented that radical formation and reactions are present in the early stages of the Maillard reaction (*18-21*). Consequently, radical mechanisms may be a key parameter in Maillard-type reactions in relation to formation of flavor compounds (i.e. aroma-actives), in addition to the traditionally focused on parameters such as pH, a_w, temperature, sugar-type, and nitrogen source. Wang (*22*) studied the affects of chlorogenic acid and hydroxycinnamic acids on Maillard reaction model systems and concluded that these phenolic compounds influence aroma development. The affect of strong antioxidant molecules on these radical pathways and subsequent reactions may have important implications on flavor formation. Interestingly, radical mechanisms have been associated with the formation of browning products (*23,24*).

The addition of strong antioxidant molecules could also limit Maillard-type reaction products via the reduction of lipid oxidation derived carbonyl compounds which for many thermally processed food products is an important part of flavor development (*25*). Furthermore, oxidation reactions are ubiquitous in numerous proposed Maillard-type aroma formation pathways which could be inhibited by the presents of a strong antioxidant compounds (i.e. oxidation of dihydropyrazine to the corresponding pyrazine compound – single electron oxidation). Conversely, flavonoids themselves can function as reducing agents and may undergo oxidation reactions to form potentially reactive carbonyl compounds (Figure 2), a basic precursor of Maillard chemistry.

The purpose of this manuscript was to investigate the influence of two common flavonoids, epicatechin and epigallocatechin gallate, on the formation of aroma compounds in model Maillard reaction systems and two food products.

Materials and Methods

Materials.

The following chemicals D-fructose and D-glucose, L-alanine, L-glycine, L-leucine, L-isoleucine, n-dodecane were obtained from the Sigma-Aldrich Co. (St. Louis, MO); *n*-Octane-d_{18} (99% Isotopic) from (Alfa Aesar, Ward Hill, MA; potassium phosphate, disodium phosphate from EMD Chemicals (Gibbstown, NJ); and (-)-epicatechin, (-)-epigallocatechin gallate from Zhejiang Yixin Pharmaceutical Co. (Zhejiang, China) and were ≥ 98% purity. Dichloromethane solvent was obtained from EMD Chemicals (OmniSolv grade). The following ingredients shortening, brown sugar, honey, oats, wheat germ were obtained from the local market.

Model Maillard Reaction System

Each of the two reaction mixtures (shown in Table I) were placed in a 600 mL Parr reactor (model 4563) and heated under constant stirring (set to 50% of maximum speed). Epicatechin (EC) was used for all treatment reaction mixtures; while epigallocatechin gallate (EGCG) was only used once for comparison. The temperature/time experimental conditions for the two reaction mixtures were as follows: 125°C for 30 minutes at pH of 6.0, 7.0, 8.2; 125°C for 15 minutes at pH 7.0; 100°C for 60 minutes at pH 7.0; 150°C for 30 minutes at pH 7.0. The reaction mixtures were cooled to 25°C with an internal cooling coil and immediately 300ml of the reaction mixture (pH was with 0.2 units of target) was extracted 3x with 30ml of dichloromethane which contained 2.75 μL of n-dodecane/L of solvent as the internal standard. The extracts were pooled and dried with anhydrous sodium sulfate, filtered and concentrated to approximately 500 μL using a 30 cm packed vacuum jacketed distilling column for further analysis.

Table 1 – Composition of Model Maillard Reaction Mixture

Reactants[a]	Control (M)	Treatment (M)
Alanine	0.01	0.01
Glycine	0.01	0.01
Leucine	0.01	0.01
isoleucine	0.01	0.01
Glucose	0.02	0.02
Fructose	0.02	0.02
EP or EGCG	none	0.01

a = all reactants were placed in 350ml of buffered (0.1 M phosphate) deionized water

Model Food Systems

Granola bar - two basic granola bar model systems (formulation outlined in Table II) were prepared as follows: the shortening, brown sugar and honey were thoroughly mixed in a stainless steal bowl using a hand mixer. The oats and germ were subsequently added and thoroughly mixed with a hand mixer. The mixture was then compressed onto a greased baking pan (control). The same procedure was used for the epicatechin enriched sample (treatment); with the exception that epicatechin was first dispersed in the honey. A depth of 1 cm was maintained with pre-marked toothpick. The samples were baked for 10 minutes at 191°C in a gravity air oven.

The granola bar samples were ground separately with Tekmar A-10 analytical mill. 40.0 g of each sample were analytically weighed into Erlenmeyer flasks. 70 mL of diethyl ether were added to the flasks for 10 minutes before decanting the liquid into a separate flask. This step was repeated two more times for each sample. n-Octane-d_{18} was added as the internal standard (1 μL/L solvent). The high vacuum distillation (<10^{-3} Pa) was performed on the ether extract. During this procedure the extract was slowly added to a high vacuum chamber and the volatiles were flashed off and trapped by a liquid nitrogen cold finger. The contents of the trap were collected, dried with anhydrous sodium sulfate and then concentrated (spinning band distillation) to approximately 500 μL for analysis.

Table II – Granola Bar Model Formulation

Control		*Treatment*	
Ingredient	*%*	*Ingredient*	*%*
shortening	15.4	shortening	15.3
brown sugar	15.4	brown sugar	15.3
honey	7.7	honey	7.7
oats	46.2	oats	46.1
wheat germ	15.4	wheat germ	15.3
		epicatechin	0.3

Roasted Cocoa – 350 g of non-roasted Ivory Coast cocoa nibs (Blommer Chocolate Co. East Greenville, PA) were ground with a Tekmar A-10 analytical mill. The ground material was sectioned into 3 separate 100 g samples. One of these samples was enriched with 0.3% (w/w) epicatechin and another with 0.6% (w/w) and the for third sample nothing was added (control). The samples were individually placed in a 600 mL beaker and roasted for 30 minutes at 145°C in a gravity air oven.

Aroma isolates were prepared by the solid-phase microextraction/headspace technique (HS-SPME). Two grams of each roasted cocoa sample was placed into a 20 mL headspace vial and sealed with poly (tetrafluoroethylene)/butyl rubber septa (MicroLiter Analytical Supplies, Inc., Suwanee, GA). HS-SPME conditions were as follows: sample incubation temperature at 45°C, sample incubation time was 60 min with agitation, extraction time was 1 min using a 50/30 μm DVB/Carboxen/PDMS fiber (Supelco, Bellefonte, PA).

Gas Chromatography/Mass Spectrometry

The extracts/isolates were analyzed by an Agilent 6890 gas chromatograph (Agilent Technologies, Palo Alto, CA) utilizing a HP5972 mass spectrometer detector (EI-mode) equipped with a liquid autosampler (model A200SE, CTC Analytics, Carrboro, NC). All analyses were performed on both a DB-5ms and DB-Wax capillary column (30 m x 0.25 mm i.d. with a 0.25 μm film thickness). The analysis parameters were as follows: inlet temperature was 200°C (250°C for cocoa analysis), column flow was constant at 0.8ml/min (H_2), and MS temperature = 175°C. The temperature program was 35°C for 2 minutes, ramped at 3°C to 250°C and held for 4 minutes for DB-5ms analysis and 35°C for 2 minutes, ramped at 5°C to 230°C and held for 6 minutes for DB-Wax analysis. The extracts/isolates were injected in split mode (15:1) for the model Maillard reactions and in splitless mode for the granola bar and cocoa analysis. For the cocoa analysis a 0.75 mm i.d. inlet liner was also used and the samples were desorbed manually (desorption time was 1 minute). Compounds were identified by their mass spectra, and linear retention indices. Each set of samples (reaction systems, granola bar and cocoa) were analyzed within a single day minimize an potential drift in the detector response (confirmed by internal standard).

Results and Discussion

The influence of epicatechin (EC) on the generation of Maillard-type aroma compounds for a 4 amino acid-2 hexose reaction mixture heated at 125°C for 30 minutes at pH 6.0, 7.0 and 8.2 are illustrated in Figures 3-5. Overall EC had a distinct inhibitory effect on the overall formation of aroma compound (various chemical classes) at all three pH conditions, however the level of inhibition appeared to decrease with increasing pH (lowest for the alkaline reaction mixture). For example, 2,5-dimethylpyrazine at pH 6.0 was not detected in the treatment (signal to noise in the control was > 100), while at pH 7.0 the peak area in the EC reaction mixture was only 18 percent of the control peak area (or 5.6x higher in the control) and at pH 8.2 the peak area in the EC reaction mixture was 70 percent of the control peak area (1.43 x higher in the control). Possibly the observed correlation with the level of EC inhibition on Maillard-type aroma compound formation and pH is related to its antioxidant activity. The two most acidic hydroxyl groups of EC have a pka values of 8.64 and 9.41 (*13*). At a pH of 8.2, EC would be partially doubly ionized (negatively charged) and consequently reduce its antioxidant activity.

The effect of reaction time on EC inhibition on aroma generation is shown in Figure 6. The formation of all thirteen compounds reported were inhibited to a greater extent by EC at the shorter reaction time (15 versus 30 minutes). This

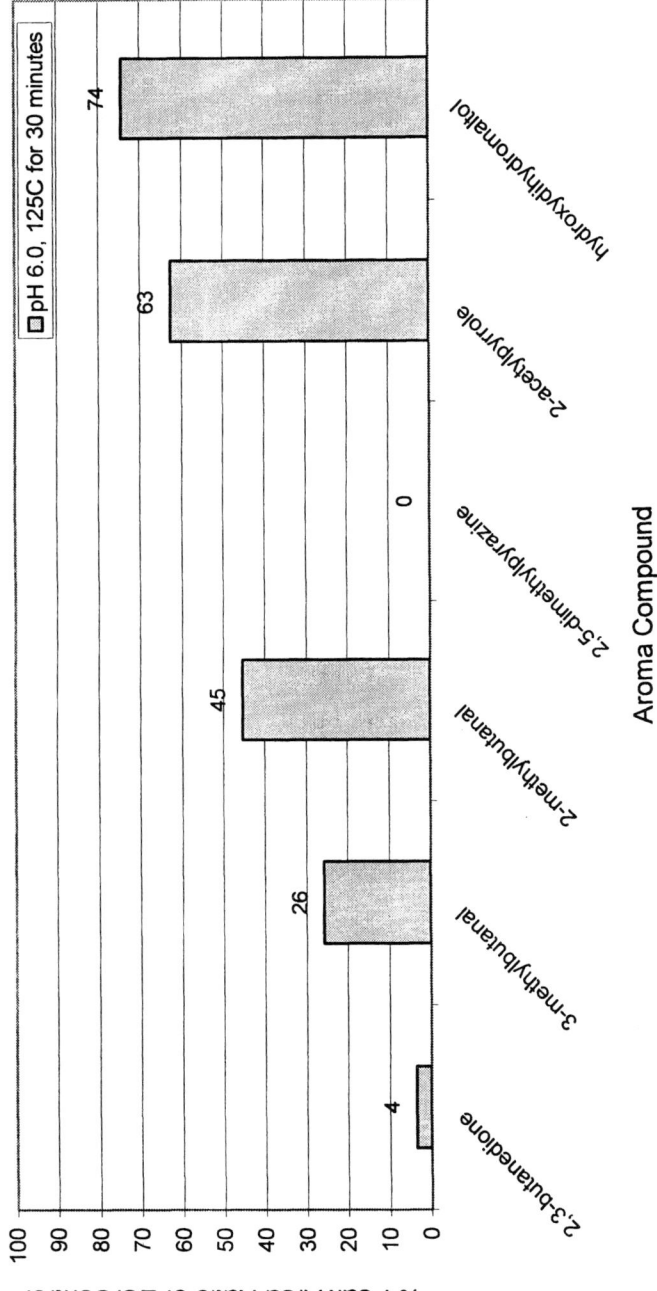

Figure 3 – Relative change in the aroma compounds generated in a model Maillard reaction systems with and without epicatechin at pH 6.0, 125ºC for 30 minutes; TIC Peak Area of Epicatechin (EC) Treatment/TIC Peak Area of Control * 100; peak area adjusted by internal standard (dodecane)

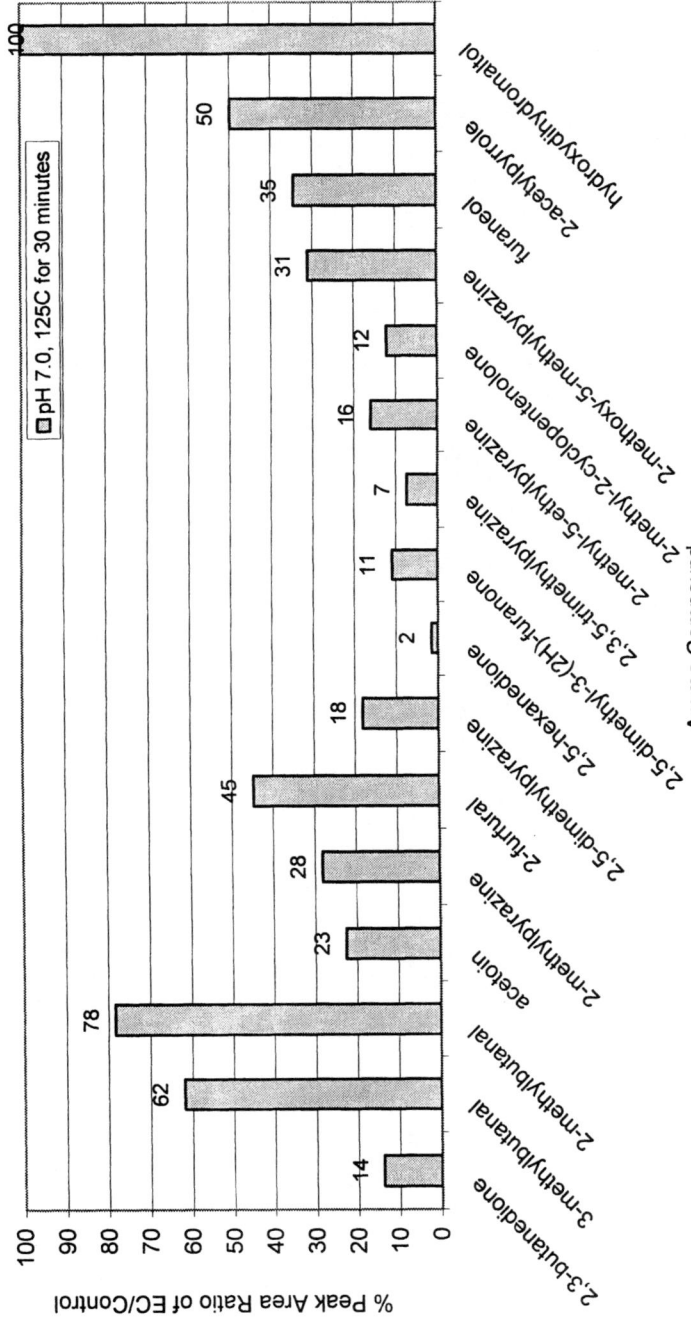

Figure 4 – Relative change in the aroma compounds generated in a model Maillard reaction systems with and without epicatechin at pH 7.0, 125°C for 30 minutes; TIC Peak Area of Epicatechin (EC) Treatment/TIC Peak Area of Control * 100; peak area adjusted by internal standard (dodecane)

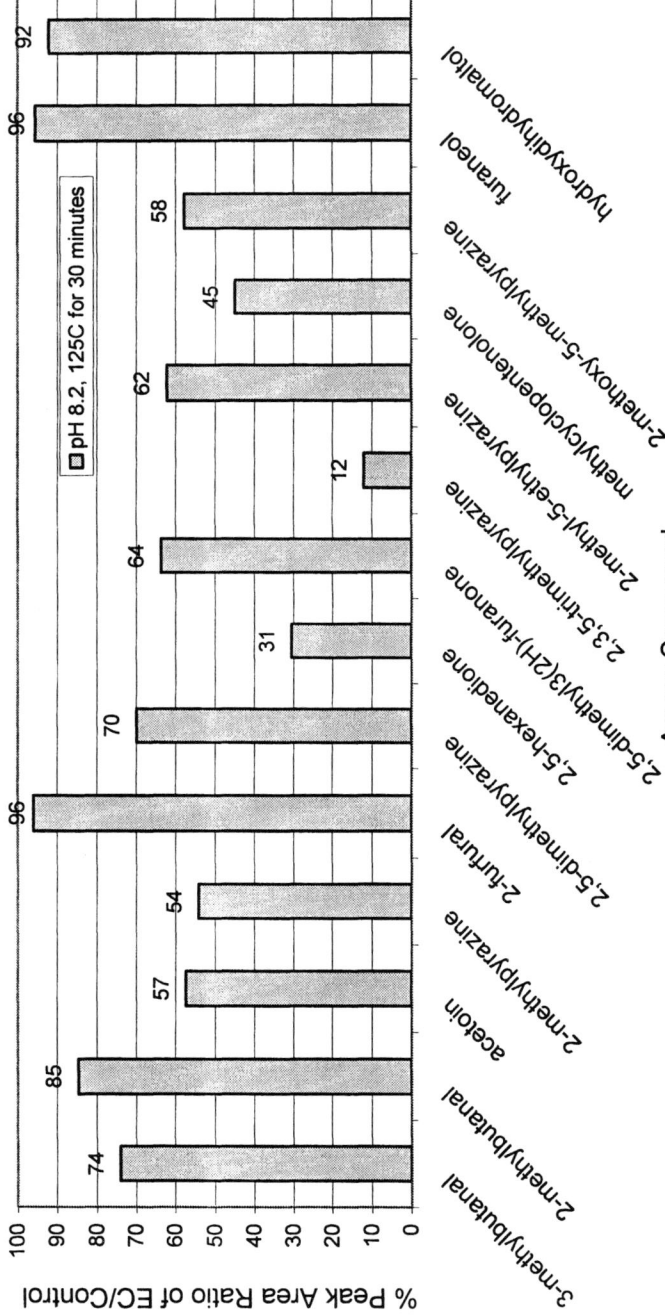

Figure 5 – Relative change in the aroma compounds generated in a model Maillard reaction systems with and without epicatechin at pH 8.2, 125°C for 30 minutes; TIC Peak Area of Epicatechin (EC) Treatment/TIC Peak Area of Control * 100; peak area adjusted by internal standard (dodecane)

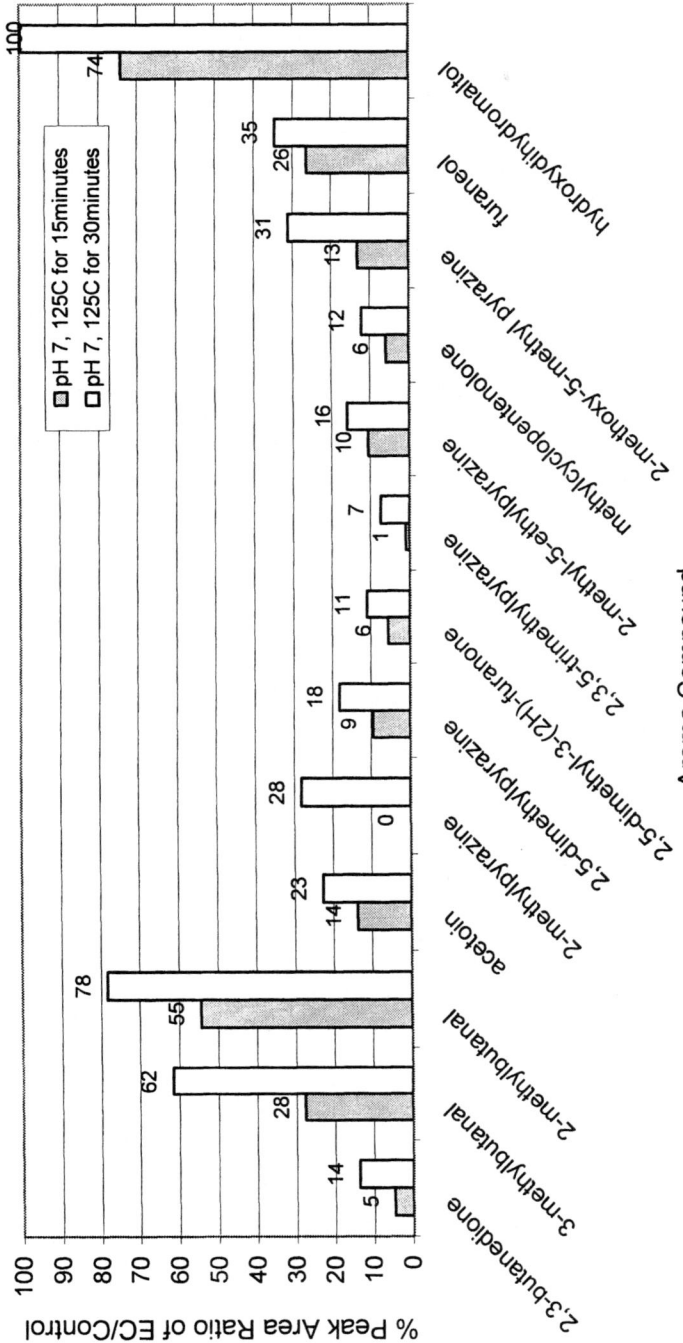

Figure 6 – Relative change in the aroma compounds generated in a model Maillard reaction systems with and without epicatechin at pH 7.0, 125°C for 15 and 30 minutes; TIC Peak Area of Epicatechin (EC) Treatment/TIC Peak Area of Control * 100; peak area adjusted by internal standard (dodecane)

suggests that either EC as a reactant was becoming rate limiting (expended) or that perhaps the reaction mechanisms are time depend (i.e. ionic versus radical). Previous work by Cammerer and Krob (*18*) reported that for a glucose:glycine reaction mixture the ionic portion of the reaction mechanism (versus radical portion) increased over time (up to 60 minutes under reflux conditions at pH 7.0); while for a fructose:glycine reaction mixture the ionic portion decreased with time.

The reaction temperature was also found to influence the affect of EC on aroma generation (see Figures 7 & 8). At higher temperatures (150°C – see Fig. 7) EC addition actually promoted the formation of three compounds (2,3-butanedione, 3-methylbutanal, and 2-methylbutanal) and although it did inhibit the formation of other compounds, the level was much lower in comparison to the 125°C reaction. The increase in strecker aldehydes may indicate that under these reaction conditions, EC readily formed the oxidized phenolic structure (quinone-type – see Figure 2) and therefore provided an source of reactive vicinal carbonyl compounds (precursor of Strecker aldehydes). This could also explain the increase in 2,3-butanedione (vicinal carbonyl) due to competition from the oxidized phenolic compounds which could reduce the utilization of 2,3-butanedione by further reactions. However at lower temperatures (Figure 8 – 100°C conditions), EC dramatically inhibited the formation of aroma compounds.

Furthermore, the choice of flavonoid compound also influenced the magnitude of the inhibitory affect on aroma generation in a Maillard reaction model (see Figure 9). ECGC has approximately twice the antioxidant strength as EC based on the trolox equivalent antioxidant activity (TEAC) assay [26]. Of the fourteen compounds listed in Figure 9, the formation of 12 compounds was inhibited to a greater extent with the stronger antioxidant compound, ECGC.

The inhibitory affect of EC on aroma formation illustrated by the model Maillard reaction system analyses was also in agreement with the affect of EC addition on aroma formation during thermal processing in two food systems (see Figures 10 & 11). Addition of 0.3% EC to a model granola bar food product (prior to baking) resulted in a decreased formation of Maillard-type aroma compounds in the baked product. 2,4-decadienal is a lipid oxidation product (not a Maillard reaction product) but was included to further illustrate additional affects of flavonoids on aroma development. For the cocoa analysis (Figure 11), EC reduced the formation of two previously identified key aroma in milk chocolate and cocoa (*27*). Cocoa beans are an excellent source of flavanols. The intrinsic levels of flavanols in cocoa prior to roasting are dependent upon many parameters, such as environmental and genetic factors, as well as handling and processing practices (e.g. fermentation conditions, harvesting conditions, etc.). As a result, bean phenolic content (as well as composition) may provide a link between cocoa roasting and flavor quality. The addition of 0.6% EC had similar

155

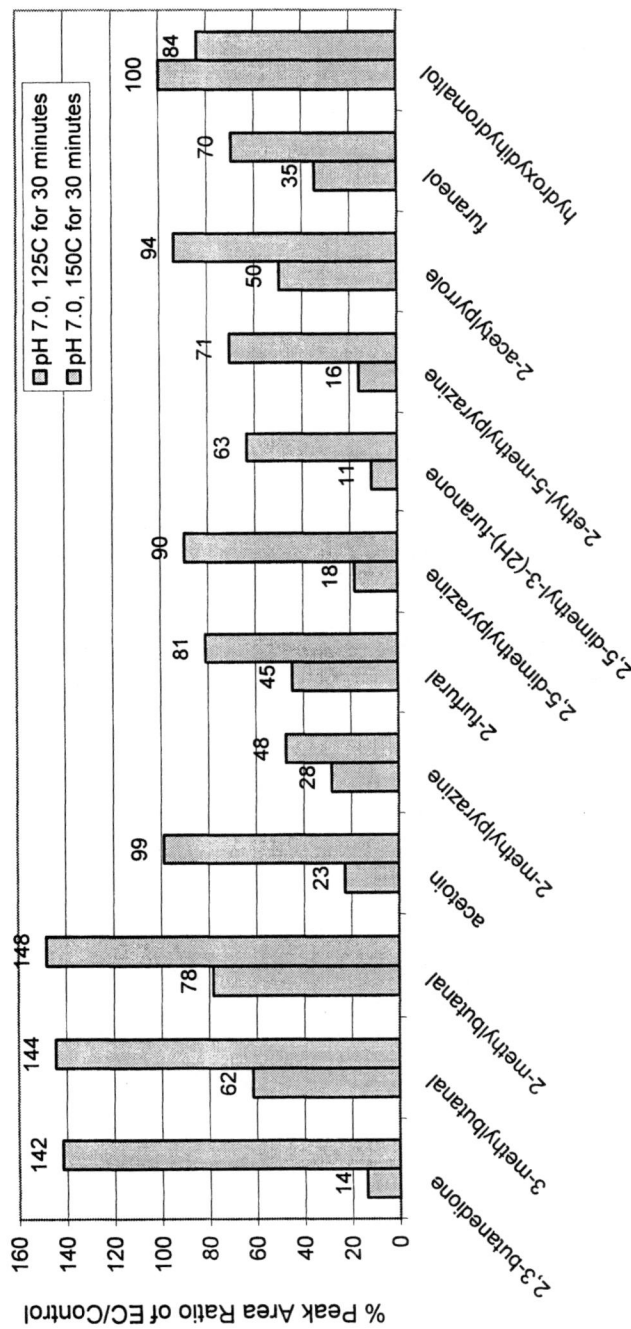

Figure 7 – Relative change in the aroma compounds generated in a model Maillard reaction systems with and without epicatechin at pH 7.0, 125°C and 150°C for 30 minutes; TIC Peak Area of Epicatechin (EC) Treatment/TIC Peak Area of Control * 100; peak area adjusted by internal standard (dodecane)

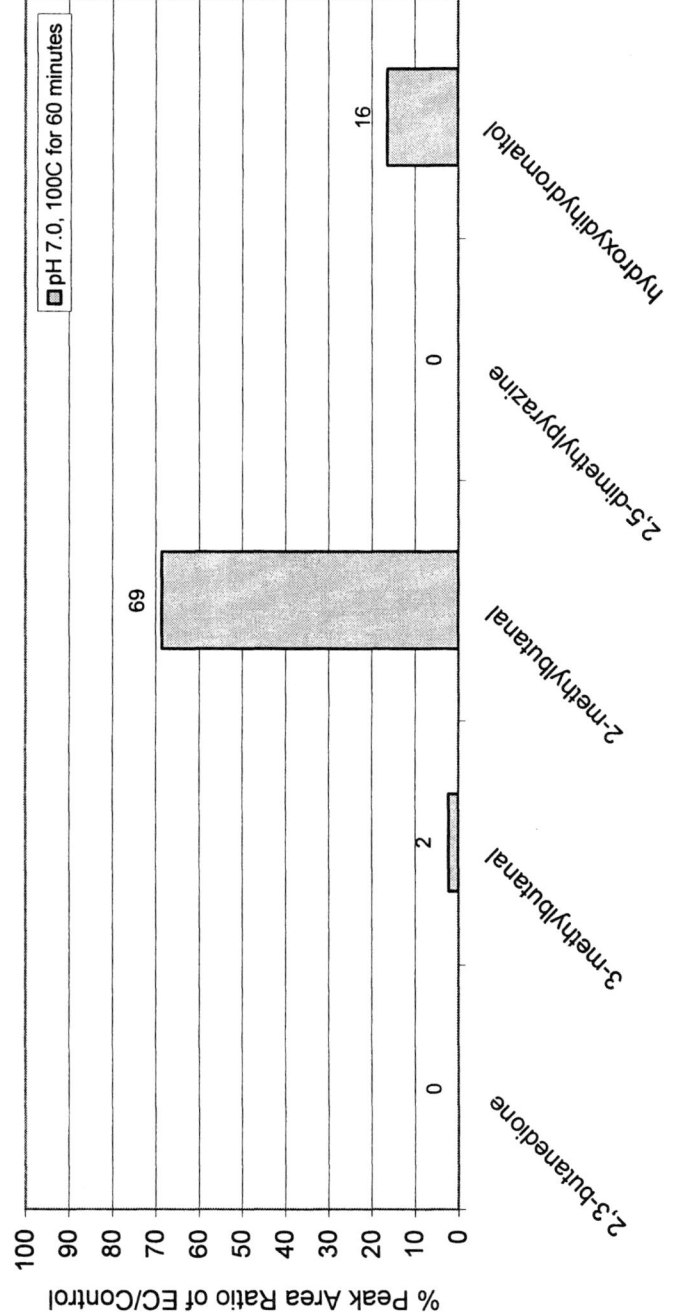

Figure 8 – Relative change in the aroma compounds generated in a model Maillard reaction systems with and without epicatechin at pH 7.0, 100°C for 60 minutes; TIC Peak Area of Epicatechin (EC) Treatment/TIC Peak Area of Control*100; peak area adjusted by internal standard (dodecane)

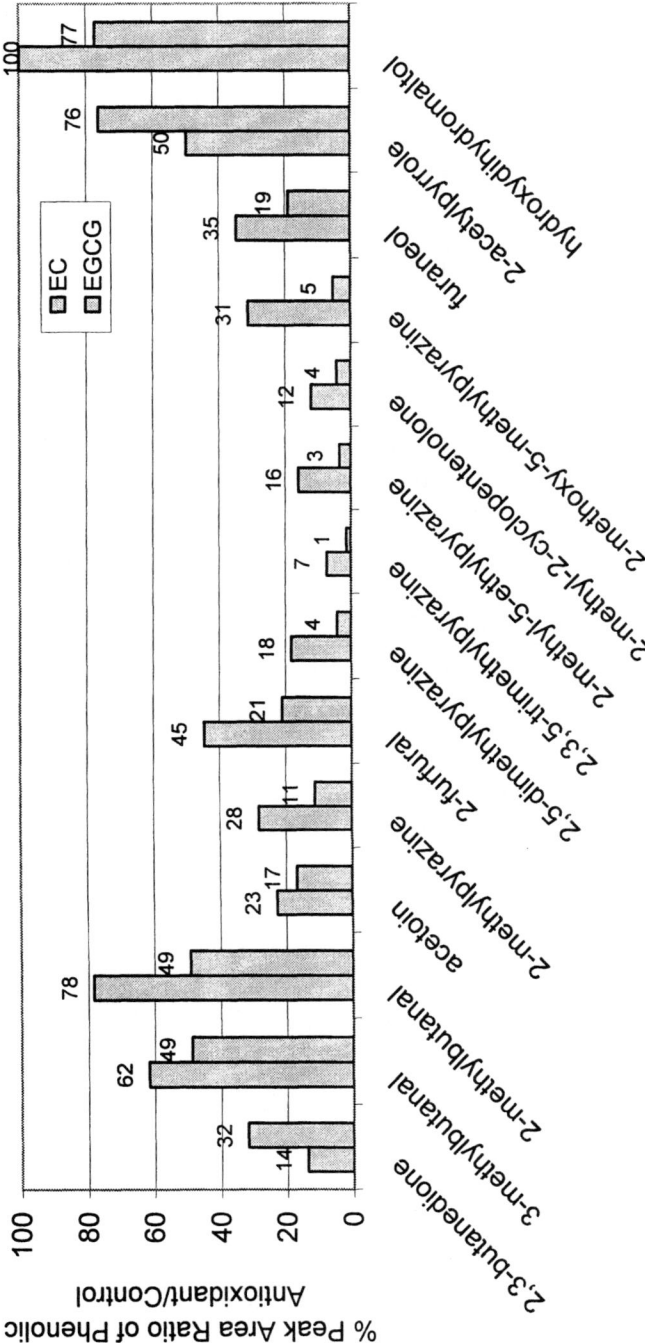

Figure 9 – Relative change in the aroma compounds generated in a model Maillard reaction systems with and without epicatechin (EC) or epigallocatechin gallate (EGCG) at pH 7.0, 125°C for 30 minutes; TIC Peak Area of Phenolic Antioxidant Treatment/TIC Peak Area of Control * 100; peak area adjusted by internal standard (dodecane)

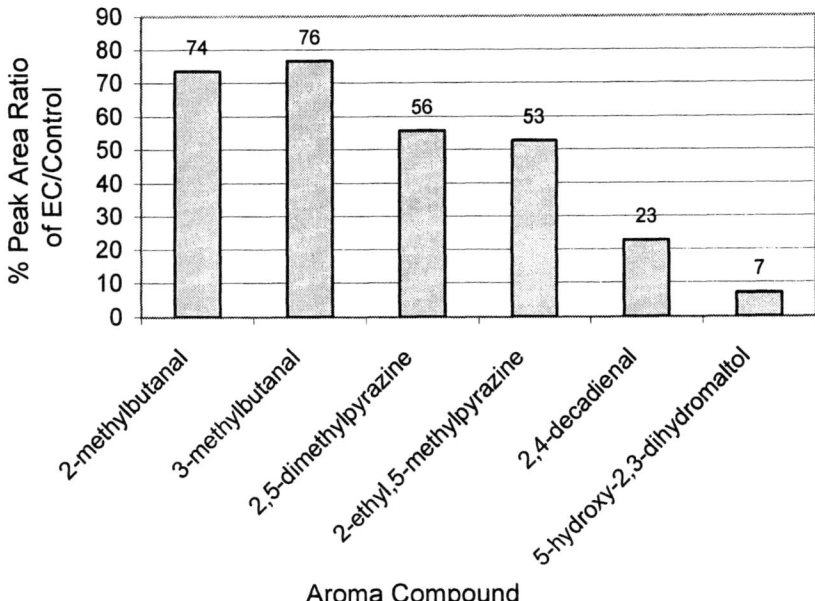

Figure 10 - Relative change in the aroma compound generated in a granola bar model with and without epicatechin at during the baking process

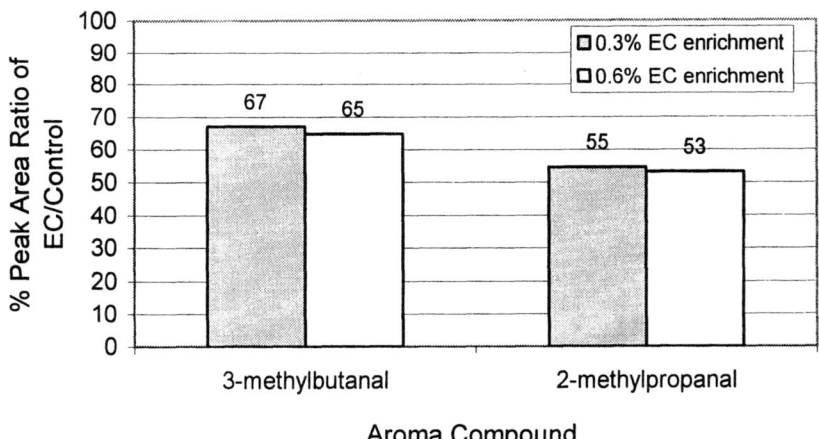

Figure 11 - Relative change in the aroma compound generated in a cocoa nibs enriched and not enriched with epicatechin at during the roasting process

affects on inhibiting aroma formation as the 0.3% treatment, indicating a zero-order reaction rate for the reactant at these concentrations.

Conclusion

Addition of flavonoids (i.e. EC, EGCG) to model Maillard reaction model system distinctly altered aroma development. Furthermore, addition of flavonoids to food products/commodities was also shown to influence the aroma generation during thermal treatment. Ultimately, little is known about how flavonoids behave as reactants or how enrichment would alter the aroma composition and subsequently the flavor properties. Although radical pathways are not part of the classical Hodge sequence, they may be a key parameter in Maillard-type reactions in relation to formation of flavor compounds.

REFERENCES:

(1) Kochian, L.; Garvin, D. *Nutr. Rev.* **1999**, *57*, S13-18.
(2) Farnham, M.; Simom, P.; Stomel, J. *Nutr. Rev.* **1999**, *57*, S19-26.
(3) Craig, W. J. *J. Am. Diet. Assoc.* **1997**, *97(suppl)*, S199-204.
(4) Potter, J. P. "Food, nutrition and the prevention of cancer: a global perspective," World Cancer Research Fund, 1997.
(5) Rodes, M. J. C. *Proc. Nutr. Soc.* **1996**, *55*, 371-384.
(6) Zhang, Y.; Talalay, P.; Cho, C.-G.; Posner, G. H. *Proc. Natl. Acad. Sci. USA* **1992**, *89*, 2399-2403.
(7) Bravo, L. *Nutr. Rev.* **1998**, *56*, 317-333.
(8) Glanz, K.; Basil, M.; Maiback, E.; Goldberg, J.; Snyder, D. *J. Am. Diet. Assoc.* **1998**, *98*, 1118-1126.
(9) Fenwick, G. R.; Griffiths, N. M.; Heaney, R. K. *J. Sci. Food Agric.* **1983**, *34*, 73-80.
(10) Rouseff, R. L. In *Bitterness in foods and beverages*; Rouseff, R. L., Ed.; Elsevier: Amsterdam, 1990; Vol. 25.
(11) Kielhorn, S.; Thorngate, J. *Food Qual. Prefer.* **1999**, *10*, 109-116.
(12) Drewnowski, A. *Annu. Rev. Nutr.* **1997**, *17*, 237-53.
(13) Jovanovic, S.; SteenKen, S.; Tosic, M.; Marjanovic, B.; Simic, M. *J. Am. Chem. Soc.* **1994**, *116*, 4846-4851.
(14) Schramm, D.; Wang, J.; Holt, R.; Ensunsa, J.; Gonsalves, J.; Lazarus, S.; Schmitz, H.; German, J.; Keen, C. *Am. J. Clin. Nutr.* **2001**, *73*, 36-40.
(15) Holt, R.; Schramm, D.; Keen, C.; Lazarus, S.; Schmitz, H. *JAMA* **2002**, *287*, 2212-2213.

(16) Reicks, M.; Randall, J.; Haynes, B. *J. Am. Diet Assoc* **1994**, *94*, 1309-1311.
(17) Hodge, J. E. *Food Chem.* **1952**, *1*, 928-943.
(18) Cammerer, B.; Kroh, L. *Food Chem.* **1996**, *57*, 217-221.
(19) Namiki, M.; Hayashi, T. *Prog. Food Nutr. Sci.* **1981**, *5*, 81-91.
(20) Kroh, L.; Zeise, S.; Stober, R.; Westphal, G. *Z. Lebensm. Unters. Forsch* **1989**, *188*, 115-117.
(21) Zeise, S.; Klein, L.; Kroh, L.; Stober, R. *Proc. Eur. Food Chem.* **1991**, *6*, 280-285.
(22) Wang, Y., Ph.D. Dissertation, Rutgers State University, 2000.
(23) Hayashi, T.; Namiki, M. *Agr. Biol. Chem.* **1981**, *45*, 933-939.
(24) Hofmann, T.; Bors, W.; Strettmaier, K. *J. Agric. Food Chem.* **1999**, *47*, 379-390.
(25) Ho, C.; Hartman, T. Lipids in Food Flavors, Denver, Colorado, 1993.
(26) Rice-Evans, C.; Miller, N.; Paganga, G. *Free Radic. Biol. Med.* **1996**, *20*, 933-956.
(27) Schnermann, P.; Schieberle, P. *J. Agric. Food Chem.* **1997**, *45*, 867-872.

Chapter 14

Capsaicinoid Oxidation by Peroxidases: Kinetic, Structural, and Physiological Considerations

Douglas C. Goodwin, Kimberley A. Laband, and Kristen M. Hertwig

Department of Chemistry and Program in Cell and Molecular Biosciences, Auburn University, Auburn, AL 36849–5312

>Capsaicinoids, biosynthesized exclusively in *Capsicum* fruits or "hot" peppers, are *o*-methoxyphenols that have substantial chemopreventive potential. Daily, *Capsicum* fruits are consumed by vast segments of the world's population. The levels of capsaicinoids in these fruits (and hence their desirability for consumption) are intimately linked with the activity of peroxidases, but several questions surrounding the kinetics of the process have gone unanswered. We applied novel transient- and steady-state methods to address the reduction of peroxidase compounds I and II by capsaicinoids and to obtain steady-state kinetic parameters for capsaicinoid oxidation. Comparison of transient-state and steady-state data identify electron transfer from capsaicinoids to compound II as rate-determining. Ascorbate rapidly reduces capsaicinoid radicals, a property essential for the kinetic studies. However, evidence also suggests that ascorbate is an important factor in capsaicinoid content of *Capsicum* fruits. The 4-substituent of *o*-methoxyphenols has a dramatic influence on rates of oxidation by peroxidases. However, our results suggest that kinetic parameters for oxidation of other chemopreventive *o*-methoxyphenols are likely to be similar to the capsaicinoids.

As the reports of this symposium firmly establish, evidence continues to accumulate suggesting that naturally-occuring phenolic and polyphenolic compounds have substantial health benefits. The *o*-methoxyphenols represent an important class of these increasingly prominent phytochemicals. Indeed, recent reports indicate that gingerol (the principle pungent compound in ginger) (*1,2*) and curcumin (a key component of tumeric) (*3-6*) show promise as chemopreventive agents through a variety of mechanisms (*7*). It is also important to note that these compounds impart the desirable or distinctive characteristics to the agricultural/food products that include them, ensuring their consumption in abundance by large sectors of the world's population.

In this sense, the capsaicinoids, produced only by plants of the *Capsicum* genus, are a very important group of *o*-methoxyphenols. They are the major contributors to the pungency of the *Capsicum* fruits (i.e., "hot" peppers). The levels of capsaicinoids in *Capsicum* fruits vary greatly among cultivars, some producing levels as high as 1% (*8*). Over one-quarter of the world's population consumes *Capsicum* fruits or related products on a daily basis, and the primary driving force behind the desirability of these fruits is their capsaicinoid content (*9*). As with gingerol and curcumin, capsaicinoids (Figure 1A) are effective antioxidant and anti-inflammatory compounds with strong potential for prevention of tumor promotion (*7,10-12*). Clearly, the factors contributing to the levels of capsaicinoids in *Capsicum* fruits have important agricultural, cultural, and biomedical implications.

Capsaicinoids and Peroxidase-Catalyzed Oxidation

Peroxidases occupy a central position in capsaicinoid catabolism. Capsaicinoid degradation has been correlated with a rise in peroxidase activity during *Capsicum* fruit maturation (*13*). Plant peroxidases have been shown to oxidize capsaicin to the lignin-like dimers, 5,5'-dicapsaicin and 4'-*O*-5 dicapsaicin ether, and further oxidation of these compounds by peroxidases leads to formation of high molecular weight polymers (*14*).

The typical peroxidase catalytic cycle (Figure 1B) is initiated by the reaction between ferric peroxidase and H_2O_2. In the reaction, the heme is oxidized by two electrons to form the ferryl-oxo porphyrin radical intermediate known as compound I. The porphyrin radical of compound I is then reduced by one electron by an exogenous electron donor (i.e., reducing substrate) to produce the ferryl-oxo intermediate compound II and one equivalent of substrate free radical.

Compound II is then reduced by a second equivalent of exogenous electron donor yielding the ferric peroxidase and a second equivalent of substrate radical (*15*).

Though capsaicinoids are well-known as reducing substrates for plant peroxidases (*16-20*), several kinetic aspects of the process have remained poorly defined. Until recently, the reactions of peroxidase compound I and compound II with capsaicinoids had not been directly observed nor the rate constants determined. Furthermore, the formation of a broad range of polymeric products (each with its own unique absorption characteristics) had hampered the assignment of steady-state kinetic parameters for capsaicinoid oxidation by peroxidases.

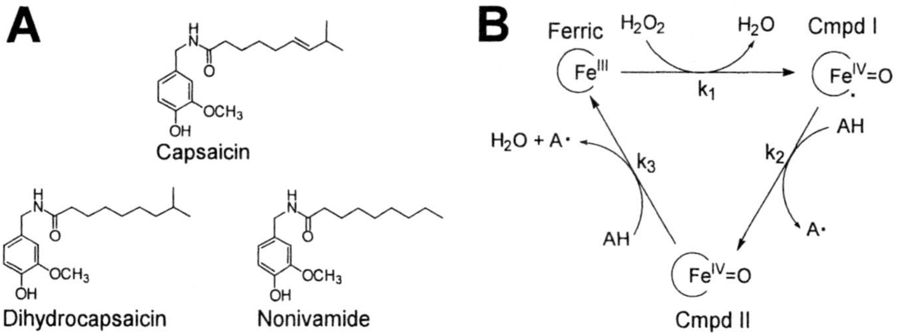

Figure 1. Structures of some of the capsaicinoids and their proposed interaction with the peroxidase catalytic cycle. Panel A shows the structures of the two most naturally abundant capsaicinoids, capsaicin and dihydrocapsaicin. Nonivamide is often referred to as "synthetic capsaicin" but was recently identified as a natural product. Panel B shows the peroxidase catalytic cycle. The capsaicinoids are proposed to be reducing substrates (AH) for peroxidase compounds I and II.

We addressed these difficulties by applying novel, ascorbate-based transient-state and steady-state kinetic approaches to reactions between the archetypal plant peroxidase (horseradish peroxidase) and typical capsaicinoids (e.g., capsaicin and nonivamide) (*21*). Our studies not only have provided valuable kinetic information on the metabolism of an important class of compounds, but they also point to a crucial role for ascorbate in the preservation of capsaicinoids in *Capsicum* fruits. Extension of our techniques to other *o*-

methoxyphenols has revealed the importance of the 4-substituent in the ability of peroxidases to descriminate between prospective electron donors. With this information it is possible to better predict the interplay between peroxidases and other agriculturally and biomedically important o-methoxyphenols in the systems that produce them.

Transient-State Kinetics of Capsainoid Oxidation

Compound I Reduction

We applied sequential-mixing stopped-flow methods to evaluate the reduction of peroxidase compound I by capsaicinoids. Briefly, compound I can be formed in a delay line by the rapid mixing of ferric peroxidase with 0.9 equivalent H_2O_2. Compound I so formed can then be reacted with an electron donor and conversion to compound II monitored by diode array or other detection method (*15,21*). Reduction of compound I to compound II in the presence of capsaicinoids is clearly indicated by a shift in the absorption maximum of the Soret band (403 nm to 419 nm), an increase in Soret intensity, an isosbestic point at 398 nm, and new minor absorption peaks at 527 nm and 555 nm (Figure 2A) (*15*).

The reactions between peroxidase compound I and capsaicinoids are very rapid. The need to retain sufficient signal intensity for stopped-flow studies requires that final reactions contain ~ 1 µM peroxidase. For such rapid reactions, it is difficult to maintain pseudo-first-order conditions with respect to reducing substrate and keep the rate of the reaction within the limits imposed by the 2 ms dead-time typical of stopped-flow instruments. To address this problem we have used ascorbate as a reductant of substrate radicals (*22*). Because rate constants for ascorbate reaction with these radicals typically far exceed those governing generation of the radical by the peroxidase, the concentration of the reducing substrate remains constant, even if it is equimolar or submolar with respect to the peroxidase (*21-24*).

Consistent with these principles, inclusion of a constant concentration of ascorbate (50 µM) produces single-exponential reduction of compound I at all capsaicinoid concentrations (*21*). The effect of capsaicinoid concentration on k_{obs} values obtained in these experiments is linear (Figure 2B), yielding rate constants of $(1.5 \pm 0.1) \times 10^7$ $M^{-1}s^{-1}$ for capsaicin and $(1.8 \pm 0.2) \times 10^7$ $M^{-1}s^{-1}$ for nonivamide. In each case, y-intercepts equal zero, suggesting that compound I reduction by these compounds is irreversible.

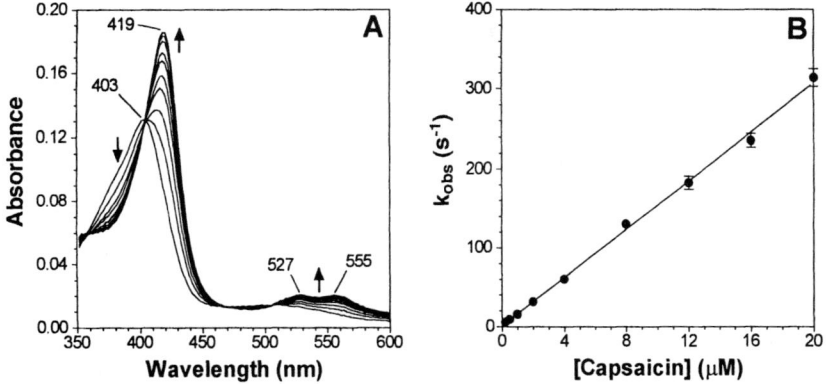

Figure 2. Reduction of peroxidase compound I by capsaicin. For panel A, 1.8 µM compound I was reacted with 2 µM capsaicin in the presence of 50 µM ascorbate, and spectra were recorded over 0.4 s. For panel B, 1 µM peroxidase compound I was reacted with variable concentrations of capsaicin in the presence of 50 µM ascorbate and the reaction monitored at 411 nm. All reactions were carried out at pH 7.0 and 25°C. (Panel A reproduced from reference 21. Copyright 2003 Elsevier Inc.)

Compound II Reduction

Our sequential-mixing stopped-flow experiments also clearly show reduction of peroxidase compound II to its ferric state (Figure 3A). Following compound II formation in the presence of 10 µM capsaicinoid (~40 ms), transition to a new absorption spectrum (λ_{max} at 403 and 485 nm) is observed with isosbestic points at 410, 453, and 521 nm.

The k_{obs} values for compound II reduction are linearly dependent on the concentration of capsaicinoid (Figure 3B), yielding rate constants of $(3.9 \pm 0.1) \times 10^5$ $M^{-1}s^{-1}$ and $(4.6 \pm 0.2) \times 10^5$ $M^{-1}s^{-1}$ for capsaicin and nonivamide, respectively. As with compound I reduction, y-intercepts are equal to zero, suggesting that compound II reduction by the capsaicinoids is also irreversible. Not surprisingly, capsaicinoids have limited solubility in neutral aqueous solutions. As such, kinetic data for concentrations of capsaicin or nonivamide exceeding 200 µM are typically inaccessible.

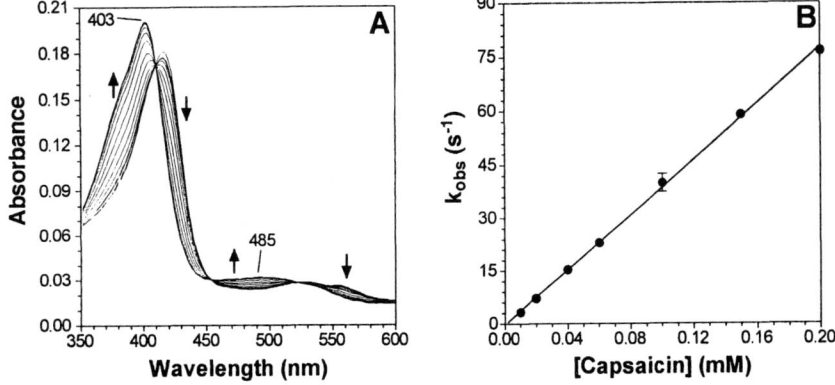

Figure 3. Reduction of peroxidase compound II by capsaicin. For panel B, 1.8 µM compound II was reacted with 10 µM capsaicin and spectra were recorded over 2 s. For panel B, 1 µM peroxidase compound II was reacted with variable concentrations of capsaicin and the reaction monitored at 428 nm. All reactions were carried out at pH 7.0 and 25°C. (Panel A reproduced from reference 21. Copyright 2003 Elsevier, Inc.)

Steady-State Kinetics of Capsaicinoid Oxidation

Measurement of steady-state kinetic parameters for oxidation of capsaicinoids by peroxidases has been plagued by difficulties involving the polymerization of the oxidation products (25). Multiple polymer products are generated including the dimeric species, 5-5' dicapsaicinoid and dicapsaicinoid ether. Additionally, larger polymeric products have been detected as well as capsaicinoid-protein copolymers (14). Each of these species has unique absorption characteristics.

Standard Spectrophotometric Methods

Peroxidase-catalyzed oxidation of capsaicinoids under steady-state conditions is typified by an increase in absorbance across the UV-visible spectrum (Figure 4A). Increases in capsaicinoid concentration lead to increases in rates of oxidation (Figure 4B). However, only relative values can be ascertained due to the multiple products generated and their individual contributions to the overall increase in absorbance. Consequently, it is difficult to assign the kinetic parameters of the oxidation process.

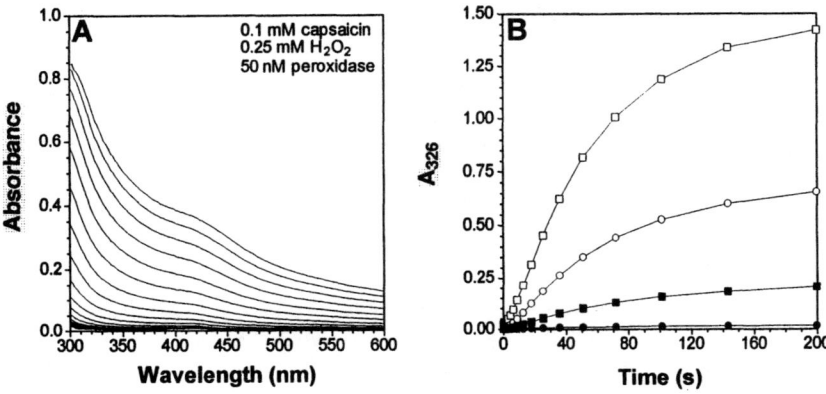

Figure 4. Capsaicin oxidation by peroxidase under steady-state conditions. Spectra collected during oxidation of 0.1 mM capsaicin are shown in Panel A. The effect of capsaicin concentration on the rate of its oxidation by peroxidase is shown in Panel B. Reactions were composed as shown in panel A except that capsaicin concentrations were 0.01 mM (●), 0.04 mM (■), 0.1 mM (○), and 0.2 mM (□). All reactions were carried out at pH 7.0 and 25 °C. (Reproduced from reference 21. Copyright 2003 Elsevier, Inc.)

Ascorbate-Dependent Chronometric Method

Given these problems, we turned to a chronometric method that uses the radical scavenging properties of ascorbate (23). In the presence of ascorbate, capsaicinoid radicals are reduced to the parent compound as rapidly as they are generated by peroxidase/H_2O_2. Thus, the accumulation of stable oxidation products (i.e., dimeric and polymeric species) is prevented, and a pronounced lag is observed (Figure 5A). Ascorbate consumption is directly related to the rate of capsaicinoid radical generation, and this rate is directly dependent on the rate of peroxidase turnover. Therefore, the length of the lag phase can be used as a measure of the initial rate of capsaicinoid oxidation by peroxidase under steady-state conditions.

In order for this method to be used effectively, three criteria must be met: 1) The concentration of capsaicinoid must remain constant over the course of the lag phase, 2) nearly all the ascorbate must be consumed before the stable capsaicinoid oxidation products begin to accumlate, and 3) the rate of ascorbate consumption must be independent of ascorbate concentration. All three requirements are met as evident in Figure 5. The *maximum* rate of accumulation of stable oxidation products is the same in the presence or absence of ascorbate (Figure 5A and 5B). This confirms that the concentration of capsaicinoid remains constant during the lag phase. When monitoring the reaction at 285 nm (in order to observe ascorbate consumption), the same minimum absorption value is obtained in the presence (t = 82 s) or absence (t = 0 s) of ascorbate (Figure 5B). This confirms that very nearly all ascorbate is consumed before any stable capsaicinoid oxidation products accumulate. Finally, the linear decrease in absorbance at 285 nm in the presence of ascorbate (Figure 5B) confirms that the rate of ascorbate consumption is independent of its concentration. Together, these data demonstrate that this method is effective for evaluating the steady-state kinetics of capsaicinoid oxidation by peroxidases.

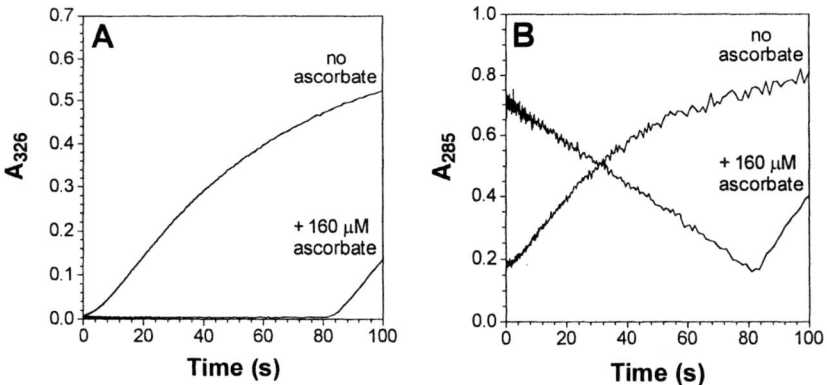

Figure 5. Effect of ascorbate on the accumulation of capsaicin oxidation products measured at 326 nm (A) and 285 nm (B). Reactions contained 50 nM peroxidase, 0.25 mM H_2O_2, 0.1 mM capsaicin, 100 mM phosphate, pH 7.0, and ascorbate as indicated. All reactions were performed at 25 °C. (Reproduced from reference 21. Copyright 2003 Elsevier, Inc.)

At a constant concentration of capsaicinoid, increasing concentrations of ascorbate produce progressive increases in the duration of the lag phase (τ) (Figure 6-inset). A series of τ values can be used to calculate an initial rate (v_o) according to the equation: $v_o = 2[\text{ascorbate}]/\tau$. To account for side reactions between H_2O_2 and ascorbate, the following equation is used: $v_o = \{2[\text{ascorbate}](1-\tau k[H_2O_2])\}/\tau$, where k is the rate constant (1.03 $M^{-1}s^{-1}$) for the reaction between H_2O_2 and ascorbate (23). Within the limits of solubility of the capsaicinoids, $v_o/[E]_T$ values are linear with increasing concentrations of nonivamide (Figure 6) and capsaicin, yielding apparent second-order rate constants of $(6.6 \pm 0.7) \times 10^5$ $M^{-1}s^{-1}$ and $(7.6 \pm 0.7) \times 10^5$ $M^{-1}s^{-1}$, respectively.

Our application of novel transient-state and steady-state kinetic methods has provided clarity on the kinetics of interplay between plant peroxidases and capsaicinoids. We clearly show that capsaicinoids are rapidly oxidized by peroxidase compounds I and II and have determined rate constants for the reactions. Reliable steady-state kinetic parameters have also been assigned to the overall process. Comparison of these two sets of data provide strong evidence that electron transfer from the capsaicinoid to the ferryl-oxo heme of compound II is rate-determining for peroxidase-catalyzed capsaicinoid oxidation. Thus, we have shed light on the kinetics of a major catabolic step involving an agriculturally and biomedically important group of phytochemicals.

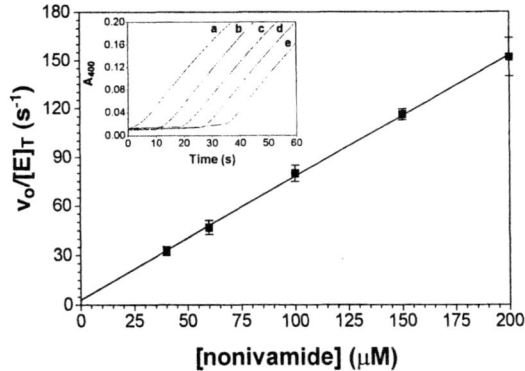

Figure 6. Apparent second-order rate constant determination for nonivamide oxidation by peroxidase. The inset shows the set of ascorbate-induced lag times used to determine the $v_o/[E]_T$ value for 150 μM nonivamide. Reactions also contained 50 nM peroxidase, 0.25 mM H2O2, 100 mM phosphate buffer, pH 7.0, and 0 μM (a), 40 μM (b), 60 μM (c), 80 μM (d), or 100 μM ascorbate (e). All reactions were carried out at 25 °C. (Reproduced from reference 21. Copyright 2003 Elsevier, Inc.)

Physiological Considerations: Capsaicinoid Oxidation and Ascorbate

Ascorbate has proven to be an effective tool for kinetic evaluation of peroxidases (*21-23*). That ascorbate was effective in kinetic studies involving capsaicinoids demonstrates that it reacts very rapidly with capsaicinoid radicals, likely at near diffusion-controlled rates. This may have important implications for the metabolism of capsaicinoids *in vivo*. Ascorbate prevents the accumulation of the stable capsaicinoid oxidation products (dimers, polymers, etc.), in essence blocking the catabolism of capsaicinoids. Consequently, ascorbate levels could profoundly influence the levels of capsaicinoids observed in *Capsicum* fruits.

There is significant evidence to support this hypothesis. First, both constituents are present together with *Capsicum* fruits. Capsainoids are known to accumulate primarily in the placental tissue of the fruit (*26*), and ascorbate has also been identified in this tissue (*27,28*).

Second, factors that increase ascorbate levels tend also to increase capsaicinoid levels. It is well known that increased light exposure increases production of ascorbate in many plant species. In *Capsicum* fruits, ascorbate *and* capsaicinoid content are enhanced by light exposure (*26,27,29-31*). Indeed, fruit from the upper (i.e., more light-exposed) portions of *Capsicum* plants have higher levels of ascorbate and capsaicinoids than fruit harvested from the lower portions of the same plants (*27,31*).

Similarly, mineral fertilizers increase the levels of both constituents in *Capsicum* fruits (*27,32,33*). Consistent with the chemistry described here, these fertilizers are also known to decrease the formation of lignin-like products (*33*). Because ascorbate reduces capsaicinoid and other phenoxyl radicals preventing the accumulation of lignin-like polymers, it is reasonable to suggest that the increase in capsaicinoids under these conditions arises due to the ascorbate-dependent inhibition of peroxidase-catalyzed capsaicinoid polymerization.

Together, these data suggest that the abundance of capsaicinoids in *Capsicum* fruits depends, at least in part, on the complex interplay between three factors: 1) production by biosynthetic pathways, 2) consumption by peroxidase-catalyzed oxidation, and 3) preservation against oxidation by ascorbate. Each of these represent potential control points for manipulating the levels of these very important compounds in *Capsicum* fruits.

Substituent Structure and Capsaicinoid Oxidation

As mentioned previously, capsaicinoids are only one group among many *o*-methoxyphenols with strong chemopreventive potential. Peroxidase compounds I and II are known to accept a very structurally diverse range of agents as electron donors. Thus, plant peroxidases are likely to have significant involvement in the metabolism of many of these compounds. Nevertheless, it is striking to note that vanillylamine, another o-methoxyphenol, is a very poor reducing substrate in comparison to the capsaicinoids (*19,21*). This suggests that rates of the o-methoxyphenol oxidation by peroxidases are influenced greatly by the nature of the 4-substituent. To explore this possibility, we used transient-state kinetic methods (as described for the capsaicinoids) (*21*) to determine rate constants for compound I and compound II reduction by various o-methoxyphenols.

Rate constants for compound I reduction ranged from $(8.0 \pm 0.9) \times 10^7$ $M^{-1}s^{-1}$ for ferulic acid to as low as $(3.2 \pm 0.2) \times 10^4$ $M^{-1}s^{-1}$ for 4-hydroxy-3-methoxymandelic acid (HMMA) (Table I). Although the rate constants for compound II reduction by each o-methoxyphenol are smaller (typically by about an order of magnitude) (Table I), the same trend in reactivity is followed for reduction of compounds I and II.

Interestingly, rate constants measured for vanillin, capsaicin, nonivamide, and guaiacol were all very similar to one another, providing strong evidence that steric constraints imposed by the 4-substituent have very little influence on selection of the o-methoxyphenols as peroxidase reducing substrates. This is consistent with recent investigations on steady-state oxidation of o-diphenols (*23*) as well as crystallographic studies using the ferri-cyano complex of HRP as an analog of peroxidase compounds I and II (*34*). Ferulic acid binds to Fe(III)-CN HRP such that the propenoate substituent is pointed directly into the surrounding medium (*34*). In this conformation, the protein imposes no steric restrictions on the 4- substituent.

The basis for the very rapid oxidation of ferulic acid in comparison to the other compounds may be explained by its relatively low reduction potential (*16*). It is known that the thermodynamic driving force for compound I reduction is an important determinant of rate, consistent with the Marcus Theory of electron transfer (*35*). Increased conjugation due to the propenate side chain is expected to lower the energy of the ferulic acid radical. The other o-methoxyphenols tested lack this structure, and hence, are expected to be oxidized to less stable radicals.

The poor reactivity of vanillylamine, HMMA, and to a lesser extent, vanillic acid is not explained by differences in thermodynamic driving force. It is likely, though not satisfactorily established, that the charged side chains of these poor substrates lead to alternative binding modes that are far less productive with respect to electron transfer. One possibility is that these charged electron donors bind to the peroxidase more like benzhydroxamic acid (*36*) than ferulic acid (*34*). Under such circumstances the phenolic hydroxyl moiety would be oriented away from the heme group leading to suboptimal electron transfer.

Table I. Rate Constants for Peroxidase Compound I and Compound II Reduction by *o*-Methoxyphenols

Compound	Substituent	$k_2\ (M^{-1}s^{-1})^a$	$k_3\ (M^{-1}s^{-1})^a$
Ferulic acid	–CH=CH–COO⁻	$(8.0 \pm 0.9) \times 10^7$	$(1.75 \pm 0.04) \times 10^7$
Vanillin	–CHO	$(2.2 \pm 0.1) \times 10^7$	$(1.6 \pm 0.2) \times 10^6$
Nonivamide[b]	See Figure 1	$(1.8 \pm 0.2) \times 10^7$	$(4.6 \pm 0.1) \times 10^5$
Capsaicin[b]	See Figure 1	$(1.5 \pm 0.1) \times 10^7$	$(3.9 \pm 0.1) \times 10^5$
Guaiacol	–H	$(9 \pm 1) \times 10^6$	$(3.6 \pm 0.4) \times 10^5$
Vanillic Acid	–COO⁻	$(2.6 \pm 0.2) \times 10^6$	$(7.0 \pm 0.2) \times 10^4$
Vanillylamine	–NH$_3^+$	$(1.46 \pm 0.04) \times 10^5$	$(3.3 \pm 0.9) \times 10^3$
HMMA	–CH(OH)–COO⁻	$(3.2 \pm 0.2) \times 10^4$	$(8 \pm 4) \times 10^3$

[a] All rate constants determined at pH 7.0 and 25°C.
[b] Reference (*21*).

Conclusion

Capsaicinoids, naturally present in *Capsicum* fruits, are greatly desired because they impart a unique flavor to the foods that include them. Thus, capsaicinoids are consumed by billions of individuals on a daily basis. Without question the factors that influence the levels of capsaicinoids in *Capsicum* fruits are of great agricultural, economic, and cultural interest. Even greater interest in the capsaicinoids (and other *o*-methoxyphenols) has been spurred by recent data suggesting that these compounds have chemopreventive and other health benefits.

The peroxidases play a key role in the catabolism of the capsaicinoids, and therefore, their levels in *Capsicum* fruits. We applied novel, ascorbate-based kinetic methods to address unresolved questions surrounding the kinetics of peroxidase-catalyzed capsaicinoid oxidation. Peroxidase compound I and compound II rapidly oxidize capsaicinoids with rate constants approaching 2×10^7 and 5×10^5 M^{-1}s^{-1}, respectively. Careful measurement of steady-state kinetic parameters reveal an apparent second order rate constant for the overall process of approaching 8×10^5 M^{-1}s^{-1}, implicating electron transfer to compound II as rate-limiting.

Ascorbate was a useful tool for obtaining accurate kinetic data, but our data and that of others suggests that ascorbate may be an important determinant of capsaicinoid content of *Capsicum* fruits. This suggests at least three potential control points for manipulating capsaicinoid abundance: increased biosynthesis, decreased peroxidase-dependent oxidation, and/or increased levels of ascorbate.

Finally, we have carried out additional studies evaluating the effect of *o*-methoxyphenol structures on rates of peroxidase-catalyzed oxidation. Our results indicate that the nature of the 4-substituent plays an important part in the selection of electron donors as peroxidase substrates. The data suggest that other agriculturally and biomedically prominent *o*-methoxyphenols will be efficient peroxidase substrates. Thus, the data we have accumulated on the capsaicinoids may have broad application to optimizing the health benefits to be derived from plant-derived *o*-methoxyphenols.

References

1. Lee, E.; Surh, Y. J. *Cancer Lett.* **1998**, *134*, 163-168.
2. Park, K. K.; Chun, K. S.; Lee, J. M.; Lee, S. S.; Surh, Y. J. *Cancer Lett.* **1998**, *129*, 139-144.
3. Chun, K. S.; Keum, Y. S.; Han, S. S.; Song, Y. S.; Kim, S. H.; Surh, Y. J. *Carcinogenesis* **2003**, *24*, 1515-1524.
4. Brouet, I.; Ohshima, H. *Biochem. Biophys. Res. Commun.* **1995**, *206*, 533-540.
5. Kim, M. S.; Kang, H. J.; Moon, A. *Arch. Pharm. Res.* **2001**, *24*, 349-354.
6. Shim, J. S.; Kim, J. H.; Cho, H. Y.; Yum, Y. N.; Kim, S. H.; Park, H. J.; Shim, B. S.; Choi, S. H.; Kwon, H. J. *Chem. Biol.* **2003**, *10*, 695-704.
7. Surh, Y. *Mutat. Res.* **1999**, *428*, 305-327.
8. Surh, Y. J.; Lee, S. S. *Food Chem. Toxicol.* **1996**, *34*, 313-316.
9. Szallasi, A.; Blumberg, P. M. *Pharmacol. Rev.* **1999**, *51*, 159-212.
10. Macho, A.; Sancho, R.; Minassi, A.; Appendino, G.; Lawen, A.; Munoz, E. *Free Radic. Res.* **2003**, *37*, 611-619.
11. Jung, M. Y.; Kang, H. J.; Moon, A. *Cancer Lett.* **2001**, *165*, 139-145.
12. Surh, Y. J. *Food Chem. Toxicol.* **2002**, *40*, 1091-1097.
13. Contreras-Padilla, M.; Yahia, E. M. *J. Agric. Food Chem.* **1998**, *46*, 2075-2079.
14. Bernal, M. A.; Ros Barcelo, A. *J. Agric. Food Chem.* **1996**, *44*, 3085-3089.
15. Dunford, H. B.; Stillman, J. S. *Coord. Chem. Rev.* **1976**, *19*, 187-251.
16. Boersch, A.; Callingham, B. A.; Lembeck, F.; Sharman, D. F. *Biochem. Pharmacol.* **1991**, *41*, 1863-1869.
17. Bernal, M. A.; Calderon, A. A.; Pedreno, M. A.; Munoz, R.; Ros Barcelo, A.; Merino de Caceres, F. *J. Agric. Food Chem.* **1993**, *41*, 1041-1044.
18. Bernal, M. A.; Calderon, A. A.; Pedreno, M. A.; Munoz, R.; Ros Barcelo, A.; Merino de Caceres, F. *J. Food Science* **1993**, *58*, 611-613.
19. Bernal, M. A.; Calderon, A. A.; Ferrer, M. A.; Merino de Caceres, F.; Ros Barcelo, A. *J. Agric. Food Chem.* **1995**, *43*, 352-355.

20. Pomar, F.; Bernal, M. A.; Diaz, J.; Merino, F. *Phytochemistry* **1997**, *46*, 1313-1317.
21. Goodwin, D. C.; Hertwig, K. M. *Arch. Biochem. Biophys.* **2003**, *417*, 18-26.
22. Goodwin, D. C.; Yamazaki, I.; Aust, S. D.; Grover, T. A. *Anal. Biochem.* **1995**, *231*, 333-338.
23. Rodriguez-Lopez, J. N.; Gilabert, M. A.; Tudela, J.; Thorneley, R. N.; Garcia-Canovas, F. *Biochemistry* **2000**, *39*, 13201-13209.
24. Schuler, R. H. *Radiat. Res.* **1977**, *69*, 417-433.
25. Calderon, A. A.; Munoz, R.; Morales, M.; Ros Barcelo, A. *Phytochem. Analysis* **1992**, *3*, 238-240.
26. Suzuki, T.; Iwai, K. In *The Alkaloids;* Brossi, A., Ed.; Academic Press, Inc., Orlando, FL, 1984; Vol. XXIII, pp. 227-299.
27. Mozafar, A. *Plant Vitamins: Agronomic, Physiological, and Nutritional Aspects*, CRC Press, Inc., Boca Raton, FL, 1994.
28. Govindarajan, V. S. *CRC Crit. Rev. Food Sci. Nutr.* **1984**, *22*, 109-176.
29. Iwai, K.; Lee, K.-R.; Kobashi, M.; Suzuki, T. *Agric. Biol. Chem.* **1977**, *41*, 1873-1876.
30. Iwai, K.; Suzuki, T.; Lee, K.-R.; Kobashi, M.; Oka, S. *Agric. Biol. Chem.* **1977**, *41*, 1877-1882.
31. Estrada, B.; Bernal, M. A.; Diaz, J.; Pomar, F.; Merino, F. *J. Agric. Food Chem.* **2002**, *50*, 1188-1191.
32. Dass, R. C.; Mishra, S. N. *Plant Sci.* **1972**, *4*, 78.
33. Estrada, B.; Pomar, F.; Diaz, J.; Merino, F.; Bernal, M. A. *J. Hort. Sci. Biotechnol.* **1998**, *73*, 493-497.
34. Henriksen, A.; Smith, A. T.; Gajhede, M. *J. Biol. Chem.* **1999**, *274*, 35005-35011.
35. Candeias, L. P.; Folkes, L. K.; Wardman, P. *Biochemistry* **1997**, *36*, 7081-7085.
36. Henriksen, A.; Schuller, D. J.; Meno, K.; Welinder, K. G.; Smith, A. T.; Gajhede, M. *Biochemistry* **1998**, *37*, 8054-8060.

Synthesis, Production, and Mechanism of Action

Chapter 15

Artepillin C Isoprenomics: Facile Total Synthesis and Discovery of Amphiphilic Antioxidant

Yoshihiro Uto*, Shutaro Ae, Hideko Nagasawa, and Hitoshi Hori*

Department of Biological Science and Technology, Faculty of Engineering, The University of Tokushima, Minamijosanjima-cho 2, Tokushima 770–8506, Japan

We examined the facile synthesis and antioxidant property of artepillin C, a medicinally active component of Brazilian propolis. The yield of o,o'-diprenylation of p-iodophenol, as a key step, was improved to 52%, when the reaction was run in toluene. Moreover, we succeeded in the acetylation of o,o'-diprenyl-p-iodophenol 2 resulting in good yield of 89%, following the Mizoroki-Heck reaction. Consequently, the total yield of artepillin C was improved to 33%. Next we demonstrated artepillin C was as potent as α-tocopherol for scavenging the stable free radical diphenyl-2-picrylhydrazyl (DPPH) with an $IC_{0.200}$ of 14.7 ± 0.6 µM. It also strongly inhibited iron-induced lipid peroxidation in rat liver mitochondria (RLM) with an IC_{50} of 0.72 ± 0.03 µM. We propose that artepillin C is a new amphiphilic antioxidant, meaning that it functions both in water and in lipid.

Artepillin C [3-{4-hydroxy-3,5-di(3-methyl-2-butenyl)phenyl}-2(*E*)-propenoic acid] is a diprenylated *p*-hydroxycinnamic acid (*p*-coumaric acid) derivative isolated from *Artemisia patustris* (*1*), *Baccharis punctulate* (*2*), *Artemisia capillaries* (*3*) and also isolated as a major constituent (>5%) (*4*) from Brazilian propolis. Recent reports indicate that artepillin C has important medicinal activities, such as antimicrobial (*4*), antitumor (*5*), apoptosis-inducing (*6*), immunomodulating (*7*), and antioxidative (*8*). These data suggest that artepillin C may be one of the important active principles of Brazilian propolis. Therefore, we recognized the importance of medicinal chemistry and the gene associated with biosynthesis of isoprenoid molecules and we termed it "isoprenomics."

We previously reported the first total synthesis of artepillin C even though total yield was very low (7%) (*9*). The problem is the inefficiency of *o,o'*-diprenylation of *p*-halophenols as a key step following the Mizoroki-Heck reaction. And also, purification of *o,o'*-diprenylated product in the presence of many prenylated byproducts is difficult. We need to solve these problems in order to establish a facile preparation of artepillin C and available synthesis tools for many prenylated natural products having medicinal activities.

Hayashi *et al.* reported that artepillin C showed a potent inhibition activity against peroxidation of linoleic acid in a micell solution (*8*). They also described artepillin C as one of the major antioxidants in Brazilian propolis. We agree with this description, however, we consider that the antioxidative mode of action of artepillin C is still important. In general, radical species are generated in an aqueous layer and then enter the lipid bilayer to attack the methylene group of a 1,4-pentadiene moiety in a biosystem. Accordingly, water- and lipid-soluble antioxidants, that are amphiphilic antioxidant, are particularly effective antioxidants *in vivo*.

In this paper we discuss the regioselective prenylation of *p*-iodophenol to synthesis of artepillin C. In addition, we have investigated the antioxidative mode of action of artepillin C compared with α-tocopherol, a typical intracellular lipid peroxidation inhibitor and butylated hydroxytoluene (BHT), a synthetic antioxidant in food.

Materials and Methods

Materials

p-Iodophenol, prenyl bromide and tri-*o*-tolylphosphine ((*o*-tol)$_3$P) were obtained from Aldrich Chemical Company (Milwakee, WS). DL-α-tocopherol (Figure 1), 2,6-di-*t*-butyl-4-methylphenol (BHT, Figure 1), 1,1-diphenyl-2-picrylhydrazyl (DPPH), adenosine 5'-diphosphate disodium salt (ADP), iron (II) sulfate (FeSO$_4$) heptahydrate, sodium hydride (NaH), 4-(dimethylamino)pyridine

(DMAP), acetyl chloride (AcCl), methyl acrylate and palladium diacetate (Pd(OAc)$_2$) were purchased from Wako Pure Chemical Industries Ltd. (Osaka, Japan). Rotenone, 2-(N-morpholino)ethanesulfonic acid (MES) and triethylamine (Et$_3$N) were obtained from Sigma Chemical Company (St. Louis, MO). Toluene was distilled under nitrogen from CaH$_2$.

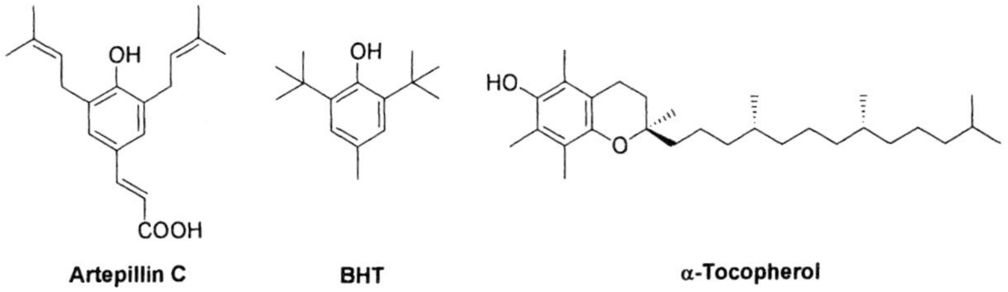

Figure 1. Chemical structure of artepillin C, BHT, and α-tocopherol.

Synthesis

Prenylation of p-iodophenol 1

p-Iodophenol **1** (44 mg, 0.20 mmol) was dissolved in dry toluene (1 mL), and then NaH (1.2 or 2.2 per mol of *p*-iodophenol) was added. After the reaction mixture was stirred at room temperature for 1 h, prenyl bromide (2.0 or 2.2 per mol of *p*-iodophenol) was added. Stirring, at room temperature, was continued for 3 h, after which the mixture was poured into ice water and then acidified with 2 M CH$_3$COOH. The mixture was extracted with Et$_2$O and washed with saturated aqueous NaHCO$_3$ followed by saturated aqueous NaCl. The Et$_2$O layer was dried (anhydrous MgSO$_4$) and evaporated under reduced pressure. Residues were purified by column chromatography on silica gel (hexanes/Et$_2$O, 20:1) to provide the prenylated products (**2–5**) previously identified (*9*).

Acetylation of o,o'-diprenylated-p-iodophenol 6

o,o'-Diprenylated-*p*-iodophenol **2** (526 mg, 1.48 mmol) and DMAP (217 mg, 1.78 mmol) were dissolved in dry CH$_2$Cl$_2$ (3 mL), and then AcCl (232 mg, 2.96 mmol) was added dropwise. The mixture was refluxed for 4 h, cooled to room temperature and evaporated. The residue was chromatographed over silica gel eluated by hexanes-Et2O (10:1) to afford **6** (552 mg, 1.39 mmol, 94%). **6**: light yellow oil; ^{1}H NMR (400 MHz, CDCl$_3$) δ 7.37 (s, 2H, H_3,H_5-Ar), 5.17 (t, J = 7.1 Hz, 2H, Ar(CH$_2$CH=C(CH$_3$)$_2$)$_2$), 3.12 (d, J = 7.1 Hz, 4H, Ar(CH_2CH=C(CH$_3$)$_2$)

$_2$), 2.29 (s, 3H, C(O)CH_3), 1.74 (s, 6H, C(CH_3)$_2$), 1.67 (s, 6H, C(CH_3)$_2$); EI-MS *m/z* 398 (M$^+$).

Mizoroki-Heck coupling of 7

A solution of **6** (212 mg, 0.53 mmol), methyl acrylate (229 mg, 2.67 mmol), Et$_3$N (109 mg, 1.08 mmol), (*o*-tol)$_3$P (17 mg, 0.054 mmol), and Pd(OAc)$_2$ (6 mg, 0.027 mmol) in dry toluene (2 ml) was refluxed for 17 h. The mixture was cooled to room temperature, diluted with Et$_2$O, and filtrated through Celite. The filtrate was evaporated under reduced pressure. Residues were purified by column chromatography on silica gel (hexanes/Et$_2$O, 5:1) to give 168 mg (0.47 mmol, 89%) of product **7** as a white solid: ^{1}H NMR (400 MHz, CDCl$_3$) 7.63 (d, 1H, *J* = 16.1 Hz, ArC*H*=), 7.23 (s, 2H, *H*$_3$,*H*$_5$-Ar), 6.35 (d, 1H, *J* = 16.1 Hz, C*H*CO$_2$CH$_3$), 5.22 (t, 2H, *J* = 7.3 Hz, Ar(CH$_2$C*H*=C(CH$_3$)$_2$)$_2$), 3.80 (s, 3H, CHCO$_2$C*H*$_3$), 3.19 (d, 4H, *J* = 7.1 Hz, Ar(C*H*$_2$CH=C(CH$_3$)$_2$)$_2$), 2.32 (s, 3H, C(O)C*H*$_3$), 1.76 (s, 6H, C(C*H*$_3$)$_2$), 1.69 (s, 6H, C(C*H*$_3$)$_2$); EI-MS *m/z* 356 (M$^+$).

Hydrolysis of 7 to synthesis of artepillin C

KOH (3 mL of a 5% aqueous solution) was added to a solution of **8** (187 mg, 0.52 mmol) in MeOH (5 ml). The mixture was refluxed for 3 h, cooled to 0 °C, and acidified with 1N HCl. The MeOH was evaporated under reduced pressure, and the aqueous residue was extracted with Et$_2$O. Extracts were washed with saturated aqueous NaHCO$_3$ and followed with saturated aqueous NaCl. The Et$_2$O layer was dried (anhydrous MgSO$_4$) and evaporated under reduced pressure. Residues were purified by column chromatography on silica gel (CH$_2$Cl$_2$/MeOH, 10:1) to give 119 mg (0.40 mmol, 76%) of artepillin C, previously identified (*9*).

DPPH Radical Scavenging Activity

Free radical scavenging activity was determined by using DPPH at 517 nm according to the method of Blois with some modifications (*10*). 3 μL of different concentrations of antioxidants being tested were added to 3.0 mL of 100 μM DPPH ethanol solution (60%) containing 40 mM MES at pH 5.5. The change in optical absorbance at 517 nm was measured for 30 min. The mean effective concentration of antioxidants tested required decreasing the absorption by 0.200

calculated from the absorbance concentration curve in the reaction with DPPH after 30 min ($IC_{0.200}$) (*11*).

Molecular Orbital (MO) Calculation

The semi-empirical MO calculation was applied using the AM1 method of Stewart and the program MOPAC2000 (Fujitsu, Japan) (*12*). The orbital energy (eV) and atomic orbital coefficients (C^2) value of the highest occupied molecular orbital (HOMO) were calculated.

Preparation of Rat Liver Mitochondria (RLM)

RLM were isolated from male Wistar rats (140–220 g) by homogenization followed by differentiated centrifugation in ice-cold medium (pH 7.4) (*13,14*). Fresh and freeze-thawed mitochondria were used for this study. The mitochondrial protein content was determined by the biuret method using bovine serum albumin as a standard (*15*).

Inhibition of RLM Lipid Peroxidation

Lipid peroxidation in RLM was performed at 25 °C and monitored as oxygen consumption with a Clark-type oxygen electrode in a total volume of 2.53 ml. Mitochondria was added at concentrations of 0.7 mg protein/ml. Peroxidation was started by addition of final concentrations of 1 mM ADP and 100 µM $FeSO_4$ in medium consisting of 175 mM KCl and 10 mM Tris-HCl buffer, pH 7.4. The amount of O_2 consumed during peroxidation was calculated assuming that the saturation concentration of O_2 at 25 °C is 258 µM. The percent inhibition with drug on RLM lipid peroxidation was calculated by the following equation:

$$\text{Inhibition}(\%) = \{1 - (R_p t / t_{inh}) \times k\} \times 100\%; \quad k = t_{inh0} / R_{p0} t_0$$

where R_p, is the rate of lipid peroxidation (nmol O/min); R_{p0} the rate of lipid peroxidation of control (nmol O/min); t the total time (min); t_0 the total time of control (min); t_{inh} the induction time (min); and t_{inh0} the induction time of control (min). The mean effective concentration, of antioxidants tested, required for 50% inhibition was defined as IC_{50} in the RLM lipid peroxidation reaction (*16,17*).

The Octanol-Water Partition Coefficient Calculation

The octanol-water partition coefficient (log$P$) of artepillin C, BHT, and α-tocopherol were calculated by the program Pallas 3.0 (CompuDrug International Inc., Arizona, USA). The calculated log$P$ was indicated by log$P$(C).

Results

Synthesis of Artepillin C

Firstly, we examined the solvent effect for the regioselective prenylation of p-iodophenol **1** to establish the preparative synthesis of artepillin C. We had previously shown that o,o'-diprenylation of p-iodophenol can be achieved in alkaline solution, but the product yield was low (the best yield was 27%) and O-prenylated compounds were produced (9). In this work, we selected a toluene as a solvent and sodium hydride as a base (Scheme 1 and Table I). When **1** was treated with 1.2 equiv of sodium hydride and 2.0 equiv of prenyl bromide, o-monoprenylated **3** was obtained selectively (60%) and o,o'-diprenylated **2** was obtained in low yield (11%, Entry 1). On the other hand, **2** was mainly obtained (52%) using 2.2 equiv of sodium hydride (Entry 2). Nevertheless, O-prenylated **4** and **5** were produced.

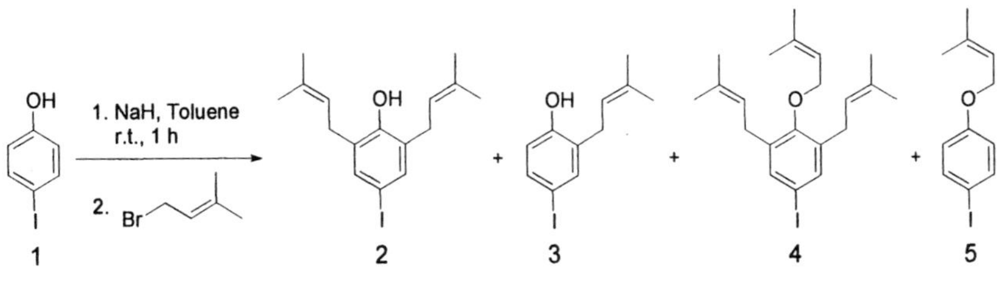

Scheme 1. Prenylation of p-iodophenol in toluene

Table I. Product Distributions in the Prenylation of p-Iodophenol

Entry	NaH (equiv)	Prenyl Bromide (equiv)	2	3	4	5	Recovered 1
1	1.2	2.0	11	60			13
2	2.2	2.2	52		29	19	

Next, we examined the protection of the phenolic hydroxyl group of **2** to improve the efficiency of Mizoroki-Heck coupling (Scheme 2). Acetylation of **2** (AcCl and DMAP) produced **6** in 94%, which was treated with methyl acrylate in the presence of Pd(OAc)$_2$, (o-tol)$_3$P, and Et$_3$N in dry toluene to give acetylartepillin C methylester **7** in 89%. Finally, **7** was hydrolyzed in methanolic potassium hydroxide to give artepillin C in 76% yield. The total yield of artepillin C was improved in 33%, compared with previous data (7%) (*9*).

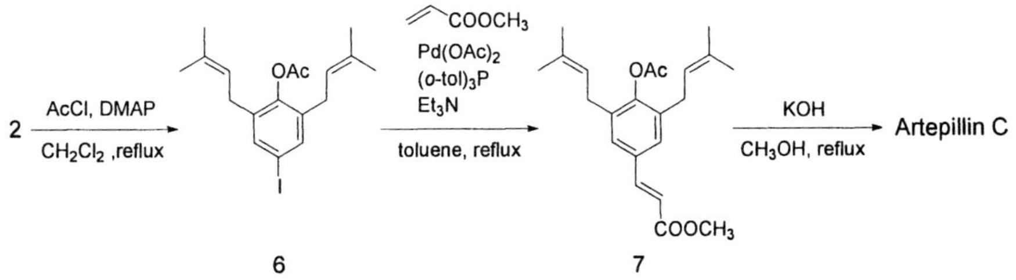

Scheme 2. Synthesis of artepillin C.

DPPH Radical Scavenging Activity

To evaluate the antioxidative properties of artepillin C, we investigated the DPPH radical scavenging activity of artepillin C, compared with BHT and α-tocopherol, an intracellular lipid peroxidation inhibitor. The degree of potency of antioxidants should be estimated by the reaction rates of radical trapping and the molar equivalent ratio of radical to antioxidant. Therefore, we monitored the absorbance decrease of DPPH radical at 517 nm (ΔA) in the presence of antioxidants. Figure 2 shows the time-course of 100 μM DPPH radical scavenging reaction with artepillin C, BHT, and α-tocopherol at 20 μM. α-Tocopherol caused very quick change in ΔA, due to scavenging of DPPH radicals, and ΔA reached a plateau level within 1 minute. Absorption change of artepillin C was similar to α-tocopherol and its attained a plateau after 5 minutes. In contrast, BHT represented a slower change in ΔA, which increased over 30 minutes.

Next we determined the $IC_{0.200}$ value, which is the final concentration of the antioxidant in a reaction solution needed to decrease by 0.200 in absorption of DPPH radical at 517 nm after 30 minutes (Table II). DPPH radical scavenging activity of artepillin C ($IC_{0.200}$ = 14.7 ± 0.6 μM) was twice as high as that of BHT ($IC_{0.200}$ = 28.5 ± 1.0 μM), but slightly lower in activity than that of α-tocopherol

($IC_{0.200}$ = 10.8 ± 0.1 µM). At the $IC_{0.200}$ value the molar equivalent ratio of DPPH to artepillin C was 1.4, and those of DPPH to α-tocopherol and BHT were 2.0 and 0.7, respectively.

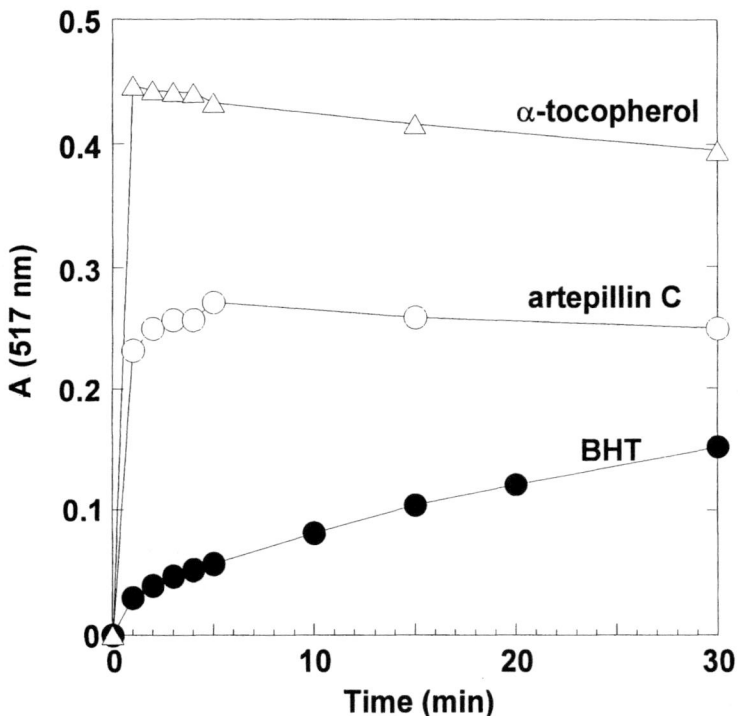

Figure 2. Time-course of DPPH radical scavenging reaction with antioxidants

Table II. $IC_{0.200}$ and HOMO Energies of Artepillin C and Antioxidants.

Compound	$IC_{0.200}$ (µM)	HOMO energy (eV)	C^2 of hydroxyl group
Artepillin C	14.7 ± 0.6	-8.911	0.098
α-Tocopherol	10.8 ± 0.1	-8.353	0.088
BHT	28.5 ± 1.0	-9.048	0.057

Note: $IC_{0.200}$ value represents the mean ± S.D. of three experiments.

To understand the differences in DPPH radical scavenging activity between artepillin C and α-tocopherol or BHT, we calculated their molecular orbital energies using the AM1 method on the MOPAC2000 program (Table II). It is known that HOMO energy will correlate with the ability of a phenol moiety to donate electrons (*18*). The HOMO energy of α-tocopherol (-8.353 eV) was the highest, while those of artepillin C and BHT were almost same (-8.911 and -9.048 eV, respectively). However, C^2 value of the hydroxyl group of artepillin C (0.098) was higher than that of BHT (0.057). These data suggest that HOMO energy of antioxidants correlated with the potency of DPPH radical scavenging activities.

Inhibitory activity of Artepillin C on RLM Lipid Peroxidation

We investigated the inhibitory activity of artepillin C on RLM lipid peroxidation. ADP and Fe^{2+} were used as starting materials for a radical reaction, and then IC_{50} value was estimated from O_2 consumption in the presence or absence of antioxidants (Table III). We found that artepillin C showed more potent inhibitory activity (IC_{50} = 0.72 ± 0.03 μM) than that of α-tocopherol (IC_{50} = 2.0 μM), a potent lipid peroxidation inhibitor. It is interesting that an inverse correlation exists between the inhibitory activity on lipid peroxidation (BHT; IC_{50} = 0.55 ± 0.03 μM > artepillin C > α-tocopherol) and the DPPH radical scavenging activity (α-tocopherol > artepillin C > BHT). Moreover, an inverse relationship also exists between the inhibitory activity on lipid peroxidation and the hydrophobic parameter log$P$(C) (α-tocopherol; log$P$(C) = 9.74 > artepillin C; log$P$(C) = 5.66 > BHT; log$P$(C) = 5.33).

Table III. Inhibitory Activity on RLM Lipid Peroxidation and log$P$(C).

Compound	IC_{50} (μM)	log$P$(C)
Artepillin C	0.72 ± 0.03	5.66
α-Tocopherol	2.0	9.74
BHT	0.55 ± 0.03	5.33

Note: IC_{50} value represents the mean ± S.D. of three experiments except for α-tocopherol of one experiment.

Discussion

In this chapter, we discuss the facile total synthesis of artepillin C, a major component of Brazilian propolis and the amphiphilic antioxidant property of artepillin C as a DPPH radical scavenger and lipid peroxidation inhibitor. The structural feature of artepillin C is a p-hydroxycinnamic acid derivative having a two-prenyl group at meta positions. It is an unique structure in natural products, therefore, our reported diprenylation of phenolic compounds at ortho positions is an exceptional result. We succeeded in regioselective o,o'-diprenylation of p-iodophenol in alkaline water, regardless of the equivalent of prenyl bromide. However, the disadvantage of the prenylation is that o-prenylation occurred preferentially as side reactions due to the fact that nucleophilicity of phenolic hydroxyl group increased in alkaline water. This side reaction can be suppressed using a less polar solvent, that is, toluene instead of water. In addition, regioselective C-prenylation of p-iodophenol can be achieved by an equivalent change in base. Using these facts we will apply the solvent-specific prenylation of phenolic compounds to the synthesis of many prenylated natural or synthetic products for isoprenomics.

This DPPH radical scavenging experiment is widely used in food chemistry and provides available information on the reactivity of antioxidant with a stable free radical, independent from any enzyme activity. Nevertheless, there is no consistency in the experimental conditions and evaluation methods. In this paper, DPPH radical scavenging activity of antioxidant was expressed by $IC_{0.200}$. Evidently, artepillin C is an effective free radical scavenger with potency similar to α-tocopherol. The molar equivalent ratio of DPPH to artepillin C was 1.4. In addition, artepillin C shows an efficient radical scavenging activity against cumylperoxyl radical in an aprotic medium, which is comparable to that of (+)-catechin (19). These data indicate that artepillin C acted as a direct free radical scavenger in water/ethanol solutions.

In the results of HOMO energy calculation of antioxidants, HOMO energy of artepillin C and antioxidants were correlated with the potency of DPPH radical scavenging activities. However, the HOMO energy only related to the ability of phenol moiety to donate electrons. Therefore, we should consider the stability of phenoxy radical produced by hydrogen abstraction to explain the DPPH radical scavenging activity of artepillin C in detail.

It is well known that mitochondria will readily subject to lipid peroxidation by reactive free radicals because of the presence in the mitochondrion of a high concentration of unsaturated lipid (20). Consequently, rat liver mitochondria are widely used as a preparation to evaluate the inhibitory activities on lipid peroxidation. Artepillin C was very effective in inhibiting the ferrous ion-stimulated lipid peroxidation of rat liver mitochondria. Its potency was nearly same as BHT but was about three times more active than α-tocopherol, a potent lipid peroxidation inhibitor. In contrast to the result of inhibitory activity of lipid peroxidation, DPPH radical scavenging activity of artepillin C was little weaker than that of α-tocopherol. This contradiction canl be explained because artepillin C, contains a polar carboxylate group that substantially deprotonized at physiological pH, and may be localized at an outer membrane surface and be favorable as a protection against ferrous ion-stimulated lipid peroxidation. Another explanation may be that α-tocopherol act as a pro-oxidant under *in vitro* experimental conditions (21).

In conclusion, we demonstrate the facile total synthesis of artepillin C, a medicinal active component of Brazilian propolis. Total yield of artepillin C was improved 33%. Moreover, our established regioselective prenylation of *p*-iodophenol may be applied to prenylated natural or synthetic products for isoprenomics. Artepillin C could be a powerful amphiphilic antioxidant against water-soluble DPPH radical and RLM lipid peroxidation.

References

1. Jakupovic, J.; Warning, U.; Bohlmann, F.; King, R. M. *Revista Latinoamericana de Quimica* **1987**, *18*, 75.
2. Huneck, S.; Zdero, C.; Bohlmann, F. *Phytochem.* **1986**, *25*, 883.
3. Okuno, I.; Uchida, K.; Nakamura, M.; Sakurawi, K. *Chem. Pharm. Bull.* **1988**, *36*, 769.
4. Aga, H.; Shibuya, T.; Sugimoto, T.; Nakajima, S.; Kurimoto, M. *Biosci. Biotech. Biochem.* **1994**, *58*, 945.
5. Kimoto, T.; Arai, S.; Aga, M.; Hanaya, T.; Kohguchi, M.; Nomura, Y.; Kurimoto, M. *Gan To Kagaku Ryoho* **1996**, *23*, 1855.
6. Matsuno, T.; Jung, S. K.; Matsumoto, Y.; Saito, M.; Morikawa, J. *Anticancer Res.* **1997**, *17*, 3565.
7. Kimoto, T.; Arai, S.; Kohguchi, M.; Aga, M.; Nomura, Y.; Micallef, M. J.; Kurimoto, M.; Mito, K. *Cancer. Detect. Prev.* **1998**, *22*, 506.
8. Hayashi, K.; Komura, S.; Isaji, N.; Ohishi, N.; Yagi, K. *Chem. Pharm. Bull.* **1999**, *47*, 1521.
9. Uto, Y.; Hirata, A.; Fujita, T.; Takubo, S.; Nagasawa, H.; Hori, H. *J. Org. Chem.* **2002**, *67*, 2355.
10. Blois, M. S. *Nature* **1958**, *181*, 1199.
11. Kubo, K.; Yoshitake, I.; Kumada, Y.; Shuto, K.; Nakamizo, N. *Arch. Int. Pharmacodyn.* **1984**, *272*, 283.
12. Stewart, J. J. *J. Comput. Aided. Mol. Des.* **1990**, *4*, 1.
13. Lash, L. H.; Sall, J. M. *Method in Toxicology*, San Diego, **1993**, 2.
14. Myers, D. K.; Slater, E. C. *J. Biochem.* **1957**, *67*, 558.
15. Gornall, A. G.; Bardawill, C. J.; David, M. M. *J. Biol. Chem.* **1949**, *177*, 751.
16. Niki, E.; Saito, T.; Kawakami, A.; Kamiya, Y. *J. Biol. Chem.* **1984**, *259*, 4177.

17. Hori, H.; Ishibashi, M.; Mohamad, S. B.; Nagasawa, H.; Uto, Y.; Sakamaki, H.; Pan, N.; Ohkura, K.; Nishibe, S. *Adv. Exp. Med. Biol.* **1999**, *471*, 395.
18. Weber, V.; Coudert, P.; Rubat, C.; Duroux, E.; Vallee-Goyet, D.; Gardette, D.; Bria, M.; Albuisson, E.; Leal, F.; Gramain, J.; Couquelet, J.; Madesclaire, M. *Bioorg. Med. Chem.* **2002**, *10*, 1647.
19. Nakanishi, I.; Uto, Y.; Ohkubo, K.; Miyazaki, K.; Yakumaru, H.; Urano, S.; Okuda, H.; Ueda, J.; Ozawa, T.; Fukuhara, K.; Fukuzumi, S.; Nagasawa, H.; Hori, H.; Ikota, N. *Org. Biomol. Chem.* **2003**, *1*, 1452.
20. Richardson, T.; Tappel, A. L.; Gruger, E. H. Jr.; *Arch. Biochem. Biophys.* **1961**, *94*, 1.
21. Bowry, V. W.; Stocker, R. *J. Am. Chem. Soc.* **1993**, *115*, 6029.

Chapter 16

Production of Theaflavins and Theasinensins during Tea Fermentation

Takashi Tanaka, Chie Mine, Sayaka Watarumi, Yosuke Matsuo, and Isao Kouno

Department of Pharmaceutical Sciences, Graduate School of Biomedical Sciences, Nagasaki University, Nagasaki 852–8521, Japan

Theaflavins are the most important phenolic pigments in black tea. Upon enzymatic synthesis of theaflavin from epicatechin and epigallocatechin, trapping of the o-quinone intermediates with glutathione demonstrated that enzymes rapidly oxidized epicatechin to epicatechin quinone, and the epicatechin quinone oxidized epigallocatechin in turn. Subsequent coupling between o-quinones of epicatechin and epigallocatechin then formed theaflavin. Four oxidation products of theaflavin were also isolated. On the other hand, theasinensins A and D, other major black tea polyphenols, were produced by the oxidation-reduction dismutation of dehydrotheasinensin A, which was in turn produced by the dimerization of the o-quinone of epigallocatechin 3-O-gallate.

Black tea and green tea are made from the same tea leaf.. In the manufacture of black tea, four major tea catechins in the fresh leaves are enzymatically oxidized during the fermentation process to produce many oxidation products including theaflavins and thearubigins, the characteristic pigments of black tea. On the other hand, in the green tea production, the enzymes are inactivated by heating immediately after harvesting; therefore, composition of the polyphenols is similar to that of fresh leaves and relatively simple. Recently, the health benefits of polyphenols in foods and beverages have been attracting the interest of scientists, and some epidemiological and biochemical studies suggested the daily intake of polyphenols may reduce the risk of cancer and heart disease (*1,2*). Figure 1 shows the increase in scientific research publications on wine, coffee, green tea and black tea derived from the SciFinder® database. Since a Japanese national project on functional foods (1984-1986) and sensational reports on "the French paradox" of red wine, scientific publications on wine and green tea have been increasing more and more. However, publications on black tea have not increased proportionally, even though black tea accounts for almost 80 % of the world's tea production and is the most important source of dietary polyphenol in the world. In addition, recent studies on the biological activities of black tea polyphenols, especially those of theaflavins, suggested that black tea may also have beneficial health effects (*3,4*). One of the reasons for the difficulty in studying black tea is its complex polyphenol composition compared to that of green tea. Despite chemical studies from the 1950's, a large number of black tea polyphenols remain chemically unknown.

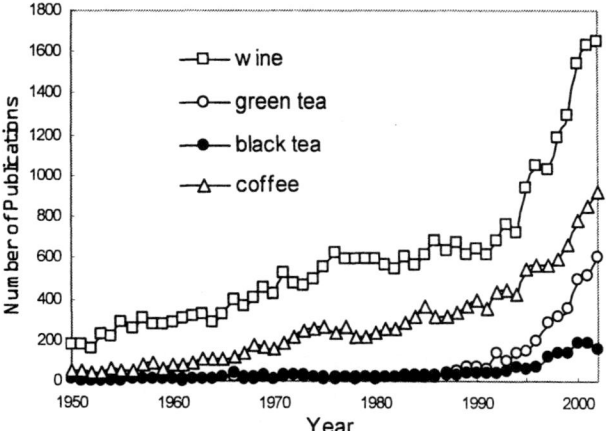

Figure 1. Increase in scientific research publications of beverages

In the course of our chemical studies on plant polyphenols, we have studied the mechanism of production of black tea polyphenols by isolating catechin oxidation products produced by model fermentation experiments. We also confirmed the presence of some new oxidation products obtained by model fermentation of commercial black tea. Oxidation of tea catechins is important also from the viewpoint of their biological activities, because tea catechins are easily oxidized even under cell culture conditions. Here, we introduce our recent results on synthesis and decomposition of theaflavins and production of theasinensins during tea fermentation. Theasinensins are major black tea polyphenols (1-1.5 % in commercial black tea, which is comparable to the concentration of theaflavins) (5), and their production is related to the antioxidative activities of epigallocatechin 3-O-gallate, which is the most important tea catechin in green tea.

Evidence for a Coupled Oxidation Mechanism of Theaflavin Synthesis

Theaflavin (**3**) is produced by oxidative coupling between (-)-epicatechin (**1**) and (-)-epigallocatechin (**2**). Takino *et al.* (*6*) first suggested a possible mechanism in which epicatechin quinone (**1a**) was coupled with epigallocatechin quinone (**2a**). In our model oxidation experiment, pure **1**, **2** or a mixture of **1** and **2** was oxidized with banana or Japanese pear fruits homogenates (*7*). Both homogenates demonstrated high enzyme activity, producing **3** from **1** and **2** while not producing any interfering products. When **1** and **2** were treated separately with banana homogenate, the rate of decrease of **1** was faster than that of **2**. However, when the mixture of **1** and **2** was treated in the same manner, **2** decreased more rapidly compared to **1**. This was because the enzymes preferentially oxidized **1** to **1a** and **2** was then oxidized by **1a** resulting in regeneration of **1** (Figure 2).

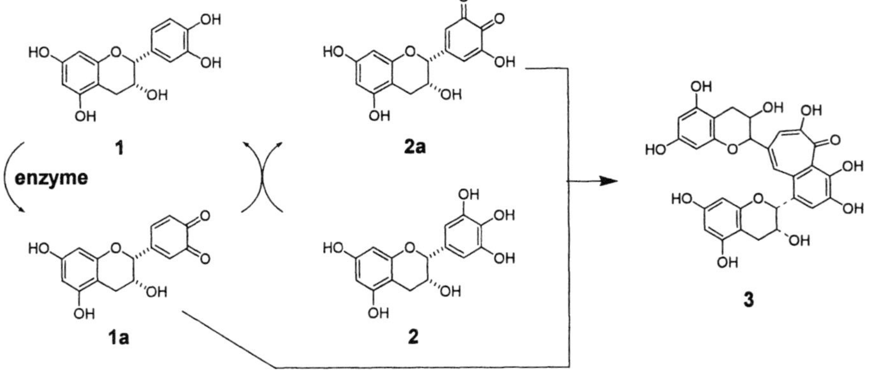

Figure 2. Coupled oxidation mechanism for theaflavin synthesis

This is referred to as a coupled oxidation mechanism (*8,9*). In order to confirm this mechanism; glutathione was added to the reaction mixture to trap the quinone intermediates. As expected, theaflavin was not produced and **1** was completely converted into two glutathione adducts (**1b** and **1c**); however less than 30% of **2** was changed to a glutathione adduct **2b**. This result confirmed the rapid turnover between **1** and **1a** during theaflavin synthesis.

Oxidation Products of Theaflavin

Theaflavin (**3**) and its galloyl esters are important reddish-yellow pigments and their concentration is closely related to the quality of black tea; however, these are not stable compounds. In our model fermentation experiments using banana fruits or fresh tea leaves (*10-12*), **3** was further oxidized to give theanaphthoquinone (**4**), bistheaflavin A (**5**), and dehydrotheaflavin (**6**). Furthermore, autoxidation of **3** in phosphate buffer (pH 7.3) afforded **4** and bistheaflavin (**7**). We have presumed the chemical mechanism for production of these oxidation products from **3** (*13*). These products have not yet been detected in commercial black tea. In addition, autoxidation of galloyl esters of **3** gave a polymeric substance and did not afford galloyl esters of **7**. The presence of galloyl groups at C-3 hydroxyl groups is crucial to the theaflavin oxidation.

Oxidation Products of Epigallocatechin and its Gallate

Epigallocatechin (**2**) and its galloyl ester (**8**) represent over 70% of tea leaf catechins; therefore, their oxidation products are also important in black tea. In our model oxidation experiments, the ten oxidation products listed below (**9-18**) were obtained. Among these compounds, theasinensins (**9-12**) and oolongtheanins (**13, 14**) were already known as black tea constituents (*5,14,15*). Recently, we found that two new products **16** and **17** obtained by the model experiments were also present in commercial black tea. Compound **15** was a product derived from the *o*-quinone of theasinensin E (**12**) (*7*), which has a biphenyl bond with *S*-atropisomerism. Compound **18** was identified as a product, which was initially reported as a radical oxidation product of **8** with an organic radical initiator in acetonitrile (*16*).

Mechanism for Theasinensin Production

Theasinensins, major polyphenols in black tea and oolong tea (*5,15*), are simple dimers of **2** and/or **8**. These compounds could be produced by a simple phenol radical coupling similar to the mechanism demonstrated for production of ellagic acid from gallic acid (*17*). However, when fresh tea leaves were crushed at the initial stage of tea fermentation, theasinensins were not detected by HPLC analysis, even though sufficient theaflavins were produced. At this stage of fermentation, addition of acidic ethanol solution of *o*-phenylenediamine to the leaves yielded several phenazine derivatives (**19** and its desgalloyl analogs), which suggested the presence of *o*-quinones (**9a**) of theasinensin A (**9**) and its desgalloyl analogs (*18*). These phenazine derivatives were not produced from the fermented tea leaves after heating, and theasinensins were detected instead. These results strongly suggested that theasinensins are produced by reduction of the theasinensin quinones on heating and drying at the final stage of black tea production.

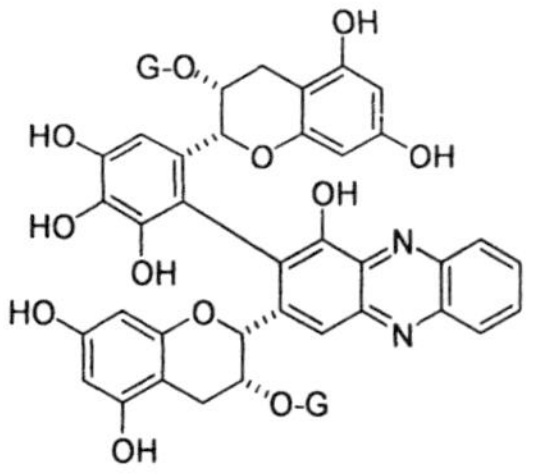

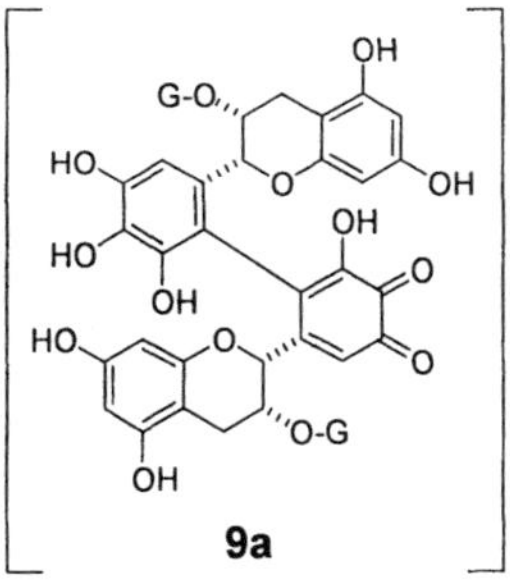

Very recently we have succeeded in isolating a theasinensin precursor **20**, equivalent to one of the hydrated forms of **9a**, by a model fermentation experiment using **8** and Japanese pear fruits homogenate, and have determined the structure based on spectroscopic analysis (*19*). The theasinensin precursor was named dehydrotheasinensin A, because its chemical reduction with ascorbic acid, cysteine methyl ester or thioglycolic acid yielded theasinensin A (**9**). It was presumed that **20** was produced by 1,4-addition between two molecules of *o*-quinone of **8** (Figure 3). Production of **19** and its desgalloyl analogs on treatment of the fermented tea leaves with phenylenediamine suggested that the dimerization of the *o*-quinones of **2** and **8** is highly stereoselective (*18*).

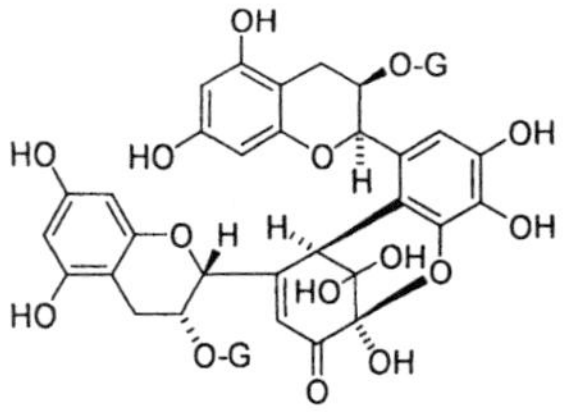

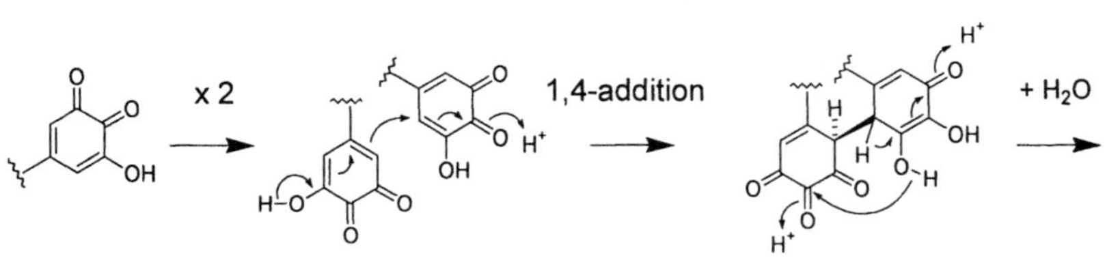

Figure 3. Mechanism for production of dehydrotheasinensin A

Dehydrotheasinensin A was stable under acidic conditions; however, it gradually decomposed under neutral conditions (pH 6.8) to give theasinensins A (**9**), D (**10**), galloyloolongtheanin (**13**), and an oxidation product **21**. In this reaction, **9** and **10** were reduction products and **13** and **21** were oxidation products of **20**; therefore, decomposition of **20** was an oxidation-reduction dismutation reaction.

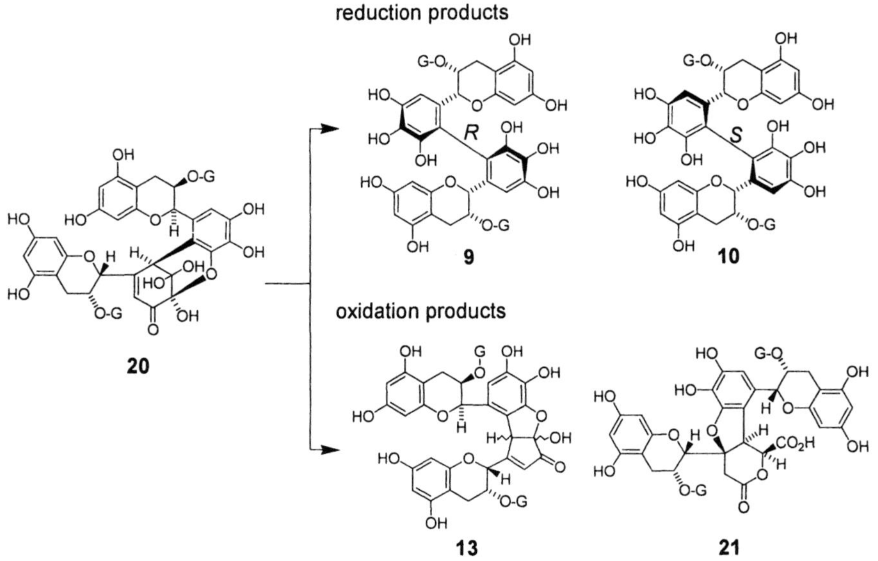

Figure 4. Oxidation-reduction dismutation of dehydrotheasinensin A

A similar reaction occurred when an aqueous solution of **20** was heated at 80°C for few minutes. In this case, only **10** and **13** were detected as the major products. As mentioned before, reduction of **20** with ascorbic acid or cysteine methyl ester yielded **9**, which has an R-biphenyl bond. In addition, inversion of the biphenyl bond of **9** was not observed under similar conditions. Therefore, the production of the S-atropisomer **10** in these dismutation reactions suggested occurrence of an inversion at the benzyl methine between a double bond and a hydrated ketone (*19*). Details of the reaction mechanism still need to be clarified.

Although biological activities of epigallocatechin 3-*O*-gallate (**8**) have been extensively studied, it is known that **8** is easily oxidized even under cell culture conditions. In fact, **8** is decomposed by non-enzymatic autoxidation in weakly basic (pH 7.4) phosphate buffer. Upon addition of *o*-phenylenediamine to the buffer solution, the phenazine derivative **19** was produced, indicating that dehydrotheasinensin A (**20**) was also produced under these conditions (*19*).

Since **20** is unstable at neutral pH, the yield of **19** was very low and it was detected only at the initial stage of the reaction.

If theasinensins are produced by simple radical coupling of the epigallocatechin radical, two electrons are abstracted per production of one molecule of theasinensin. On the other hand, when theasinensins are produced by a dismutation reaction via dehydrotheasinensin A and its analogs, a total of four electrons are abstracted per production of one molecule of theasinensin. This may be important from the viewpoint of antioxidative activities of **8**.

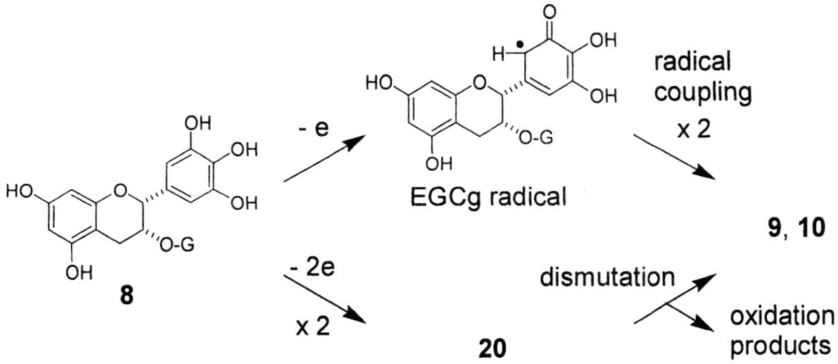

Figure 5. Two possible pathways for theasinensin production

We have proposed an oxidation-reduction dismutation mechanism for theasinensin synthesis via dehydrotheasinensin. Surprisingly, at the end of the 1950's, Roberts, who was a pioneer of black tea chemistry, had already presumed that the dimer quinone intermediate **9a** might be produced during tea fermentation and pointed out the possibility of oxidation-reduction dismutation of dimer quinines (20,21). He could not show any chemical evidence at that time and his proposal has long been ignored.

Conclusion

In our model fermentation experiments, some new oxidation products of tea catechins were isolated and it was demonstrated that some are also present in commercial black tea. However, many other unknown products including thearubigins are produced during tea fermentation. Thearubigins are other major black tea pigments originally named by Roberts 50 years ago, and are believed to be complex mixtures of catechin oxidation products with higher molecular weights. The total amounts of these unknown oxidation products still account for the majority of black tea polyphenols.

References

1. Frei, B.; Higdon, J. V. *J. Nutr.*, **2003**, *133*, 3275S-3284S.
2. Lambert, J. D.; Yang, C. S. *J. Nutr.* **2003**, 133, 3262S-3267S.
3. Lu, J.; Ho, C. T.; Ghai, G.; Chen, K. Y. *Cancer Res.* **2000**, 60, 6465-6471.
4. Feng, Q.; Torii, Y.; Uchida, K.; Nakamura, Y.; Hara, Y.; Osawa, T. *J. Agric. Food Chem.* **2002**, *50*, 213-220.
5. Hashimoto, F.; Nonaka, G.; Nishioka, I. *Chem. Pharm. Bull.* **1992**, *40*, 1383-1389.
6. Takino, Y.; Imagawa, H.; Horikawa, H.; Tanaka, A. *Agric. Biol. Chem.* **1964**, *28*, 64-71.
7. Tanaka, T.; Mine, C.; Inoue, K.; Matsuda, M.; Kouno, I. *J. Agric. Food Chem.* **2002**, *50*, 2142-2148.
8. Roberts, E. A. H. *Chem. Ind.* **1957**, 1354-1355.
9. Robertson, A. *Phytochemistry* **1983**, *22*, 889-896.
10. Tanaka, T.; Betsumiya, Y.; Mine, C.; Kouno, I. *Chem. Commun*, **2000**, 1365-1366.
11. Tanaka, T.; Inoue, K., Betsumiya, Y.; Mine, C.; Kouno, I. *J. Agric. Food Chem.* **2001**, *49*, 5785-5789.
12. Tanaka, T.; Mine, C.; Kouno, I. *Tetrahedron*, **2002**, *58*, 8851-8856.
13. Tanaka, T.; Kouno, I., *Food Sci. Technol. Res.* **2003**, *9*, 128-133.
14. Nonaka, G.; Kawahara, O.; Nishioka, I. *Chem. Pharm. Bull.* **1983**, *31*, 3906-3914.
15. Hashimoto, F.; Nonaka, G.; Nishioka, I. *Chem. Pharm. Bull.* **1988**, *36*, 1676-1684..
16. Valcic, S.; Muders, A.; Jacobsen, N.E. ; Liebler, D.C. ; Timmermann, B.N. *Chem. Res. Toxicol.* **1999**, *12*, 382-386.
17. Yoshida, T.; Mori, K.; Hatano, T.; Okumura, T.; Uehara, I.; Komagoe, K.; Fujita, Y.; Okuda, T. *Chem. Pharm. Bull.* **1989**, *37*, 1919-1921.
18. Tanaka, T.; Mine, C.; Watarumi, S.; Fujioka, T.; Mihashi, K.; Zhang, Y.-J.; Kouno, I. *J. Nat. Prod.* **2002**, *65*, 1582-1587.
19. Tanaka, T.; Watarumi, S.; Matsuo, Y.; Kamei, M.; Kouno, I. *Tetrahedron* **2003**, *59*, 7939-7947.
20. Roberts, E.A.H. *Chem. Ind.* **1957**, 1355-1356.
21. Roberts, E. A. H.; Myers, M. *J. Sci. Food Agric.* **1959**, *10*, 167-172.

Chapter 17

New Momentum on the Action Mechanisms of Black Tea Polyphenols, the Theaflavins

Jen-Kun Lin[1], Min-Hsiung Pan[1], Yu-Chih Liang[1], Shoui-Yn Lin-Shiau[2], and Chi-Tang Ho[3]

Institutes of [1]Biochemistry and [2]Toxicology, College of Medicine, National Taiwan University, Taipei, Taiwan
[3]Department of Food Science, Rutgers, The State University of New Jersey, 65 Dudley Road, New Brunswick, NJ 08901

The biological functions of both black and green teas have been attributed to their polyphenols including theaflavins and catechins. Tea polyphenols exhibit a wide range of biological properties, namely antioxidative effects, inhibition of extracellular mitotic signals through blocking growth receptors signalings, inhibition of cell cycle at G1 phase through cyclin-dependent kinase suppression, suppression of iNOS through inhibiting the activation of IKK and NFκB and induction of apoptosis in cancer cells through releasing cytochrome c and activating caspase cascades. Tea polyphenols suppress the fatty acid synthase (FAS) in human breast carcinoma MCF-7 cells. Tea and tea polyphenols may induce hypolipidemic and anti-obesity effects through suppressing FAS expression. The biological activities of theaflavins and catechins have been critically reviewed. It seems that green tea contains higher levels of catechins, while black tea theaflavins show higher biological activities in several systems assessed.

© 2005 American Chemical Society

Tea is one of the most popular beverages consumed worldwide. It is the brew prepared from the leaves of the plant *Camellia sinenesis*. Freshly harvested tea leaves require processing to convert them into green, oolong and black teas. Black and green teas are the two main types, defined by their respective manufacturing processes. Green tea is consumed mostly in Asian countries such as China and Japan, while black tea is more popular in North America and Europe (*1*). Oolong tea is an intermediate variant (partially fermented product) between green and black tea. Its production is confined to southern part of China and Taiwan. It is estimated that 80% of world-produced tea is consumed as black tea, while 16% and 2% as green and oolong tea, respectively.

During the past decade, the research of tea sciences was stimulated by the discovery of the health promoting effects of tea. The published research papers on tea science have increased tremendously (Table I). The data base of Entrez-Pubmed, National Library of Medicine, indicates that published papers dealing with tea number 8710. Among these papers 1381 (15.6%) deal with green tea while 554 (6.3%) deal with black tea. Based on the fact that black tea is heavily consumed (80%) by most people in the world, it is reasonabe to suggest that more effort be focused on black tea research in the future, especially, since recent investigations indicate that black tea is equally effective, even more effective in some cases, in cancer chemoprevention when compared with green tea.

In this reveiew, the biological activities of black and green teas closely associated with their cancer chemoprevention are discussed. In most cases, epigallocatechin gallae (EGCG) is considered the representative catechin of green tea, while TF3 (theaflavin-3,3'-digallate) is recognized as the representative theaflavin of black tea.

Antioxidative Effects of Black Tea and Catechin

Human red blood cells (RBC) were taken as the model and oxidative damage was induced by a variety of inducers, such as phenylhydrazine, Cu^{2+}-ascorbic acid and xanthine/xanthine oxidase systems (*2*). Lipid peroxidation of pure erythrocyte membrane and of whole RBC could be completely prevented by black tea extract (Table II). Similarly black tea provided total protection against degradation of membrane proteins. Black tea extract, in comparison to free catechins, seemed to be the better protecting agent against various types of oxidative stress (*2*).

Apparently, conversion of catechins to partially polymerized products, such as thaflavins or thearuigins, during the fermentation process for making black tea has no deleterious effect on its reactive oxygen species (ROS) scavenging

Table I. Summary on the Published Papers Describing Tea and Tea Polyphenols

Category	Number of published paper
Tea	8710
Green tea	1381
Black tea	554
Green tea only	175
Black tea only	92
Oolong tea	103
Oolong tea only	12
Puerh tea	1
Green tea polyphenols	362
Black tea polyphenols	150
EGCG	604
Theaflavins	75
Thearubigins	24

As of January 2004. Data from Entrez Pubmed, National Library of Medicine

Table II. Protection against Lipid Peroxidation by Catechin and Black Tea

System	Malondialdehyde equivalent production (nmol/mg protein)	% of control
Experiment A		
1. Membrane (control)	1.28 +/- 0.03	100
2. 1 + Phenylhydrazine (1 mM)	4.00 +/- 0.21	352
3. 2 + Catechin (5 mM)	1.56 +/-0.09	121
4. 2 + Black tea extract* (50 µl)	1.26 +/- 0.07	99
Experiment B		
1. Red blood cell suspension	0.045 +/- 0.07	100
2. 1 + Cu^{2+} + Ascorbic acid (1 mM)	0.168 +/- 0.01	372
3. 2 + Catechin (5 mM)	0.053 +/-0.04	111
4. 2 + Black tea ectract (50 µl)	0.046 +/-0.07	102

* 50 µl black tea extract was estimated to be equivalent to 200 µg of catechin and epicatechin taken together; Data from reference 2.

properties. In fact, black tea appears to be a better scavenger of various ROS than free catechins on a weight basis (2,3).

Inhibition of Xanthine Oxidase and ROS by Tea Polyphenols

The inhibitory effects of black tea polyphenols, namely theaflavin (TF1), theaflavin-3-gallate + theaflavin-3'-gallate (TF2) and theaflavin-3,3'-digallate (TF3), green tea polyphenol (-)-epigallocatechin-3-gallate (EGCG), and simple gallates including propyl gallate (PG) and gallic acid (GA) on xanthine oxidase (XO) were investigated (4). Theaflavins and EGCG inhibit XO to produce uric acid and also act as scavengers of superoxide. TF3 acts as a competitive inhibitor and is the most potent inhibitor of XO among these compounds. Tea polyphenols and PG all have potent inhibitory effects (>50%) on phorbol myristate acetate (PMA)-induced superoxide production at 20-50 µM in HL-60 cells. GA showed no inhibition under the same conditions. The superoxide scavenging abilities of these compounds are as follows: EGCG>TF2>TF1>GA>TF3>PG. Meanwhile, the order of hydrogen peroxide scavenging ability was TF2>TF3>TF1>EGCG>PG>GA. The experimental data demonstrated that the antioxidative activity of tea polyphenols and PG was due not only to their ability to scavenge superoxides and hydrogen peroxide but also to their ability to block XO and related oxidative signal transducers (4).

Antimutagenic Effects of Black and Green Tea Extracts

Black and green tea extracts were made and tested as inhibitors of the mutagenicity caused by the food mutagen 2-amino-1-methyl-6-phenylimidazo[4,5-b]pyridine (PhIP) in a *Salmonella typhimurium* TA98 assay containing S9 fraction from rat livers induced with α-naphthoflavone and Phenobarbital (5). Extracts of both black and green teas were good inhibitors of mutagnicity. The polyphenols of black tea were more potent inhibitors of mutagenicity than the polyphenols of green tea (Table III). These findings suggest that black tea may have similar or better health-promoting properties than those reported previously for green tea (5).

Another study was undertaken to compare the antimutagenicity of aqueous extracts, at the concentrations used for human consumption, from green tea, black tea and decaffeinated black tea. All these types of tea gave rise to strong and concentration-dependent suppression of mutagenicity of the tested premutagens, including Glu-P-1, benzo(a)pyrene (BP), and nitroso-pyrrolidine in the presence of an activation system (6). In order to compare the ability of the three teas to scavenge electrophiles, generated metabolically, the Ames procedure was modified. Carcinogen, appropriate activation system and fresh bacterial culture were incubated in a shaking waterbath at 37°C for 20 min. Microsomal activation was terminated by the addition of menadione, aqueous tea extracts were added and a further 20 min preincubation was carried out at 37°C. Incorporation of aqueous extracts from three teas caused a concentration-dependent decrease in the mutagenic response of the three model mutagens, BP being the most susceptible. Furthermore, green tea was clearly the least effective when BP served as the model carcinogen (Table IV).

Table III. Inhibition of PhIP Mutagenicity by Tea and Tea Polyphenols

System		Mutagenicity (revertants/plate)	IC_{50} ($\mu g/mL$)
Control system (PhIP)	10µM	1000	--
+ Black tea extract	600 µg/mL	200	400
+ Green tea extract	600 µg/mL	200	300
+ Black tea polyphenol	500 µg/mL	180	160
+ Green teapolyphenol	500 µg/mL	320	280

Data from reference 5.

Table IV. Inhibition of Black and Green Tea Extracts on the Microsome-Derived Genotoxic Species of BP

First preincubation	Second preincubation	Black tea	Green tea
Blank	--	129 ± 13	129 ± 13
BP	--	724 ± 12	724 ± 12
BP and Menadione	--	142 ± 18	142 ± 18
BP	Menadione	578 ± 37	578 ± 37
BP	Menadione + tea ext (0.25 mL)	214 ± 22	448 ± 30
BP	Menadione + tea ext (0.50 mL)	189 ± 17	419 ± 15
BP	Menadione + tea ext (0.75 mL)	186 ± 13	386 ± 19
BP	Menadione + tea ext (1.0 mL)	172 ± 13	228 ± 25

Results are expressed as histidine revertants/plate and mean +/- SD of triplicates; A 2.5% tea infusion was prepared by adding boiling water (400 mL) to the tea (10 g) in a prewarmed thermos flask, leaving to stand for 10 min and then filtering through cotton wool, 1 mL tea extract was estimated to contain 0.025 g tea; Data from reference 6.

In order to compare the ability of the three types of tea to scavenge genotoxic electrophiles, their antimutagenic potential was investigated against the direct-acting mutagen 9-aminoacridine. Inhibition of the mutagenicity of 9-aminoacridine was evident with all three types of tea, but green tea was clearly less potent (6).

Anti-carcinogenic Effects of Black and Green Teas

Administration of a water extract of black tea and green tea leaves as the sole source of drinking fluid inhibited ultraviolet light (UVB)-induced carcinogenesis in SKH-1 mice previously initiated with 7,12-dimethyl-benz[a]antracene (DMBA) (7,8). UVB-induced formation of skin tumors was markedly inhibited by oral administration of 0.63 or 1.25 % black tea and green tea as the sole source of drinking fluid two weeks prior to and during 31 weeks of UVB treatment. Administration of each tea preparation not only inhibited the number of tumors, but tumor size was also markely decreased, as illustrated in Table V (7). It is summarized by the authors that administration of black tea was comparable to green tea as an inhibitor of UVB-induced carcinogenesis in DMBA-initiated SKH-1 mice. However, published data (7) indicates a higher anti-carcingenic effect for black tea than green tea (Table V).

Table V. Inhibitory Effects of Black Tea and Green Tea on Tumorigenesis by UVB light in DMBA initiated SKH-1 Mice

Group	Drinking fluid(%)	Time of appearance of first tumor (wks)	% of mice with tumor	Tumor volume /mouse (mm^3)
D	Water(Control)	7	93	243 ± 72
E	Black tea (0.63)	13	60	12 ± 6
F	Black tea (1.25)	19	35	10 ± 6
G	Green tea (0.63)	13	80	22 ± 9
H	Green tea (1.25)	21	53	16 ± 9

Female SKH-1 mice (7-8 weeks old) were treated topically with 200 nmol DMBA. One week later the mice were treated with gradually increasing concentrations of tea in the drinking water for 1 week and full strength tea (0.63 or 1.25%) for an additional week prior to and during treatment with UVB (30 mJ/cm^2) twice weekly for 31 weeks. Each value represents the mean or mean +/- SE from 28-30 mice.

Data from reference 7.

Modulation of Tumor Promotion Signaling Proteins by TF3 and EGCG

The action mechanisms of anti-carcinogenesis or cancer chemoprevention by tea polyphenols have been investigated intensively (*9-12*). Different target enzymes and transducing proteins at different cellular compartments have been affected by tea polyphenols, TF3 and EGCG (Table VI). EGFR tyrosine kinase on the plasma membrane is significantly inhibited by EGCG and TF3 (*15,16*). The membrane bound XO is inhibited by TF1, TF2 and TF3 and EGCG (*4*). Several cytosolic enzymes and transducing proteins such as PI3K, Akt, Raf, MEK, MRK, JNK, p38 kinase and IKK have been shown to be inhibited by EGCG and theaflavins (*11,12,17*). On the other hand, cytosolic caspases 3 and 9 were activated by various tea polyphenols (*17*).

Table VI. Effects of TF3 and EGCG on Some Enzyme Systems

Enzyme system	TF3	EGCG	Reference
Xanthine oxidase (IC_{50}, μM)	4.5	12.5	(*4*)
Superoxide dismutase in HL-60 (IC_{50}, μM)	3.1	50.8	(*4*)
iNOS induced by LPS in macrophage (10 μM, % of inhibition)	95	65	(13)
NFκB activation by LPS in macrophage (10 μM, % of inhibition)	95	80	(*13*)
IκB phosphorylation induced by LPS in macrophage (10 μM, % of inhibition)	98	60	(*13*)
Caspase cascades in U-937 cells (25 μM, % of activation)	250	150	(*14*)
PARP cleavage in U-937 cells (25 μM, % of cleavage)	80	5	(*14*)

Some cell cycle regulating proteins in the nucleus, such as Rb, CDK2 and CDK4 were inhibited by tea polyphenols (*16*). On the other hand, the CDK inhibitors p21 and p27 were elevated by tea polyphenols. All these enzymes and transducing proteins are topologically arranged into a signal transducing network from cell membrane, through cytosolic compartment and finally reach nuclear genome, the site of gene expression.

Several lines of evidence demonstrate that tumor promotion processes are tightly associated with signal transduction pathways. Important signal transducing proteins such as PKC, AP-1, c-Jun, EGFR, Raf-1, MEK-1, ERK1/2, ELK, PI3K, Akt, etc. are affected by tea polyphenols TF3 and EGCG, as illustrated in Table VII. Based on these data, it is evident that TF3 is more active than EGCG in modulating these important signal transducing proteins. It should be emphasized that, on a molar basis, the black tea polyphenol TF3 is more active than the green tea polyphenol EGCG in modulating or inhibiting most signal transduction pathways described in Table VII.

Table VII. Modulation of Tumor Promtion Signalings by TF3 and EGCG

Signaling process	TF3	EGCG	Reference
PKC induced by PMA in NIH3T3 cells (40 μM, % of inhibition)	95	49	18
AP-1 binding in NIH3T3 cells (40 μM, % of inhibition)	90	56	18
c-Jun expression in NIH3T3 cells (20 μM, % of inhibition)	80	60	18
EGFR binding in A431 cells (10 μM, % of inhibition)	90	66	15
EGFR autophosphorylation in A431 cells (10 μM, % of inhibition)	100	80	15
Raf-1 expression in 30.7b Ras 12 cells (20 μM, % of inhibition)	87	0	12
MEK-1 expression in 30.7b Ras 12 cells (20 μM, % of inhibition)	45	30	12
ERK1/2 phosphorylation in Ras 12 cells (20 μM, % of inhibition)	50	30	12
Elk-1 phosphorylation , in vitro assay (20 μM, % of inhibition)	38	28	12
PI3K expression induced by UVB in JB6 cells (20 μM, % of inhibition)	80	60	11

Hypolipidemic Effects of TF3 and EGCG

Fatty acid synthase (FAS) is a key enzyme in lipogenesis. FAS is overexpressed in malignant human breast carcinoma MCF-7 cells and its expression is further enhanced by the epidermal growth factor (EGF). The EGF-induced expression of FAS was inhibited by black and green tea extracts. The expression of FAS was also inhibited by the tea polyphenols TF3 and EGCG at both protein and mRNA levels (Table VIII), which may lead to the inhibition of cell lipogenesis and proliferation (19).

Table VIII. Inhibition of EGF-induced Expression of FAS by TF3 and EGCG in MCF-7 Cells

Compound	Concentration (μM)	Inhibition of FAS Expression(%)	
		Protein level	RNA level
TF3	5	31	15
	10	54	40
	20	87	57
	40	100	--
EGCG	10	49	44
	20	60	69
	60	62	99

Data from reference 19.

Both TF3 and EGCG inhibit the activation of Akt and block the binding of Sp-1 to its target site. Furthermoe, the EGF-induced biosynthesis of lipids, including free fatty acid, free cholesterol, esterified cholesterol and triacylglycerol was significantly suppressed by TF3 and EGCG (Table IX). We have demonstrated that the tea polyphenols TF3 and EGCG showed profound inhibitory effects on FAS through down regulation of the PI3K/Akt/Sp-1 signaling pathway (19). Several FAS inhibitors have been shown to be effective antitumor agents (20, 21). FAS is highly expressed in carcinoma, adenoma and in regenerative epithelium and intestinal metaplasia of the stomach (22). In addition, FAS is also expressed at a markedly elevated level in subsets of the human breast, ovarian, endometrial and prostate carcinomas (23).

Numerous studies have demonstrated the striking similarity between the tumor promotion process in cancer induction and lipogenesis in proliferation tissues. These findings have pointed out the pivotal role FAS might play in the process of carcinogenesis and the modulation of FAS expression may be an effective approach in cancer chemoprevention.

The central theme for multiple stage mouse skin and colon carcinogenesis has been focused on the alterations in signal transduction pathways that modulating the processes of tumor promotion and progression (24). Recent studies on cancer therapy directed at specific, frequently molecular alterations in signal pathways of cancer cells has been validated through the clinical development and regulatory approval of agents such as Herceptin for the treatment of advanced breast cancer and Gleevec for chronic meylogenic leukemia and gastrointestinal stromal tumor (25). Herceptin and Gleevec are the first examples of gene-based cancer drugs, and represent the most significant development toward a new era of target-directed therapies through modulating receptor tyrosine kinase activity. These agents not only prolong life and improve its quality, they also provide clinical evaluation of the emerging field of molecular oncology, especially therapies targeting kinase enzymes that play a critical role in tumorigenesis. As early as 1997, we found that tea polyphenols TF3 and EGCG suppress the proliferation of cancer cells through inhibiting the binding of EGF to its receptor and blocking the autophosphorylation of EGF receptor (15). These findings provide the important molecular basis for cancer chemoprevention of tea polyphenol (9).

Table IX. Suppression of Lipid Biosynthesis in MCF-7 Cells by TF3 and EGCG

Compound	Concentration	Lipid biosynthesis (%)			
		TG	FFA	CE	FC
1. Basal (as control, 100%)		100	100	100	100
2. 1 + EGF	100 ng/mL	123	194	116	207
3. 2 + TF3	10 μM	70	92	91	93
4. 2 + TF3	20 μM	83	82	56	80
5. 2 + EGCG	10 μM	77	106	100	103
6. 2 + EGCG	20 μM	96	132	114	148

Data from reference 19.
Abbreviations are: TG, triacylglycerol; FFA, free fatty acid; CE, cholesterol ester; and FC, free cholesterol. % of inhibition can be easily obtained by the difference between the EGF system (2) and the tested tea polyphenol system (3-6).

It has been demonstrated that overexpression of epidermal growth factor receptor (EGFR) commonly occurs in many types of cancer, such as head and neck, non-small cell lung, largngeal, esophageal, gastric, pancreatic, colon, renal cell, bladder, breast, ovarian, cervical, prostate and papillary thyroid cancers, melanoma, and gliomas, and correlates with a poor clinical outcome (26). Signaling through EGFR activates pathways that stimulate many of the properties associated with cancer cells namely proliferation, migration, stromal invasion, tumor neovascularization and resistance to cell apoptosis-inducing signals (27). Based on the aforementioned findings, it is proposed that the suppression of EGFR function may be considered the first pivotal step in the action mechanisms of anti-carcinogenesis or cancer chemoprevention of tea polyphenols (9,28).

Overall Consideration on the Health Promotion of Black and Green Teas

On the basis of present findings, this is an interesting issue that deserves further elaboration. Both black tea and green tea exhibit profound inhibitory effects on cellular mutation and carcinogenesis (Tables IV and V). Both teas showed strong antioxidative activity (Table III) and hypolipidemic effects (Table IX). Both teas and their tea polyphenols suppress the enzymes and signal transducing proteins that functioning in various signal transduction pathways (Tables VIII and IX).

Most of these data regarding the biochemical and pharmacological effects of tea polyphenols are collected from the experiments designed on the basis of their molar concentrations. When extrapolating these data to the original black tea and green tea, the concentrations of their tea polyphenols should be considered. It has been demonstrated that different concentrations of catechins, theaflavins and thearubigins exsist in black, green and oolong teas (Table X). It is apparent that green tea contains a high concentration of catechins, while black tea contains some catechins and large amounts of theaflavins and thearubigins. Black tea also contains an appreciable amount of highly polymerized substances (7.27-7.66%). The chemical structures of these polymers are completely unknown and this has hampered direct assessment of health promoting activities of black tea. For the time being, it is a fair conclusion that black tea and green tea are comparable in inhibiting skin carcinogenesis as stated by the Conney group (7,8). We would like to follow this statement and point out that black tea and green tea are comparable in promoting human health if people are daily tea drinkers. Furthermore, recent published data (Tables III-IX) demonstrated that black tea polyphenol (TF3) might be more effective than green tea polyphenol (EGCG) in several biological systems. Further investigations are required to provide more *in vivo* data to assess the health promoting actions of these two groups of tea polyphenols.

Table X. Concentrations(%) of Tea Polyphenols in Different Teas

Tea polyphenol	Black tea	Green tea	Oolong tea
(-)-Epicatechin	--	0.74-1.00	0.21-0.33
(-)-Epicatechin 3-gallate	0.29-0.42	1.67-2.47	0.99-1.66
(-)-Epigallocatechin	--	2.60-3.36	0.92-1.08
(-)-Epigallocatechin 3-gallate	0.39-0.60	7.00-7.53	2.93-3.75
Theaflavins	0.98-2.12	--	--
Thearubigins	7.63-8.03	--	--
Highly polymerized substances	7.27-7.66	--	--
Total tea polyphenols	15.56-18.83	12.01-14.36	5.03-6.82

Data from reference *1*.

Acknowledgements

This study was supported by the National Science Council NSC92-2311-B002-022 and NSC 92-2320-B-002-192.

References

1. Ho, C.-T.; Zhu, N. In: *Cafeinated Beverages: Health Benefits, Physiological Effects and Chemistry*; Parliament, T.H.; Ho, C.-T.; Schieberle, P.; Eds., ACS Symp. Ser. 754; Americal Chemical Society, Washington, D.C., **2000**; pp 316-326.
2. Halder, J.; Bhaduri, A.N. *Biochem. Biophys. Res. Commun.* **1998**, *244*, 903-907.
3. Bokudava, M.A.; Skobeleva, N.I. *Crit. Rev. Food Sci. Nutr.* **1980**, *12*, 303-370.
4. Lin, J.K.; Chen, P.C., Ho, C.-T.; Lin-Shiau, S.Y. *J. Agric. Food Chem.* **2000**, *48*, 2738-2743.
5. Apostolides, Z.; Balentine, D.A.; Harbowy, M.E.; Weisburger, J.H. *Mutat. Res.* **1996**, *359*, 159-163.
6. Bu-Abbas, A.; Nunez, X.; Clifford, M.N.; Walker, R.; Ioannides, C. *Mutagenesis*, **1996**, *11*, 597-603.
7. Wang, Z.Y.; Huang, M.T.; Lon, Y.R.; Xie, J.G.; Reuhl, K.R.; Neumark, H.L.; Ho, C.-T.; Yang, C.S.; Conney, A.H. *Cancer Res.* **1994**, *54*, 3428-3435.
8. Lou, Y.R.; Lu, Y.P.; Xie, J.G.; Huang, M.T.; Conney, A.H. *Nutr. Cancer*, **1999**, *33*, 146-153.
9. Lin, J.K.; Liang, Y.C.; Lin-Shiau, S.Y. *Biochem. Pharmacol.* **1999**, *58*, 911-915.
10. Lin, J.K.; Liang, Y.C. *Proc. Natl. Sci. Council, ROC (B)*, **2000**, *24*, 1-13.
11. Nomura, M.; Kaji, A.; He, Z.; Ma, W. Y.; Miyamoto, K.; Yang, C. S.; Dong, Z. *J. Biol. Chem.* **2001**, *276*, 46624-46631.
12. Chung, J.Y.; Park, J.O.; Phyu, H.; Dong, Z.; Yang, C.S. *FASEB J.* **2001**, *15*, 2022-2024.
13. Lin, Y.L.; Tsai, S.H.; Lin-Shiau, S.Y.; Ho, C.-T.; Lin, J.K. *Eur. J. Pharmacol.* **1999**, *367*, 379-388.
14. Pan, M.H.; Liang, Y.C.; Lin-Shiau, S.Y.; Zhu, N.Q.; Ho, C.-T.; Lin, J.K. *J. Agric. Food Chem.* **2000**, *48*, 6337-6346.
15. Liang, Y.C.; Lin-Shiau, S.Y.; Chen, C.F.; Lin, J.K. *J. Cell. Biochem.* **1997**, *67*, 55-65.
16. Liang, Y.C.; Lin-Shiau, S.Y.; Chen, C.F.; Lin, J.K. *J. Cell. Biochem.* **1999**, *75*, 1-12.

17. Pan, M.H.; Lin-Shiau, S.Y.; Ho, C.-T.; Lin, J.H.; Lin, J.K. *Biochem. Pharmacol.* **2000**, *59*, 357-367.
18. Chen, Y.C.; Liang, Y.C.; Lin-Shiau, S.Y.; Ho, C.-T.; Lin, J.K. *J. Agric. Food Chem.* **1999**, *47*, 1416-1421.
19. Yeh, C.W.; Chen, W.J.; Chiang, C.T.; Lin-Shiau, S.Y.; Lin, J.K. *The Pharmacogenomics J.* **2003**, *3*, 267-276.
20. Kuhajda, F.P. *Nutrition,* **2000**, 16, 202-208.
21. Kuhajda, F.P.; Pizer, E.S.; Li,, J.N.; Manii, N.S.; Frehywot, G.L., Townsend, C.A, *Proc. Natl. Acad. Sci. USA* **2000**, *97*, 3450-3454.
22. Kusakabe, T.; Nashimoto, A.; Honma, K., Suzuki, T. *Histopathology* **2002**, *40*, 71-79.
23. Pizer, E.S.; Lax, S.F.; Kuhajda, F.P.; Pasternack, G.R.; Kuman, R.J. *Cancer* **1998**, *83*, 528-571.
24. Zoumpourtis, V.; Solakidi, S.; Papathoma, A.; Papaevangelious, D. *Carcinogenesis* **2003**, *24*, 1159-1165.
25. Shawver, L.K.; Slamon, D.; Ullrich, A. *Cancer Cell* **2002**, *1*, 117-123.
26. Nicholson, R.I.; Gee, J.M.; Harper, M.E. *Eur. J. Cancer* **2001**, *37* (Suppl. 4), S9-15.
27. Schlessinger, J. *Cell* **2000**, *103*, 211-225.
28. Lin, J.K. *Arch. Pharm. Res.* **2002**, *25*, 561-571.

Health Effects

Chapter 18

Biotransformation and Bioavailability of Tea Polyphenols: Implications for Cancer Prevention Research

Joshua D. Lambert, Jungil Hong, Mao-Jung Lee, Shengmin Sang, Xiaofeng Meng, Hong Lu, and Chung S. Yang

Department of Chemical Biology, Ernest Mario School of Pharmacy, Rutgers, The State University of New Jersey, Piscataway, NJ 08854

Consumption of tea (*Camellia sinensis*) has been suggested to prevent cancer, heart disease, and other diseases. Animal studies have shown that tea and tea constituents inhibit carcinogenesis at a number of organ sites including the skin, lung, oral cavity, esophagus, stomach, liver, intestine, colon, and prostate. A number of potential cancer prevention mechanisms for the tea polyphenols have been proposed based mainly on studies with human cancer cells. These include protection from oxidative stress, induction of oxidative stress, inhibition of enzymes (MAP kinases, cyclin-dependent kinases, telomerase, etc.), and inhibition of growth factor-related cell signaling (epidermal growth factor and others). Whereas some studies report effects of epigallocatechin-3-gallate (EGCG) at submicromolar levels, most experiments require concentrations of greater than 10 or 20 µM to demonstrate the effect. In humans, mice, and rats, tea polyphenols undergo glucuronidation, sulfation, methylation, and ring fission. Recent reports also suggest that EGCG and other catechins may be substrates for active efflux. The peak plasma concentrations of EGCG, epigallocatechin (EGC), and epicatechin (EC) following oral administration of green tea are 0.04 –1 µM, 0.3–5 µM, and 0.1–2.5 µM, respectively. The plasma levels of theaflavins are much lower (~ 2 nM). The present chapter reviews the literature concerning the biotransformation and bioavailability of tea polyphenols. It is intended to serve as a guide for designing future bioavailability experiments and for interpreting mechanistic data regarding the actions of tea polyphenols *in vitro* and *in vivo*.

Tea [*Camellia sinensis* (Theaceae)] is second only to water in terms of worldwide popularity as a beverage. Consumption of tea has been suggested to have many health benefits, including the prevention of cancer and heart disease (*1*).

Green, black, and Oolong tea are the three major commercial types of tea and differ in how they are produced and in their chemical composition. Green tea is prepared by pan-frying or steaming fresh leaves to heat inactivate oxidative enzymes and then dried. By contrast, black tea is produced by crushing fresh tea leaves and allowing enzyme-mediated oxidation to occur in a process commonly known as fermentation. Green tea is chemically characterized by the presence of large amounts of polyphenolic compounds known as catechins (Figure 1). A typical cup of brewed green tea contains, by dry weight, 30–40% catechins including epicatechin (EC), epigallocatechin (EGC), epicatechin-3-gallate (ECG), and epigallocatechin-3-gallate (EGCG). Through fermentation, most of the catechins are converted to oligomeric theaflavins and polymeric thearubigins in black tea (Figure 1). The resulting brewed black tea contains 3–10% catechins, 2–6% theaflavins, and >20% thearubigins (*2*).

Both green and black tea and their constituents have been extensively studied *in vitro* and in animal models of carcinogenesis (*3*). Whereas these compounds have been shown to inhibit tumor formation in a number of models of carcinogenesis, the epidemiological data of cancer prevention remains mixed. Likewise, the primary cancer preventive mechanisms of tea in animal models remain unclear. EGCG, the most widely studied catechin, has been shown to inhibit enzymes (topoisomerase, matrix metalloproteinases, and telomerase), growth factor signaling (epidermal growth factor, vascular endothealial growth factor) and transcription factors (AP-1, NFκB) (*3,4,5*). None of these mechanisms have been firmly established in animals or humans.

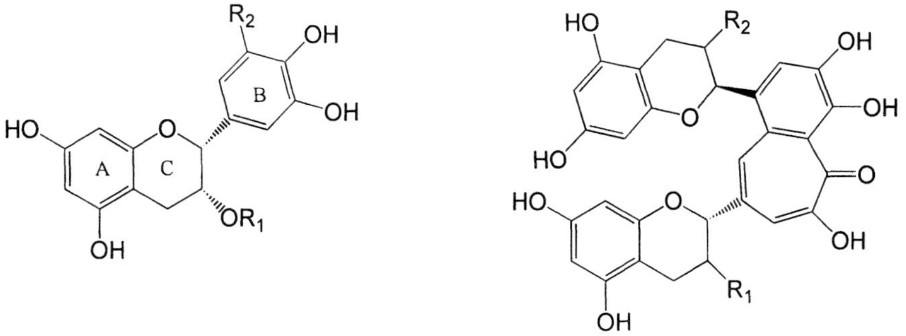

Epicatechin: $R_1 = R_2 = H$

Epigallocatechin: $R_1 = H$; $R_2 = OH$

Epicatechin-3-gallate: $R_1 = Galloyl$; $R_2 = H$

Epigallocatechin-3-gallate: $R_1 = Galloyl$; $R_2 = OH$

Theaflavin: $R_1 = R_2 = OH$

Theaflavin-3-gallate: $R_1 = Galloyl$; $R_2 = OH$

Theaflavin-3'-gallate: $R_1 = OH$; $R_2 = Galloyl$

Theaflavin-3,3'-digallate: $R_1 = R_2 = Galloyl$

Figure 1. Structures of the tea polyphenols

The catechins have been shown to undergo considerable biotransformation and to have low bioavailability (Figure 2) (*3*). The theaflavins are even less bioavailable. This poor bioavailability confounds attempts to correlate *in vitro* findings with cancer prevention in animal models. Cell line studies typically require concentrations of compound in the 10–100 µM range. Such concentrations are typically not observed systemically. The low bioavailability of the tea polyphenols is likely due to their relatively high molecular weight and the large number of hydrogen-bond donating hydroxyl groups (*6*). These hydroxyl groups not only serve as functional handles for phase II enzymes but may also reduce the absorption of the compounds from the intestinal lumen. According to Lipinski's Rule of 5, compounds with a molecular weight greater than 500, greater than 5 hydrogen-bond donors, or 10 hydrogen-bond acceptors have poor bioavailability due to their large actual size (high molecular weight) or large apparent size (due to the formation of a large hydration shell) (*6*).

In the present article, we discuss the current literature regarding the bioavailability of the tea polyphenols and the potential impact of these data on the study of tea polyphenols as cancer preventive agents in humans.

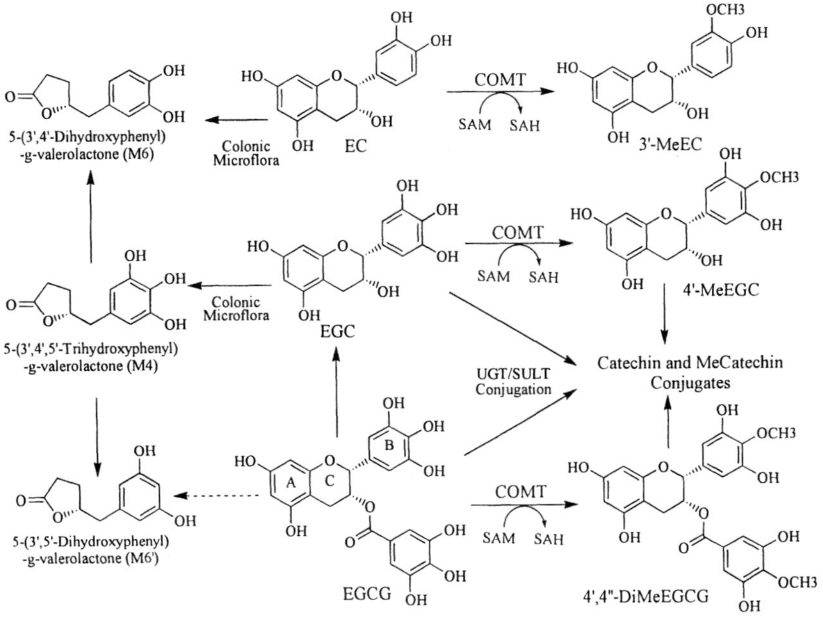

Figure 2. Biotransformation of the tea catechins.

Biotransformation

The catechins are subject to extensive biotransformation including methylation, glucuronidation, sulfation, and ring-fission metabolism (Figure 2). Recent studies on the enzymology of EGC and EGCG methylation have shown that EGC is methylated to form 4'-O-methyl-(-)-EGC and EGCG is methylated by catechol-O-methyltransferase (COMT) to form 4"-O-methyl-(-)-EGCG and 4',4"-O-dimethyl-(-)-EGCG (7). At low concentrations of EGCG, the dimethylated compound is the major product. Rat liver cytosol shows higher COMT activity toward EGCG and EGC than did human or mouse liver cytosol. Additionally, the K_m and V_{max} values are higher for EGC than for EGCG (e.g. in human liver cytosol, K_m is 4 and 0.16 µM for EGC and EGCG, respectively).

Studies of EGCG and EGC glucuronidation reveal that EGCG-4"-*O*-glucuronide is the major metabolite formed by human, mouse, and rat microsomes (*8*). Mouse small intestinal microsomes have the greatest catalytic efficiency (V_{max}/K_m) for glucuronidation followed, in decreasing order, by mouse liver, human liver, rat liver, and rat small intestine. Human UGT1A1, 1A8, and 1A9 have that highest activity toward EGCG, with the intestinal-specific UGT1A8 having the highest catalytic efficiency. EGC-3'-*O*-glucuronide is the major product formed by microsomes from all species with the liver microsomes having a higher efficiency than intestinal microsomes. Based on these studies, it appears that mice are more similar to humans than are rats in terms of enzymatic ability to glucuronidate tea catechins. While these similarities must still be confirmed *in vivo*, this information will aid in choosing the most appropriate animal model to study the potential health benefits of tea constituents.

Vaidyanathan *et al.* have shown that EC undergoes sulfation catalyzed by human and rat intestinal and liver cytosol, with the human liver being the most efficient (*9*). Further studies have revealed that sulfotransferase (SULT)1A1 is largely responsible for this activity in the liver, whereas both SULT1A1 and SULT1A3 are active in the human intestine. The catalytic efficiency for SULT1A1 and SULT1A3-mediated sulfation of EC are 5834 and 55 µL/min/mg. EGCG is also time- and concentration-dependently sulfated by human, mouse, rat liver cytosol (*10*). The rat has the greatest activity followed by the mouse and the human.

Anaerobic fermentation of EGC, EC, and ECG with human fecal microflora has been shown to result in the production of the ring fission products 5-(3',4',5'-trihydroxyphenyl)-γ-valerolactone (M4), 5-(3',4'-dihydroxyphenyl)-γ-valerolactone (M6), and 5-(3',5'-dihydroxyphenyl)-γ-valerolactone (M6') (Fig 2) (*11*). We have found these ring fission products are present in human urine and plasma approximately 3 h after oral ingestion of 20 mg/kg decaffeinated green tea (*12*). The compounds have a T_{max} of 7.5 - 13.5 h and reach peak plasma concentrations of 100–200 nM. Peak urine concentrations of 8, 4, and 8 µM have been demonstrated for M4, M6, and M6', respectively following ingestion of 200 mg EGCG. M4, M6, and M6' retain the polyphenolic character of the parent compound; have the addition of a potentially, biologically-active valerolactone structure; and may therefore have biological activities.

In animals, the phase II metabolism reactions likely compete with one another. The relative concentration of each enzyme and their activities for the tea polyphenols determine the metabolic profile *in vivo*. Since EGCG has a lower K_m for COMT than UGTs, methylation may be favored at physiological (usually low) concentrations. Indeed, EGCG is first methylated to form 4"-*O*-methyl-EGCG and then further methylated to form 4',4"-di-*O*-methyl-EGCG *in vivo* (*13*). At high doses, glucuronidation becomes more prominent, leading the formation of EGCG-4"-glucuronide in the mouse (*8*). This compound can be further methylated on the B-ring to produce different methylated metabolites. This is consistent with the observation that four mono-methylated and two di-methylated compounds are found in mouse urine after hydrolysis with β-glucuronidase and sulfatase following administration of high doses of EGCG to the mouse (*13*). The methylated compounds were found to have similar peaks heights: if conjugation had not preceeded methylation, 4"-*O*-methyl-EGCG peak would have been the predominant mono-methylated metabolite.

Active Efflux

Active efflux has been shown to limit the bioavailability and cellular accumulation of many compounds. The multidrug resistance-associated proteins (MRP) are ATP-dependent efflux transporters that are expressed in many tissues and are overexpressed in many human tumor types. MRP1 is located on the basolateral side of cells and is present in nearly all tissues. The physiological function of this protein is to transport compounds from the interior of the cells into the interstitial space (*14*). In contrast, MRP2 is located on the apical surface of the intestine, kidney, and liver, where it transports compounds from the bloodstream into the lumen, urine, and bile, respectively (*14*). Recent studies on EGCG uptake in our laboratory have shown that indomethacin (MRP inhibitor) increases the intracellular accumulation of EGCG, EGCG 4"-*O*-methyl-EGCG, or 4',4"-di-*O*-methyl-EGCG by 10-, 11-, or 3-fold in Madin-Darby canine kidney (MDCKII) cells overexpressing MRP-1 (*15*). Similarly, treatment of MRP-2 overexpressing MDCKII cells with MK-571 (MRP-2 inhibitor) results in 10-, 15-, or 12-fold increase in the intracellular levels of EGCG, 4"-*O*-methyl-EGCG, and 4',4"-di-*O*-methyl-EGCG, respectively. Treatment of PGP-overexpressing MDCKII cells with a variety of PGP inhibitors, however, results in no significant effect on the intracellular levels of EGCG or its metabolites. Treatment of HT-29 human colon cancer cells with indomethacin results in increase intracellular accumulation of EGCG and its methylated and glucuronidated metabolites (*16*). These data suggest a role for MRPs, but not PGP, in affecting the bioavailability of EGCG.

Vaidyanathan and Walle have shown that treatment of Caco-2 cells with MK-571 enhances apical to basolateral movement of EC and ECG compared to untreated control cells (*17,18*). MK-571 also reduces the efflux of EC-sulfates from the cytosol to the apical well, suggesting the EC-sulfates are also substrates for MRP-2 (*17*). Whereas we have previously suggested that the uptake of EGCG into HT-29 cells is by passive diffusion, others have demonstrated that ECG is a substrate for the monocarboxylate transporter (MCT) and that inhibition of this transporter with benzoic acid or phloretin significantly reduces uptake by Caco-2 cells (*18*).

The combined effects of MRP-1, MRP-2, and MCT on the bioavailability of the tea polyphenols remain to be determined *in vivo*. The apical location of MRP2 suggests that it acts to limit EGCG bioavailability by actively exporting EGCG in the enterocyte back into the intestinal lumen either before or after being methylated by cytosolic COMT. The remaining fraction of EGCG would then be absorbed into the portal circulation, enter the liver, and could subsequently be effluxed by MRP2 located on the canalicular membrane of the hepatocytes. In contrast, MRP1 is located on the basolateral membrane of enterocytes, hepatocytes, and other tissues. Substrates of this pump are effluxed from the interior of the cells into the blood stream or intestinal space. The role of MRP1 would be expected to increase the bioavailability of EGCG *in vivo*. The influence of MRP1 and 2 on the bioavailability of EGCG *in vivo*, however, is likely to depend on the tissue distribution of each pump. It was reported that the transcript level of MRP2 was over 10 fold higher than that of MRP1 in the human jejunum (*19*); therefore efflux of EGCG by MRP2 may be predominant in the intestine, resulting in a decrease of bioavailability. Further *in vivo* studies are needed to determine the effect of MRPs on the bioavailability of the tea polyphenols.

Pharmacokinetics

Studies of [^{3}H]-EGCG in both the rat and the mouse have shown that following a single *i.g.* dose, radioactivity is found throughout the body (*20;21*). After 24 h, 10% of the initial dose (radioactivity) is in the blood with 1% found in the prostate, heart, lung, liver, kidney and other tissues. The major route of elimination is via the feces. In the rat, 77% of an *i.v.* dose of [^{3}H]-EGCG is eliminated in the bile while only 2% is found in the urine.

Detailed pharmacokinetic and biotransformation studies of the tea catechins have been conducted in rats, mice, and humans (*22*). Following *i.g.* administration of decaffeinated green tea (200 mg/kg) to rats, plasma levels of EGCG, EGC, and EC were fit to a two-compartment model with elimination half-lives of 165, 66, and 67 min, respectively. The absolute bioavailability of EGCG, EGC, and EC after *i.g.* administration of decaffeinated green tea is 0.1, 14, and 31%, respectively. Studies with bile duct-cannulated rats have shown that after oral administration of 100 mg EGCG, 3.28% of the dose is recovered in the bile as: EGCG (2.65%), 4"-*O*-methyl-EGCG (0.25%), 3"-*O*-methyl-EGCG (0.11%), 4'-*O*-methyl-EGCG (0.11%), 3'-*O*-methyl-EGCG (0.10%), 4',4"-di-*O*-methyl-EGCG (0.06%) (*23*). With the exception of 4"-*O*-methyl-EGCG and 4',4"-di-*O*-methyl-EGCG, which are present as the sulfated form, the other metabolites and EGCG are presently largely (>58%) in the glucorondiated form with less sulfate present (<42%). Other methylated catechins including 3'- and 4'-O-methyl-EGC have also been reported following oral dosing of rodents and humans with green tea (*3*).

By comparison, the absolute bioavailability of EGCG in mice following *i.g.* administration of 75 mg/kg EGCG is 26.5%. Concentrations of EGCG in the small intestine and colon are 45 and 7.9 nmol/g following *i.g.* administration. The levels in other tissues are less than 0.1 nmol/g. Following *i.v.* administration, levels of EGCG are highest in the liver (3.6 nmol/g), lung (2.7 nmol/g), and small intestine (2.4 nmol/g). Whereas greater than 50% of plasma EGCG is present as the glucuronide, EGCG is present mainly as the free form in the tissues (*24*).

Treatment of rats with a green tea polyphenol preparation (0.6% w/v) in the drinking fluid has been shown to result in increasing plasma levels over a 14-day period with levels of EGC and EC being higher than those of EGCG (*25*). Plasma levels then decrease over the subsequent 14 days suggesting an adaptive effect. EGCG levels were found to be highest in the rat esophagus, intestine, and colon, which have direct contact with tea catechins, whereas EGCG levels are lower in the bladder, kidney, colon, lung, and prostate. When the same polyphenol preparation is given to mice, the EGCG levels in the plasma, lung, and liver are much higher than in rats. These levels appear to peak on day 4 and then decrease to less than 20% of the peak values in days 8 – 10 (*25*).

Several studies of the systemic bioavailability of orally-administered green tea and catechins in human volunteers have been conducted. Most recently, we have shown that oral administration of 20 mg green tea solids/kg body weight results in C_{max} in the plasma for EGC, EC, and EGCG of 223, 124, and 77.9 ng/mL, respectively (*26*). T_{max} is found to range from 1.3 to 1.6 h with $t_{1/2\beta}$ of 3.4, 1.7, and 2 h for EGCG, EGC, and EC, respectively. Plasma EC and EGC are present mainly in the glucuronidated and sulfated form whereas 77% of the EGCG was in the free form. EGC but not EC is also methylated (to 4'-*O*-methyl-EGC) in humans. Plasma and urine levels of 4'-*O*-methyl-EGC have been shown to exceed those of EGC by 10 and 3-fold, respectively, in some subjects (*26*). EGCG has also been found as a methylated metabolite. The maximum plasma concentration of 4',4''-di-*O*-methyl-EGCG is 20% of that of EGCG but the cumulative excretion of 4',4''-di-*O*-methyl-EGCG in the urine is 10-fold higher (140 µg) than that of EGCG (16 µg) over 24 h (*13*). In addition to methylated and conjugated metabolites, the ring-fission metabolites, M4, M6, and M6' have been detected in urine at 8, 4, and 8 µM, respectively following ingestion of 200 mg EGCG (*12,13*). In a recent report, Chow *et al.*, have demostrated that following 4 weeks of green tea polyphenol treatment at a dosing schedule of 800 mg once daily, there is an increase in the area under the plasma EGCG concentration-time curve from 95.6 to 145.6 min/(µg·min) (*27*). No significant changes are observed in the pharmacokinetics of EGCG after repeated green tea polyphenol treatment at a regimen of 400 mg twice daily. Similarly, there was no significant change in the area under the curve for EGC or EC.

There is only a single report on the detection of theaflavins in plasma following ingestion of theaflavins. Mulder et al. have shown that following ingestion of 700 mg of pure theaflavins, the peak concentration is 1 ng/mL and 4.2 ng /mL in the plasma and urine, respectively (28). There are no published reports concerning the metabolic fate of the theaflavins or the thearubigins in humans or animals. It is possible that these compounds undergo microbial degradation in the gut to yield smaller molecular weight compounds. This would explain the observed biological activities of black tea (1).

Whereas the tea polyphenols have low systemic bioavailability, high concentrations have been demonstated in the oral cavity. We have demonstrated that holding green tea solution (1.2 g green tea solids per 200 mL water) in the mouth without swallowing results in salivary concentrations of EGCG and EGC of 153 and 327 μM, respectively (29). These concentrations are 400–1000 times greater than those observed in plasma following ingestion of tea. More recently, we have reported that holding black tea extract for 2 – 5 min results in high salivary theaflavin (1–2 μM) concentrations (30). The rather high local concentrations may support the use of green tea in the prevention of oral cancer and dental caries.

Concluding Remarks

Whereas experiments with human cancer cell lines have suggested numerous mechanisms for the reported cancer preventive activity of tea polyphenols, few have been conclusively demonstrated *in vivo*. Much of the uncertainty stems from the differences in concentration used in cell line studies of tea polyphenols and those present in the plasma and tissue following consumption of tea. A detailed understanding of the factors that affect tea polyphenol bioavailability is necessary to design effective intervention trials for human diseases.

The currently available literature suggests that for some organ sites, such as the oral cavity, small intestine, and colon, the high concentrations used in tissue culture studies may be justifiable. For other tissues, this appears not to be the case. Phase II metabolism and active efflux must be considered when discussing the potential efficacy of tea at sites such as the lung and prostate. Significant work remains to fully characterize the enzymology and importance of phase II metabolism of tea catechins, as well as the importance of the MRP transporters, *in vivo*. A proposed model for the interaction of phase II metabolism and active efflux is shown in Figure 3. Tea catechins are absorbed into the enterocytes and

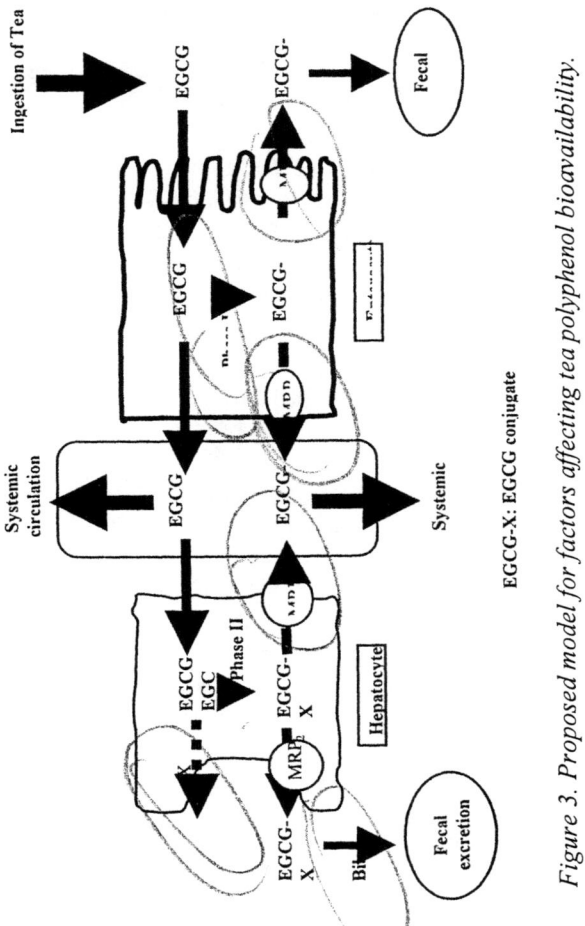

Figure 3. Proposed model for factors affecting tea polyphenol bioavailability.

EGCG-X: EGCG conjugate

subjected to glucuronidation, sulfation, or methylation. A fraction of the parent compound or the metabolite is then effluxed either into the lumen of the intestine by MRP-2 or into the portal circulation by MRP-1. Hepatic phase II enzymes may act to conjugate some portion of the remaining free compound and hepatic efflux pumps may secrete polyphenols or metabolites into the bile duct (MRP-2) or back into the bloodstream (MRP-1). In addition, intestinal microorganisms may metabolize catechins in the intestinal contents resulting in the formation of the ring-fission products. For some compounds such as ECG, active uptake into the cell by MCT must also be considered. Complete characterization of the absorption, distribution, and metabolism of the theaflavins and thearubigins remains to be done.

To summarize, further studies are needed to gain a complete understanding of the factors that affect tea polyphenol bioavailability, particularly with regard to the interplay of these factors *in vivo*. These data, coupled with carefully designed *in vitro* studies and mechanistically-oriented *in vivo* studies will allow greater understanding of the potential health benefits of tea.

References

1. Lambert, J. D.; Yang, C. S. Mechanisms of cancer prevention by tea constituents. *J. Nutr.* **2003**, *133*, 3262S-3267S.
2. Balentine, D. A.; Wiseman, S. A.; Bouwens, L. C. The chemistry of tea flavonoids. *Crit. Rev. Food Sci. Nutr.* **1997**, *37*, 693-704.
3. Yang, C. S.; Maliakal, P.; Meng, X. Inhibition of carcinogenesis by tea. *Annu. Rev. Pharmacol. Toxicol.* **2002**, *42*, 25-54.
4. Ahmad, N.; Katiyar, S. K.; Mukhtar, H. Antioxidants in chemoprevention of skin cancer. *Curr. Probl. Dermatol.* **2001**, *29*, 128-139.
5. Cherubini, A.; Beal, M. F.; Frei, B. Black tea increases the resistance of human plasma to lipid peroxidation in vitro, but not ex vivo. *Free Radic. Biol. Med.* **1999**, *27*, 381-387.
6. Lipinski, C. A.; Lombardo, F.; Dominy, B. W.; Feeney, P. J. Experimental and computational approaches to estimate solubility and permeability in drug discovery and development settings. *Adv. Drug Deliv. Rev.* **2001**, *46*, 3-26.
7. Lu, H.; Meng, X.; Yang, C. S. Enzymology of methylation of tea catechins and inhibition of catechol-O-methyltransferase by (-)-epigallocatechin gallate. *Drug Metab. Dispos.* **2003**, *31*, 572-579.
8. Lu, H.; Meng, X.; Li, C.; Sang, S.; Patten, C.; Sheng, S.; Hong, J.; Bai, N.; Winnik, B.; Ho, C. T.; Yang, C. S. Glucuronides of tea catechins:

enzymology of biosynthesis and biological activities. *Drug Metab. Dispos.* **2003**, *31*, 452-461.
9. Vaidyanathan, J. B.; Walle, T. Glucuronidation and sulfation of the tea flavonoid (-)-epicatechin by the human and rat enzymes. *Drug Metab. Dispos.* **2002**, *30*, 897-903.
10. Lu, H. Mechanistic studies on the Phase II metabolism and absorption of tea catechins. Ph.D. Dissertation, Rutgers, The State University of New Jersey, 2002.
11. Meselhy, M. R.; Nakamura, N.; Hattori, M. Biotransformation of (-)-epicatechin 3-*O*-gallate by human intestinal bacteria. *Chem. Pharm. Bull. (Tokyo)* **1997**, *45*, 888-893.
12. Li, C.; Lee, M. J.; Sheng, S.; Meng, X.; Prabhu, S.; Winnik, B.; Huang, B.; Chung, J. Y.; Yan, S.; Ho, C. T.; Yang, C. S. Structural identification of two metabolites of catechins and their kinetics in human urine and blood after tea ingestion. *Chem. Res. Toxicol.* **2000**, *13*, 177-184.
13. Meng, X.; Sang, S.; Zhu, N.; Lu, H.; Sheng, S.; Lee, M. J.; Ho, C. T.; Yang, C. S. Identification and characterization of methylated and ring-fission metabolites of tea catechins formed in humans, mice, and rats. *Chem. Res. Toxicol.* **2002**, *15*, 1042-1050.
14. Leslie, E. M.; Deeley, R. G.; Cole, S. P. Toxicological relevance of the multidrug resistance protein 1, MRP1 (ABCC1) and related transporters. *Toxicology* **2001**, *167*, 3-23.
15. Hong, J.; Lambert, J. D.; Lee, S. H.; Sinko, P. J.; Yang, C. S. Involvement of multidrug resistance-associated proteins in regulating cellular levels of (-)-epigallocatechin-3-gallate and its methyl metabolites. *Biochem. Biophys. Res. Commun.* **2003**, *310*, 222-227.
16. Hong, J.; Lu, H.; Meng, X.; Ryu, J. H.; Hara, Y.; Yang, C. S. Stability, cellular uptake, biotransformation, and efflux of tea polyphenol (-)-epigallocatechin-3-gallate in HT-29 human colon adenocarcinoma cells. *Cancer Res.* **2002**, *62*, 7241-7246.
17. Vaidyanathan, J. B.; Walle, T. Transport and metabolism of the tea flavonoid (-)-epicatechin by the human intestinal cell line Caco-2. *Pharm. Res.* **2001**, *18*, 1420-1425.
18. Vaidyanathan, J. B.; Walle, T. Cellular uptake and efflux of the tea flavonoid (-)epicatechin-3-gallate in the human intestinal cell line Caco-2. *J. Pharmacol. Exp. Ther.* **2003**, *307*, 745-752.
19. Taipalensuu, J.; Tornblom, H.; Lindberg, G.; Einarsson, C.; Sjoqvist, F.; Melhus, H.; Garberg, P.; Sjostrom, B.; Lundgren, B.; Artursson, P. Correlation of gene expression of ten drug efflux proteins of the ATP-binding cassette transporter family in normal human jejunum and in human intestinal epithelial Caco-2 cell monolayers. *J. Pharmacol. Exp. Ther.* **2001**, *299*, 164-170.

20. Kohri, T.; Matsumoto, N.; Yamakawa, M.; Suzuki, M.; Nanjo, F.; Hara, Y.; Oku, N. Metabolic fate of (-)-[4-(3)H]epigallocatechin gallate in rats after oral administration. *J. Agric. Food Chem.* **2001**, *49*, 4102-4112.
21. Suganuma, M.; Okabe, S.; Oniyama, M.; Tada, Y.; Ito, H.; Fujiki, H. Wide distribution of [3H](-)-epigallocatechin gallate, a cancer preventive tea polyphenol, in mouse tissue. *Carcinogenesis* **1998**, *19*, 1771-1776.
22. Chen, L.; Lee, M. J.; Li, H.; Yang, C. S. Absorption, distribution, elimination of tea polyphenols in rats. *Drug Metab. Dispos.* **1997**, *25*, 1045-1050.
23. Okushio, K.; Suzuki, M.; Matsumoto, N.; Nanjo, F.; Hara, Y. Identification of (-)-epicatechin metabolites and their metabolic fate in the rat. *Drug Metab. Dispos.* **1999**, *27*, 309-316.
24. Lambert, J. D.; Lee, M. J.; Lu, H.; Meng, X.; Ju, J.; Hong, J.; Seril, D. N.; Sturgill, M. G.; Yang, C. S. Epigallocatechin-3-gallate is absorbed but extensively glucuronidated following oral administration to mice. *J. Nutr.* **2003**, *133*, 4172-4177.
25. Kim, S.; Lee, M. J.; Hong, J.; Li, C.; Smith, T. J.; Yang, G. Y.; Seril, D. N.; Yang, C. S. Plasma and tissue levels of tea catechins in rats and mice during chronic consumption of green tea polyphenols. *Nutr. Cancer* **2000**, *37*, 41-48.
26. Lee, M. J.; Maliakal, P.; Chen, L.; Meng, X.; Bondoc, F. Y.; Prabhu, S.; Lambert, G.; Mohr, S.; Yang, C. S. Pharmacokinetics of tea catechins after ingestion of green tea and (-)-epigallocatechin-3-gallate by humans: formation of different metabolites and individual variability. *Cancer Epidemiol. Biomarkers Prev.* **2002**, *11*, 1025-1032.
27. Chow, H. H.; Cai, Y.; Hakim, I. A.; Crowell, J. A.; Shahi, F.; Brooks, C. A.; Dorr, R. T.; Hara, Y.; Alberts, D. S. Pharmacokinetics and safety of green tea polyphenols after multiple-dose administration of epigallocatechin gallate and polyphenon E in healthy individuals. *Clin. Cancer Res.* **2003**, *9*, 3312-3319.
28. Mulder, T. P.; van Platerink, C. J.; Wijnand Schuyl, P. J.; van Amelsvoort, J. M. Analysis of theaflavins in biological fluids using liquid chromatography-electrospray mass spectrometry. *J. Chromatogr. B Biomed. Sci. Appl.* **2001**, *760*, 271-279.
29. Yang, C. S.; Lee, M. J.; Chen, L. Human salivary tea catechin levels and catechin esterase activities: implication in human cancer prevention studies. *Cancer Epidemiol. Biomarkers Prev.* **1999**, *8*, 83-89.
30. Lee, M. J.; Lambert, J. D.; Prabhu, S.; Meng, X.; Lu, H.; Maliakal, P.; Ho, C. T.; Yang, C. S. Delivery of tea polyphenols to the oral cavity by green tea leaves and black tea extract. *Cancer Epidemiol. Biomarkers Prev.* **2004**, *in press*.

Chapter 19

Prevention of Cancer by Dietary Phytochemicals

Zigang Dong and Ann M. Bode

The Hormel Institute, University of Minnesota, 801 16th Avenue N.E., Austin, MN 55912

Diet is one of the most important factors contributing to the development of cancer. Consumption of fruits and vegetables has been linked to a decreased risk of human cancer. Many dietary substances have been isolated and tested for their chemopreventive activity. This article is a brief review of the work related to the molecular mechanisms explaining the anti-cancer effects of certain dietary substances such as tea polyphenols, resveratrol, inositol hexaphosphate and gingerol.

Introduction

Research data from epidemiological and experimental studies indicate that dietary factors are one of the most important etiological aspects of human cancer. The general public desires to find the "magic pill" or "health food" to prevent cancers without unwanted side effects. Therefore, the popular media frequently reports prevention of diseases, including cancer, by "health supplements" or "food therapy".

More and more people are using dietary supplements and herbal remedies without advice from a physician (1,2). Unfortunately, much of the information regarding the effectiveness and safety of these remedies has been gleaned from anecdotal or historical accounts, which seem to be readily available from a variety of sources. For example, advice offered to pregnant women by "medical herbalists" is readily available over the Internet and generally the advice offered is misleading and may even be dangerous (*3*). With rapid development of knowledge and techniques in biology, especially microbiology, substantial progress has been made in the study of cancer prevention (*4-8*). Hundreds of dietary factors have been identified as potential cancer preventive agents. This short review will explore the current state of knowledge regarding selected dietary phytochemicals and their beneficial effects on cancer prevention and the molecular mechanisms underlying these effects.

Signal Transduction and Carcinogenesis

Carcinogenesis is a multifactorial and multistage process in which numerous genes are affected. Many of these genes are prime targets for chemopreventive agents because they regulate intracellular, cell-surface or extracellular functions. Many cancer causing agents, especially tumor promoters, induce signaling transduction cascades (*5,9*). One of the most important protein kinase cascades is the mitogen-activated protein (MAP) kinases. The MAP kinase cascades include extracellular-signal-regulated protein kinases (ERKs), c-Jun N-terminal kinases/stress-activated protein kinases (JNKs/SAPKs) and p38 kinases (Figure 1). ERKS are believed to be strongly activated and to play a critical role in transmitting signals initiated by tumor promoters such as 12-*O*-tetradecanoyl-phorbol-13-acetate (TPA), epidermal growth factor (EGF) and platelet-derived growth factor (PDGF). On the other hand, various forms of stress activate JNKs/SAPKs and p38 kinases potently, including ultraviolet (UV) irradiation and arsenic (*5,9*).

Because different tumor promoters activate distinct MAP kinases (*10,11*), we proposed that the tumor promotion process induced by these different tumor promoters might depend on specific MAP kinase pathways. Indeed, we found that ERKs are required for TPA- or EGF-induced cell transformation in JB6 cells (*12,13*). A shortage of ERKs is responsible for resistance to AP-1 transactivation and transformation in JB6 P$^-$ cells (*12*). Blocking MAP kinase activation by dominant negative mutant (DNM) ERK1 blocks TPA-induced AP-1 transactivation in JB6 P$^+$ cells and DMN-ERK1 also blocks arsenic-induced cell transformation (*14*). On the other hand, JNKs activation is required for JB6 cell transformation induced by TNFα but not by arsenic (*15*). By using knockout mice, we showed that JNK2 is critical for tumor promotion in mouse skin carcinogenesis (*16*).

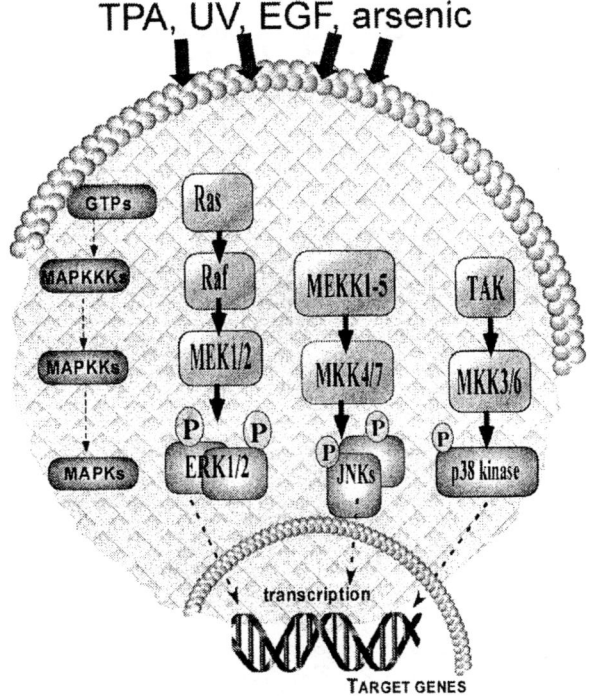

Figure 1. Schematic representation of the MAP kinase signal transduction pathways activated by tumor promoters.

MAP kinases are activated by translocation to the nucleus, where they phosphorylate a variety of target transcription factors, including activator protein-1 (AP-1) and nuclear factor kappaB (NF-κB) (7,9). AP-1 is a well-characterized transcription factor composed of homodimers and/or heterodimers of the *jun* and *fos* gene families (17). AP-1 regulates the transcription of various genes that govern cellular processes including inflammation, proliferation and apoptosis (Figure 2) (17). Blocking tumor promoter-induced AP-1 activity inhibits neoplastic cell transformation (18). By using an AP-1 luciferase transgenic mouse model, we also showed that inhibition of AP-1 activity repressed skin tumor promotion (19). Based on these results, AP-1 is a prime target for chemopreventive agents.

Figure 2. Tumor promoters induce activation and phosphorylation of MAP kinase cascades, which in stimulate AP-1 transcriptional activation resulting in cellular responses including proliferation, apoptosis and inflammation.

Tumor promoters also induce activation of the transcription factor, NF-κB. NF-κB is a stress-responsive transcription factor that is rapidly induced and functions to intensify the transcription of a variety of genes including cytokines, growth factors, COX-2 and acute response proteins (20). Its activation is strongly linked to MAP kinase signaling pathways (21). NF-κB is found in the cytosol bound to an inhibitory protein called inhibitory kappa B (IκB). Following release of IκB, NF-κB is translocated into the nucleus, where it activates gene transcription by binding to its specific DNA sequence found in certain genes, including iNOS and COX-2 (Figure 3). Initiation or acceleration of tumorigenesis is strongly associated with NF-κB activation (22), and inhibition of NF-κB blocks tumor promoter-induced malignant cell transformation (23,24). Like AP-1, NF-κB is another major target for chemopreventive agents.

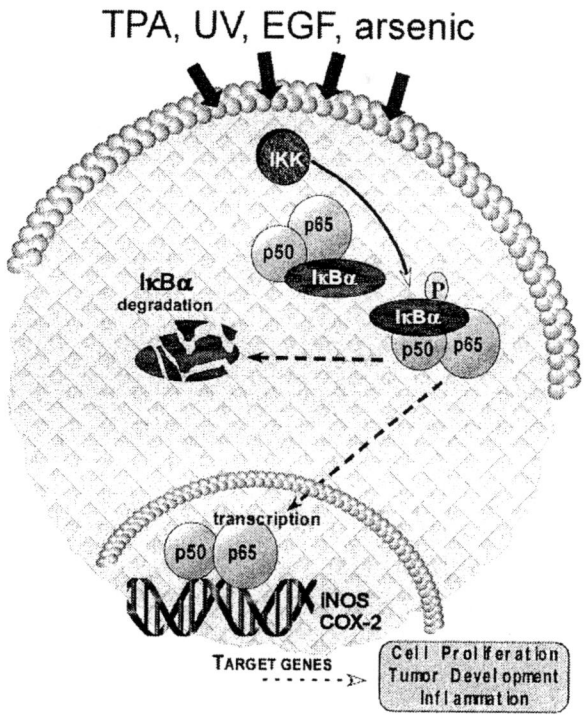

Figure 3. Tumor promoters induce activation of NFκB through IκB kinases. Activation results in separation and degradation of the inhibitory complex, resulting in translocation of active NFκB to the nucleus where it can activate transcription of target genes including iNOS and COX-2.

Tea Polyphenols

Many studies have shown that green tea, black tea, and tea polyphenol preparations have inhibitory effects on carcinogenesis in rodent models (25-28). These include cancers of the skin, lung, esophagus, stomach, liver, duodenum and small intestine, pancreas, and colorectal. Skin carcinogenesis induced by chemicals, UV light, and TPA is one of the most extensively studied systems (*29-31*).

Polyphenols are the most abundant group of compounds in tea leaves, and (-)-epigallocatechin 3-gallate (EGCG) is the best-studied tea polyphenol. A cup of green tea (2.5 g of dried green tea leaves brewed in 200 ml of water) usually contains about 90 mg of EGCG. In addition, it contains a similar or slightly smaller amount (65 mg) of (-)-epigallocatechin (EGC), about 20 mg each of (-)-epigallocatechin 3-gallate (ECG) and (-)-epigallocatechin (EC), and about 50 mg of caffeine (*25,32*). In black tea, polyphenols are reduced to about one fourth of those in green tea, and theaflavins account for 1 to 2% of the total dry matter (*25,33*). To study cellular and molecular mechanisms underlying the observed anticancer effect of tea, we again used the JB6 mouse epidermal cell line (*33*). We found that EGCG and theaflavins inhibited cell transformation and AP-1 activity induced by EGF or TPA. Furthermore, these tea compounds inhibited TPA-induced or EGF-induced c-Jun phosphorylation and JNKs activation, but not ERKs phosphorylation (*33*). We suggest that EGCG and theaflavins may exert their chemopreventive effects primarily through the inhibition of AP-1 transactivation and subsequent AP-1 DNA binding activity (*33*).

Another component important in the ultimate activation of AP-1 is the *ras* pathway. Mutation of the *ras* gene, which perpetually turns on the growth signal transduction pathway, occurs frequently in many types of cancer. We have been studying the effect of tea polyphenols on cell growth and AP-1 activation in a JB6 cell line transfected with a mutant H-*RAS* gene (*34*). Significantly, almost all of the tea polyphenols showed strong inhibition of cell growth and AP-1 activity, and ECGC and TF3 inhibited phosphorylation of p38 kinase, c-Jun and ERKs (*34*). Because the *ras* genes are activated in many animal carcinogenesis models and in human cancers, the inhibition of phosphorylation of c-Jun and ERKs may be important for the repression of cancer formation and growth.

Topical application of EGCG before UVB exposure has been shown to prevent UVB-induced skin cancer (*31*). Pretreatment of JB6 cells with EGCG or theaflavins, resulted in an inhibition of UVB-induced AP-1 activity (*35*). Further, EGCG inhibited UVB-induced AP-1 activity in mouse skin and was linked to an inhibition of UVB-induced transcriptional activation of the *c-fos* gene and UVB-induced accumulation of the c-Fos protein (*36-38*). Therefore, the inhibitory effects of EGCG on *c-fos* expression may offer a further explanation for the antitumor-promoting effects of EGCG. EGCG has also been shown to effectively block arsenite-induced AP-1 transcriptional activation and subsequent DNA binding activity (*38*). Arsenite-induced ERKs, but not p38 kinase activity, was inhibited by EGCG (*39*).

Treatment of EGCG has been reported to induce apoptosis in cancer cells but not normal cells (*40,41*). EGCG treatment was reported to inhibit cell growth, induce G0/G1-phase cell cycle arrest, and cause apoptosis in human epidermal carcinoma (A431) cells but not in normal human epidermal keratinocytes (NHEK) (*40*). The mechanism was related to a differential inhibition by EGCG of NF-κB activation with the cancer cells being more responsive to EGCG (*40*).

EGCG inhibited TPA-induced NF-κB activity and NF-κB sequence specific DNA binding in JB6 cells (*24*). The inhibition appeared to occur through a suppression of the TPA-induced phosphorylation of IκBα at Ser32. In Jurkat T cells, EGCG inhibited 20S proteasome activity, which targets IκBα for degradation, resulting in cell growth arrest and an accumulation of IκBα (*42*). The effects of tea polyphenols on signal transduction are summarized in Figure 4.

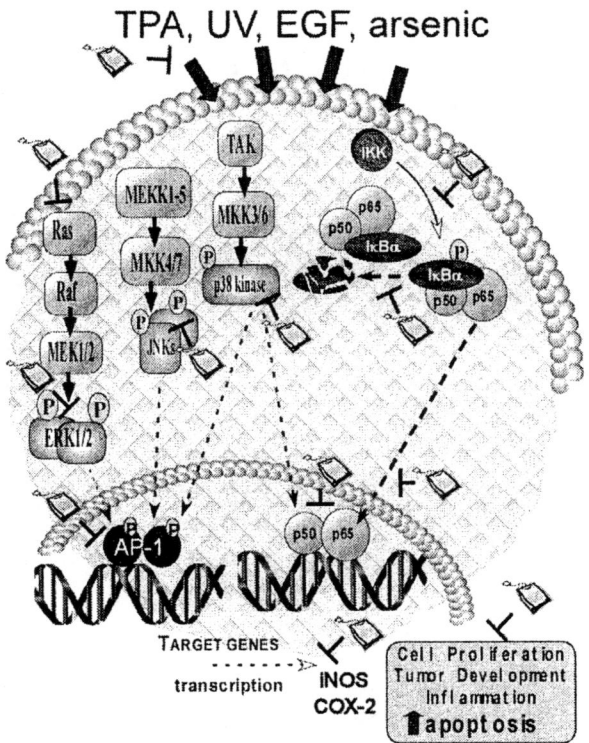

Figure 4. EGCG has inhibitory effects on AP-1 and NFκB activation resulting in inhibition of tumor development, proliferation, inflammation and increased apoptosis of cancer cells.

Inositol Hexaphosphate (IP$_6$)

IP$_6$, also known as phytic acid, is a ubiquitous compound found in the plant kingdom (*43*), but it is also a component of mammalian cells found at concentrations between 10 and 100 µM in both resting and stimulated cells (*44, 45*). Studies by Shamsuddin et al. (*46,47*) and others (*48,49*) demonstrated a striking anticarcinogenic effect of IP$_6$ and myo-inositol. IP$_6$ was shown to be both chemopreventive and chemotherapeutic in rodent colon and mammary carcinogenesis models, as well as in transplanted fibrosarcoma models (*47*). IP$_6$ also significantly inhibited human cancer growth with induction of cell differentiation at concentrations of 1-5 mM. IP$_6$ also inhibited 7,12-dimethylbenzanthracene (DMBA)-induced mouse skin carcinogenesis (*50*).

Several reports indicate the IP$_6$ may act by interfering with the cancer cell cycle. In human prostate carcinoma DU145 cells treated with IP$_6$, growth inhibition was associated with an increase in G1 arrest, increased expression of Cip1/p21 and Kip1/p27 and apoptosis (*51*). IP$_6$ has been shown to have similar effects in other cancer cell types. For example, IP$_6$ induced cell cycle arrest in human leukemia cell lines (*52*), estrogen receptor-positive and -negative human breast cancer cell lines and HT-29 human colon cancer cell lines (*53*).

We have investigated the influence of IP$_6$ on tumor promoter-induced cell transformation in JB6 cells (*54*). The results indicated that IP$_6$ markedly blocks epidermal growth factor-induced phosphatidylinositol-3 (PI-3) kinase activity in a dose-dependent manner in JB6 cells and directly *in vitro*. Blocking PI-3 kinase activity by IP$_6$ profoundly impairs EGF- or phorbol ester-induced JB6 cell transformation and ERKs activation, as well as AP-1 activation. Because PI-3 kinase is a necessary event in tumor promoter-induced cell transformation (*55*), inhibition of PI-3 kinase also inhibits skin carcinogenesis *in vivo*. These results provided the first mechanistic evidence that IP$_6$ targets and blocks PI-3 kinase activation explaining its anticancer effects (*54*). The effects of IP6 on signal transduction are summarized in Figure 5.

Resveratrol

Resveratrol (3,5,4'-trihydroxy-*trans*-stilbene) is found in more than 70 plants including peanuts, grapes and mulberries. Fresh grape skin contains about 50-100 µg of resveratrol per gram of fresh weight (*56*). The relatively high quantity of resveratrol found in grape skin is possibly due to the response of grapes to fungal attack (*57-61*). Much experimental evidence indicates that resveratrol exhibits potent anti-cancer effects in skin and other carcinogenesis models (*61,62*). Our group was the first to report that resveratrol-induced apoptosis may be involved in its chemopreventive effects (*63*).

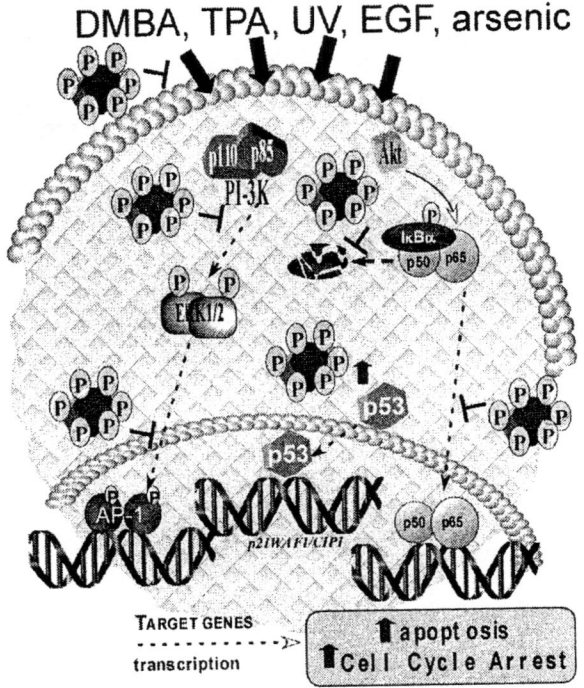

Figure 5. Inositol hexaphosphate also inhibits NFκB and AP-1 activation. In addition, it has inhibitory effects on the PI-3 kinase/Akt pathway and may also induce apoptosis and cell cycle arrest through modulation of p53.

Resveratrol suppressed tumor promoter-induced cell transformation and markedly induced apoptosis, transactivation of p53 and expression of p53 protein in the same cell line and at the same dosage (*63*). Also, resveratrol-induced apoptosis occurred only in cells expressing wild-type p53 ($p53^{+/+}$), but not in p53-deficient ($p53^{-/-}$) cells, whereas no difference in apoptosis induction was observed between normal lymphoblasts and sphingomyelinase-deficient cell lines (*63*). Further, resveratrol-induced apoptosis may be mediated through p53 phosphorylation by ERKs, JNKs and p38 activation (*64*).

Resveratrol has been reported to reduce paclitaxel-induced apoptosis in a human neuroblastoma cell line (SH-SY5Y) by blocking paclitaxel-induced phosphorylation of JNKs, Raf-1 and Bcl-2 (*65*). Apoptosis was associated with suppressed expression of iNOS and Bcl-2, reduced mitochondrial membrane potential and an activation of caspase 3 in the cancer cells (*66*). However, resveratrol has also been shown to have a major effect on the NF-κB-dependent gene expression as a result of its inhibitory effect on IκB kinase (*67*). A naturally-occurring structural analog of resveratrol, piceatannol, also effectively suppressed TNF-induced DNA binding of NF-κB in human myeloid, lymphocyte and epithelial cells. Resveratrol also induces apoptosis in human HL-60 cells (*68*).

Resveratrol inhibited both COX-2 enzyme activity and phorbol ester-induced activation (PMA) of COX-2 (*62*).

In JB6 cells, resveratrol inhibited tumor promoter TPA- and EGF-induced cell transformation in a dose-dependent manner over a range of 2.3-40 μM (*63*). The relationship between the chemical structure of resveratrol and its inhibitory effect on neoplastic cell transformation has been investigated (*69*). Two derivatives, RSVL-1 (3,5,3',4'-tetrahydroxy-*trans*-stilbene) and RSVL-2 (3,5,3',4',5'-pentahydroxy-*trans*-stilbene), were synthesized and their activities compared with that of resveratrol. Results show that RSVL-2 exhibited a more potent inhibitory effect on EGF-induced cell transformation (*69*), whereas RSVL-1 showed a reduced inhibition of cell transformation. In contrast to resveratrol, RSVL-2 appeared to exert its anticarcinogenic effect by targeting phosphatidylinositol-3 (PI-3) kinase and was significantly less toxic than the parent compound (*69*). Lu et al. (*70*) have also studied structural analogs of resveratrol and found at least one that specifically inhibits the growth of transformed WI38 cells, but has little effect on normal WI38 cells. This growth inhibition was linked to an increased expression of p53, GADD45, and Bax with corresponding suppression of Bcl2 (*70*). The effects of resveratrol on signal transduction are summarized in Figure 6.

[6]-Gingerol

Plants of the ginger (*Zingiber officinale* Roscoe, Zingiberaceae) family are one of the most highly-consumed dietary substances in the world (*8*). The oleoresin from the root of ginger contains [6]-gingerol, which is the major pharmacologically active component and a variety of other gingerols, gingerdiols, paradols and zingerones. Studies by others and us suggest that these compounds suppress proliferation of cancer cells (*71-74*). However, very little is known regarding the molecular mechanisms by which they exert their anti-tumorigenic effects. Several aspects of the chemopreventive effects of various phytochemical dietary and medicinal substances, including ginger, have been reviewed (*75*).

235

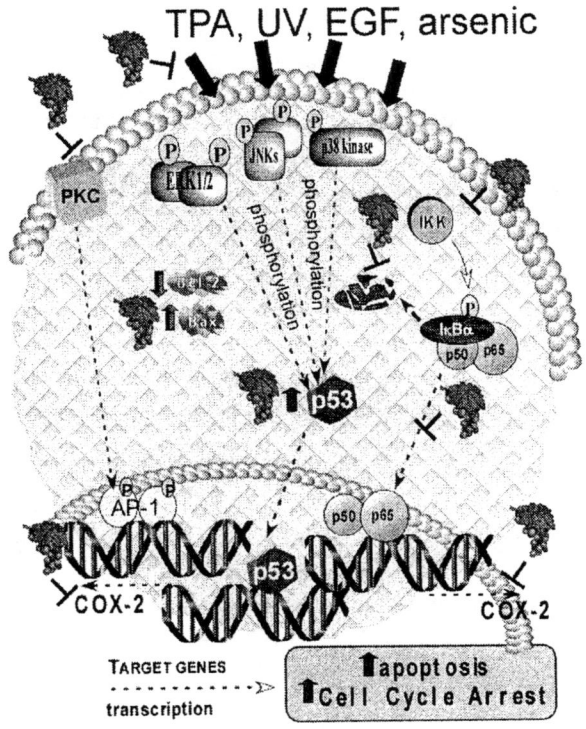

Figure 6. Resveratrol has inhibitory effects on NFκB and AP-1 activation but appears to act primarily by inducing apoptosis and cell cycle arrest mediated by p53. It may also act directly on COX-2 and PKC to exert its anticancer effects.

Earlier studies suggested that gingerol was an effective inhibitor of azoxymethane-induced intestinal carcinogenesis in rats (76). Several ginger components were shown to have effective anti-tumor promoter activity based on their ability to inhibit TPA-induced Epstein-Barr virus early antigen (EBV-EA) in Raji cells (77,78). The most common model used to study the effectiveness of ginger and compounds as anti-tumor agents in vivo is the two-stage initiation-promotion mouse skin model. In this paradigm, tumors are initiated by one application of DMBA followed by repeated topical applications of TPA beginning one to two weeks later. Ginger and its constituents have been shown to inhibit tumor promotion in the SENCAR mouse skin model (79). One study indicated that topical application of [6]-gingerol onto the shaven backs of female ICR mice reduced the incidence of DMBA-initiated/TPA-promoted skin papilloma formation and also suppressed TPA-induced epidermal ornithine decarboxylase activity and inflammation (*80*). In a similar, more recent study, Chung et al. (*81*) reported that in the DMBA/TPA skin tumor model, topical application of [6]-paradol and [6]-dehydroparadol prior to the application of TPA significantly reduced both the number of tumors per mouse and the fraction of mice with tumors (*81*). They suggested that the antitumor-promoting effect might be related to ability of these compounds to suppress TPA-induced oxidative stress.

On the other hand, evidence also suggests that ginger and other related compounds may act as chemopreventive agents by inducing programmed cell death or apoptosis (*82*). At least two recent studies suggest that these compounds suppress proliferation of human cancer cells through the induction of apoptosis (*71,72*). However, very little is known regarding the specific molecular mechanisms or pathways through which they may exert their anti-tumorigenic and apoptotic effects. We recently investigated the effect of two structurally related compounds of the ginger family, [6]-gingerol and [6]-paradol, on EGF-induced cell transformation and AP-1 activation (*74*). Our results provide the first evidence that both compounds block EGF-induced cell transformation and although [6]-gingerol inhibited AP-1 activation, both can act by inducing apoptosis (*74*). Another recent study showed that [6]-paradol and other structurally related derivatives, [10]-paradol, [3]-dehydroparadol, [6]-dehydroparadol, and [10]-dehydroparadol inhibited proliferation of KB oral squamous carcinoma cells in a time- and dose-dependent manner (*83*). [6]-Dehydroparadol (75 µM) was more potent compared to the other compounds tested and induced apoptosis through a caspase-3-dependent mechanism (*83*). Exposure of Jurkat human T-cell leukemia cells to various ginger constituents resulted in apoptosis mediated through the mitochondrial pathway (*84*). Apoptosis was accompanied by a down regulation of anti-apoptotic Bcl-2 protein and an enhancement of pro-apoptotic Bax expression, further supporting the idea that ginger compounds are potential anticancer agents. The effects of [6]-gingerol on signal transduction are summarized in Figure 7.

Conclusion

An accumulation of evidence showed that numerous substances derived from foods are linked to decreased risk of cancer. However, much of the early information is often contradictory and confusing. More recently, solid mechanistic data are rapidly being generated. Continued research efforts in this area are critically needed to identify the dietary factors having real effects and the molecular basis underlying the effects.

Large-scale animal and molecular biology studies are needed to address the bioavailability, toxicity, molecular target, signal transduction pathways, and side effects of dietary factors. Clinical trials based on clear mechanistic studies are

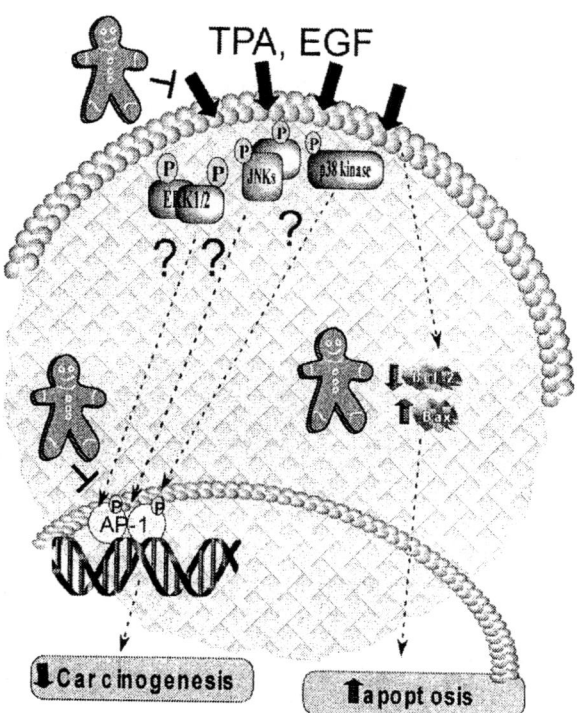

Figure 7. Much less is known about the mechanism of [6]-gingerol's anticancer effects. It appears to induce apoptosis mediated through the mitochondria and disrupts the balance anti- and pro-apoptotic proteins. It also inhibits AP-1 activation but the upstream kinases or other modulators are unclear.

also needed to assess the effectiveness of these dietary factors in the human population.

References

1. Cohen, R. J.; Ek, K.; Pan, C. X. *J. Gerontol. A Biol. Sci. Med. Sci.* **2002**, *57*, M223-227.
2. Zeilmann, C. A.; Dole, E. J.; Skipper, B. J.; McCabe, M.; Dog, T. L.; Rhyne, R. L. *Pharmacotherapy* **2003**, *23*, 526-532.
3. Ernst, E.; Schmidt, K. *Wien. Med. Wochenschr.* **2002**, *152*, 190-192.
4. Dong, Z. *Biofactors* **2000**, *12*, 17-28.
5. Bode, A. M.; Dong, Z. *Lancet Oncol.* **2000**, *1*, 181-188.
6. Dong, Z. *ACS Symposium Series 851, Food Factors for Health Promotion and Disease Prevention*, F. Shahidi, C. T. Ho, S. Watanabe and T. Osawa, American Chemical Society 2003, 40-49.
7. Bode, A. M.; Dong, Z. *Phytochemicals and Health and Disease*, 2003, in press.
8. Surh, Y. *Mutat. Res.* **1999**, *428*, 305-327.
9. Bode, A. M.; Dong, Z. *Sci. STKE.* **2003**, *2003*, RE2.
10. Huang, C.; Ma, W. Y.; Dong, Z. *Oncogene* **1999**, *18*, 2828-2835.
11. Huang, C.; Ma, W. Y.; Li, J.; Dong, Z. *Cancer Res.* **1999**, *59*, 3053-3058.
12. Huang, C.; Ma, W. Y.; Young, M. R.; Colburn, N.; Dong, Z. *Proc. Natl. Acad. Sci. U. S. A.* **1998**, *95*, 156-161.
13. Watts, R. G.; Huang, C.; Young, M. R.; Li, J. J.; Dong, Z.; Pennie, W. D.; Colburn, N. H. *Oncogene,* **1998**, *17*, 3493-3498.
14. Huang, C.; Ma, W. Y.; Li, J.; Goranson, A.; Dong, Z. *J. Biol. Chem.* **1999**, *274*, 14595-14601.
15. Huang, C.; Li, J.; Ma, W. Y.; Dong, Z. *J. Biol. Chem.* **1999**, *274*, 29672-29676.
16. Chen, N.; Nomura, M.; She, Q. B.; Ma, W. Y.; Bode, A. M.; Wang, L.; Flavell, R. A.; Dong, Z. *Cancer. Res.* **2001**, *61*, 3908-3912.
17. Angel, P.; Karin, M.. *Biochim. Biophys. Acta.* **1991**, *1072*, 129-157.
18. Dong, Z.; Birrer, M. J.; Watts, R. G.; Matrisian, L. M.; Colburn, N. H. *Proc. Natl. Acad. Sci. U. S. A.* **1994**, *91*, 609-613.
19. Huang, C.; Ma, W. Y.; Dawson, M. I.; Rincon, M.; Flavell, R. A.; Dong, Z. *Proc. Natl. Acad. Sci. U. S. A.* **1997**, *94*, 5826-5830.
20. Baldwin, A. S., Jr. *Annu. Rev. Immunol.* **1996**, *14*, 649-683.
21. Schulze-Osthoff, K.; Ferrari, D.; Riehemann, K.; Wesselborg, S. *Immunobiology* **1997**, *198*, 35-49.
22. Gilmore, T. D. *J. Clin. Invest.* **1997**, *100*, 2935-2936.
23. Li, J. J.; Westergaard, C.; Ghosh, P.; Colburn, N. H. *Cancer Res.* **1997**, *57*, 3569-3576.

24. Nomura, M.; Ma, W.; Chen, N.; Bode, A. M.; Dong, Z. *Carcinogenesis* **2000**, *21*, 1885-1890.
25. Yang, C. S.; Wang, Z. Y. *J. Natl. Cancer Inst.* **1993**, *85*, 1038-1049.
26. Yang, C. S.; Yang, G. Y.; Lee, M. L.; Chen, L. *Proceedings of the International Conference on Food Factor in Cancer Prevention*, H. Ohigashi, Springer-Verlag 1997; 113-117
27. Katiyar, S. K.; Mukhtar, H. *Int. J. Oncology*, **1996**, *8*, 221-238.
28. Dreosti, I. E.; Wargovich, M. J.; Yang, C. S. *Crit. Rev. Food Sci. Nutr.* **1997**, *37*, 761-770.
29. Yoshizawa, S.; Horiuchi, T.; Fukiji, H.; Yoshida, T.; Okuda, T.; Sugimura, T. *Photother. Res.* **1987**, *1*, 44-47.
30. Wang, Z. Y.; Khan, W. A.; Bickers, D. R.; Mukhtar, H. *Carcinogenesis* **1989**, *10*, 411-415.
31. Gensler, H. L.; Timmermann, B. N.; Valcic, S.; Wachter, G. A.; Dorr, R.; Dvorakova, K.; Alberts, D. S. *Nutr. Cancer* **1996**, *26*, 325-335.
32. Huang, M. T.; Ho, C. T.; Wang, Z. Y.; Ferraro, T.; Finnegan-Olive, T.; Lou, Y. R.; Mitchell, J. M.; Laskin, J. D.; Newmark, H.; Yang, C. S.; Conney, A. H. *Carcinogenesis*, **1992**, *13*, 947-954.
33. Dong, Z.; Ma, W.; Huang, C.; Yang, C. S. *Cancer Res.* **1997**, *57*, 4414-4419.
34. Chung, J. Y.; Huang, C.; Meng, X.; Dong, Z.; Yang, C. S. *Cancer Res.* **1999**, *59*, 4610-4617.
35. Nomura, M.; Ma, W. Y.; Huang, C.; Yang, C. S.; Bowden, G. T.; Miyamoto, K.; Dong, Z. *Mol. Carcinog.* **2000**, *28*, 148-155.
36. Barthelman, M.; Bair, W. B. 3rd; Stickland, K. K.; Chen, W.; Timmermann, B. N.; Valcic, S.; Dong, Z.; Bowden, G. T. *Carcinogenesis*, **1998**, *19*, 2201-2204.
37. Chen, W.; Borchers, A. H.; Dong, Z.; Powell, M. B.; Bowden, G. T. *J. Biol. Chem.* **1998**, *273*, 32176-32181.
38. Chen, W.; Dong, Z.; Valcic, S.; Timmermann, B. N.; Bowden, G. T. *Mol Carcinog.* **1999**, *24*, 79-84.
39. Chen, N. Y.; Ma, W. Y.; Yang, C. S.; Dong, Z. *J. Environ. Pathol. Toxicol. Oncol.* **2000**, *19*, 287-295.
40. Ahmad, N.; Feyes, D. K.; Nieminen, A. L.; Agarwal, R.Mukhtar, H. *J. Natl. Cancer Inst.* **1997**, *89*, 1881-1886.
41. Lu, J.; Ho, C. T.; Ghai, G.; Chen, K. Y. *Cancer Res.* **2000**, *60*, 6465-6471.
42. Nam, S.; Smith, D. M.; Dou, Q. P. *J. Biol. Chem.* **2001**, *276*, 13322-13330.
43. Cosgrove, D. J. *Elsevier Science Publishers*, **1980**
44. Bunce, C. M.; French, P. J.; Allen, P.; Mountford, J. C.; Moor, B.; Greaves, M. F.; Michell, R. H.; Brown, G. *Biochem J.* **1993**, *289 (Pt 3)*, 667-673.
45. Szwergold, B. S.; Graham, R. A.; Brown, T. R. *Biochem. Biophys. Res. Commun.* **1987**, *149*, 874-881.
46. Shamsuddin, A. M.; Vucenik, I.; Cole, K. E. *Life Sci.* **1997**, *61*, 343-354.

47. Vucenik, I.; Tomazic, V. J.; Fabian, D.; Shamsuddin, A. M. *Cancer Lett.* **1992**, *65*, 9-13.
48. Estensen, R. D.; Wattenberg, L. W. *Carcinogenesis* **1993**, *14*, 1975-1977.
49. Pretlow, T. O.; O'Riorda, M. A.; Pretlow, T. G. *Diet and Cancer Markers, Prevention and Treatment,* M. M. Jacobs, Plenum Press 1994; 244.
50. Gupta, K. P.; Singh, J.; Bharathi, R. *Nutr. Cancer* **2003**, *46*, 66-72.
51. Singh, R. P.; Agarwal, C.; Agarwal, R. *Carcinogenesis* **2003**, *24*, 555-563.
52. Deliliers, G. L.; Servida, F.; Fracchiolla, N. S.; Ricci, C.; Borsotti, C.; Colombo, G.; Soligo, D. *Br. J. Haematol.* **2002**, *117*, 577-587.
53. El-Sherbiny, Y. M.; Cox, M. C.; Ismail, Z. A.; Shamsuddin, A. M.; Vucenik, I. *Anticancer Res.* **2001**, *21*, 2393-2403.
54. Huang, C.; Ma, W. Y.; Hecht, S. S.; Dong, Z. *Cancer Res.* **1997**, *57*, 2873-2878.
55. Huang, C.; Ma, W. Y.; Dong, Z. *Mol. Cell Biol.* **1996**, *16*, 6427-6435.
56. Gusman, J.; Malonne, H.; Atassi, G. *Carcinogenesis* **2001**, *22*, 1111-1117.
57. Bavaresco, L.; Fregoni, C.; Cantu, E.; Trevisan, M. *Drugs Exp. Clin. Res.* **1999**, *25*, 57-63.
58. Soleas, G. J.; Diamandis, E. P.; Goldberg, D. M. *Clin. Biochem.* **1997**, *30*, 91-113.
59. Dercks, W.; Creasy, L. L. *Physiol. Mol. Plant Path.* **1989**, *34*, 189-202.
60. Dercks, W.; Creasy, L. L. *Physiol. Mol. Plant Path.* **1989**, *34*, 203-213.
61. Bode, A. M.; Dong, Z. *Phytochemicals in Health and Diseases*, 2003.
62. Jang, M.; Cai, L.; Udeani, G. O.; Slowing, K. V.; Thomas, C. F.; Beecher, C. W.; Fong, H. H.; Farnsworth, N. R.; Kinghorn, A. D.; Mehta, R. G.; Moon, R. C.; Pezzuto, J. M. *Science,* **1997**, *275*, 218-220.
63. Huang, C.; Ma, W. Y.; Goranson, A.; Dong, Z. *Carcinogenesis* **1999**, *20*, 237-242.
64. She, Q. B.; Huang, C.; Zhang, Y.; Dong, Z. *Mol. Carcinog.* **2002**, *33*, 244-250.
65. Nicolini, G.; Rigolio, R.; Scuteri, A.; Miloso, M.; Saccomanno, D.; Cavaletti, G.; Tredici, G. *Neurochem. Int.* **2003**, *42*, 419-429.
66. Billard, C.; Izard, J. C.; Roman, V.; Kern, C.; Mathiot, C.; Mentz, F.; Kolb, J. P. *Leuk. Lymphoma.* **2002**, *43*, 1991-2002.
67. Holmes-McNary, M.; Baldwin, A. S., Jr. *Cancer Res.* **2000**, *60*, 3477-3483.
68. Surh, Y. J.; Hurh, Y. J.; Kang, J. Y.; Lee, E.; Kong, G.; Lee, S. J. *Cancer Lett.* **1999**, *140*, 1-10.
69. She, Q. B.; Ma, W. Y.; Wang, M.; Kaji, A.; Ho, C. T.; Dong, Z.; *Oncogene,* **2003**, *22*, 2143-2150.
70. Lu, J.; Ho, C. H.; Ghai, G.; Chen, K. Y. *Carcinogenesis* **2001**, *22*, 321-328.
71. Lee, E.; Park, K. K.; Lee, J. M.; Chun, K. S.; Kang, J. Y.; Lee, S. S.; Surh, Y. J. *Carcinogenesis* **1998**, *19*, 1377-1381.
72. Lee, E.; Surh, Y. J. *Cancer Lett.* **1998**, *134*, 163-168.
73. Surh, Y. J.; Lee, E.; Lee, J. M. *Mutat. Res.* **1998**, *402*, 259-267.

74. Bode, A. M.; Ma, W. Y.; Surh, Y. J.; Dong, Z. *Cancer Res.* **2001**, *61*, 850-853.
75. Surh, Y. J. *Food Chem. Toxicol.* **2002**, *40*, 1091-1097.
76. Yoshimi, N.; Wang, A.; Morishita, Y.; Tanaka, T.; Sugie, S.; Kawai, K.; Yamahara, J.; Mori, H. *Jpn. J. Cancer Res.* **1992**, *83*, 1273-1278.
77. Vimala, S.; Norhanom, A. W.; Yadav, M. *Br. J. Cancer.* **1999**, *80*, 110-116.
78. Kapadia, G. J.; Azuine, M. A.; Tokuda, H.; Hang, E.; Mukainaka, T.; Nishino, H.; Sridhar, R. *Pharmacol. Res.* **2002**, *45*, 213-220.
79. Katiyar, S. K.; Agarwal, R.; Mukhtar, H. *Cancer Res.* **1996**, *56*, 1023-1030.
80. Park, K. K.; Chun, K. S.; Lee, J. M.; Lee, S. S.; Surh, Y. J. *Cancer Lett.* **1998**, *129*, 139-144.
81. Chung, W. Y.; Jung, Y. J.; Surh, Y. J.; Lee, S. S.; Park, K. K. *Mutat. Res* **2001**, *496*, 199-206.
82. Thatte, U.; Bagadey, S.; Dahanukar, S. *Cell. Mol. Biol. (Noisy-le-grand)*, **2000**, *46*, 199-214.
83. Keum, Y. S.; Kim, J.; Lee, K. H.; Park, K. K.; Surh, Y. J.; Lee, J. M.; Lee, S. S.; Yoon, J. H.; Joo, S. Y.; Cha, I. H.; Yook, J. I. *Cancer Lett.* **2002**, *177*, 41-47.
84. Miyoshi, N.; Nakamura, Y.; Ueda, Y.; Abe, M.; Ozawa, Y.; Uchida, K.; Osawa, T. *Cancer Lett.* **2003**, *199*, 113-119.

Chapter 20

Effect of Black Tea Theaflavins and Related Benzotropolone Derivatives on 12-*O*-Tetradecanoylphorbol-13-acetate-Induced Mouse Ear Inflammation and Inflammatory Mediators

Divya Ramji[1], Shengmin Sang[1,2], Yue Liu[2], Robert T. Rosen[1], Geetha Ghai[1], Chi-Tang Ho[1], Chung S. Yang[2], and Mou-Tuan Huang[2]

[1]Department of Food Science and Center for Advanced Food Technology, Rutgers, The State University of New Jersey, 65 Dudley Road, New Brunswick, NJ 08901
[2]Department of Chemical Biology, Ernest Mario School of Pharmacy, Rutgers, The State University of New Jersey, Piscataway, NJ 08854

> Tea, a popular beverage, is consumed worldwide and exists in several varieties. Extensive literature exists on the health benefits of green tea while black tea has received less attention untill recently. In this study the effect of the black tea compounds theaflavin and theaflavin 3,3'-digallate along with two synthetic catechol complexes structurally similar to theaflavin, namely, the epigallocatechin catechol (EGCCa) and epigallocatechin gallate catechol (EGCGCa), were evaluated in a 12-*O*-tetradecanoylphorbol-13-acetate (TPA)-induced mouse ear inflammatory model. Inflammation was quantified using ear punch weight and various inflammatory biomarkers such as interleukin-1 (IL-1), interleukin-6 (IL-6), prostaglandin E_2 (PGE_2) and leukotriene B4 (LTB_4). The effect of these compounds was compared to sulindac, a standard non-steroidal anti-inflammatory drug (NSAID). The black tea compounds tested here are effective inhibitors of inflammation and inflammatory mediators.

Introduction

Tea, the second largest consumed beverage in the world after water, is made from the dried and processed leaves of the plant *Camellia sinensis* (1,2). The major chemical constituents of fresh tea leaves that make up to 30% of the dry weight are the polyphenols such as flavanoids and phenolic acids (3). Catechins, the major flavanoids present in tea are flavan-3-ols that include (-)-epigallocatechin gallate (EGCG), (-)-epigallocatechin (EGC), (-)-epicatechin-3-gallate (ECG), (-)-epicatechin (EC), (-)-gallocatechin (GC) and (+)-catechin (C). The three major varieties of tea, green, oolong and black are produced with the leaves of *Camellia sinensis* using different manufacturing processes (3).

Black tea common in western countries constitutes 78% of the tea manufactured worldwide (3,4). Its unique chemistry and flavor can be attributed to its four step manufacturing process that consists of withering, rolling, fermentation, and drying. During this process the enzymes polyphenol oxidase and peroxidase interact with the catechins to form theaflavins and thearubigens (Figure 1), the products of oxidative polymerization (1,5-7). Theaflavins, the characteristic orange-red colored compounds in black tea, are a mixture that constitutes 1-2% of its dry weight (2,8) and includes theaflavin, theaflavin-3-*O*-gallate, theaflavin-3'-*O*-gallate, theaflavin-3,3'-*O*-digallate, isotheaflavins, neotheaflavins and theaflavic acids (3). The hydroxyl-substituted benzotropolone ring present in theaflavins and thearubigens is responsible for the characteristic color, astringent taste and flavor of black tea. In addition, black tea contains 3-10% unpolymerized catechins (3,9).

Tea has been reported to have several health promoting effects based on research using cell culture, animal models and clinical studies (10). The effects have been primarily attributed to its antioxidant potential which may be due to the radical scavenging or metal chelating property of polyphenols (7,11). This study was undertaken to evaluate the anti-inflammatory properties of black tea theaflavin derivatives on TPA-induced mouse ear inflammation. Theaflavin, theaflavin-3,3'-digallate and two synthetic compounds structurally similar to theaflavins were evaluated for their anti-inflammatory potential. The synthetic compounds, namely, the catechol adducts of (-)-epigallocatechin (EGCCa) and (-)-epigallocatechin gallate (EGCGCa) were made by the reaction of catechol with EGC and EGCG, leading to the formation of the benzotropolone ring similar to that present in theaflavins (Figure 2).

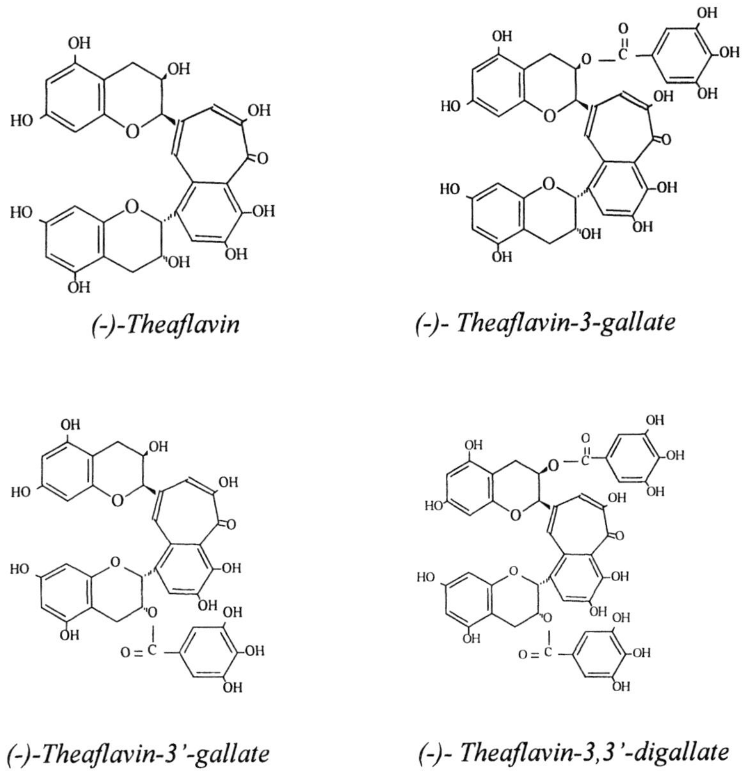

(-)-Theaflavin *(-)- Theaflavin-3-gallate*

(-)-Theaflavin-3'-gallate *(-)- Theaflavin-3,3'-digallate*

Figure 1. Structures of four major theaflavins in black tea.

Materials and Methods

Animals and Chemicals

Female CD-1 mice (3-4 weeks old) were purchased from the Charles River Breeding Laboratories (Kingston, NY). Theaflavin-related compounds were prepared in our laboratory (*12*). Acetone, 12-*O*-tetradecanoylphorbol-13-acetate, sulindac and other chemicals were purchased from Sigma Chemical Co. (St. Louis, MO). Phosphate buffered saline containing 0.4 M NaCl, 0.05% Tween-20, 0.5% bovine serum albumin, 0.1 mM phenylmethylsulfonylfluoride, 0.1 mM benzethonium, 10 mM EDTA and 20 KI aprotinin per mL was used to homogenize mouse ear tissue samples.

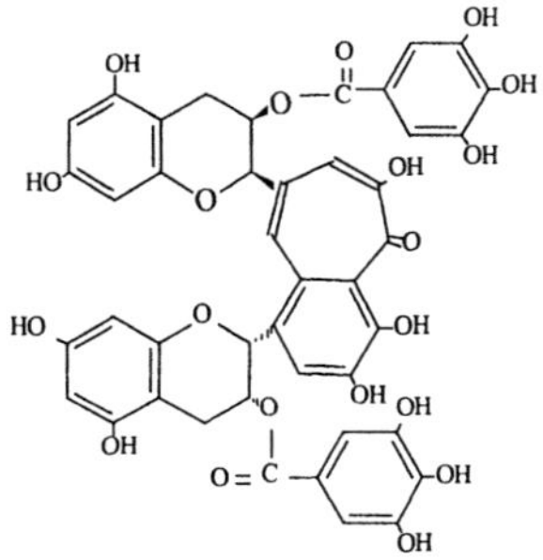

Theaflavin-3,3'-digallate

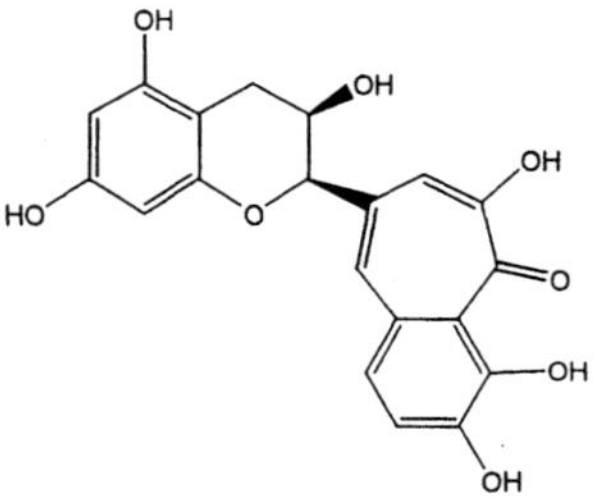

Epigallocatechin catechol
(EGCCa)

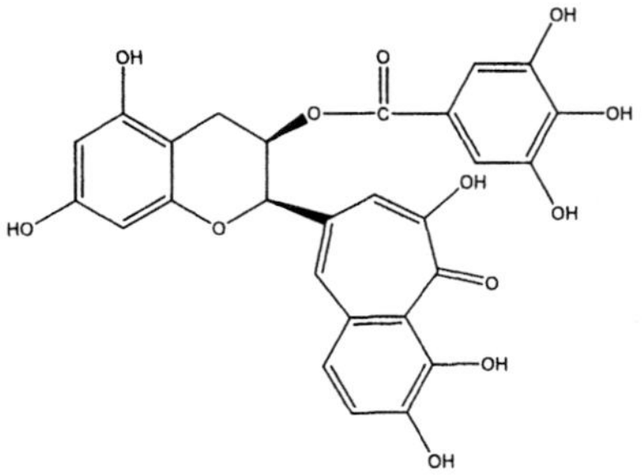

Epigallocatechin gallate catechol
(EGCGCa)

Sulindac

Figure 2. Structures of experimental compounds.

Preparation of Theaflavin Digallate and Related Compounds

ECG (1 g) and EGCG (1 g) were dissolved in a mixture of acetone-pH 5.0 phosphate-citrate buffer (1:10, v/v, 50 mL), containing 4 mg horseradish peroxidase. This mixture was stirred for 45 min with the addition of 2.0 mL of 3.13% H_2O_2 four times. The reaction mixture was extracted with ethyl acetate (50 mL × 3), concentrated and the residue subjected to Sephadex LH-20 column chromatography and eluted with acetone-water solvent system (45:55, v/v) yielding 100 mg of (-)-theaflavin 3,3'-digallate.

For the preparation of EGCCa, EGC (1 g) and catechol (1.5 g) were reacted according to the same procedure as described above, 226 mg of EGCCa was obtained.

Similarly, EGCGCa (230 mg) was obtained from the reaction of EGCG (1 g) and catechol (1.5 g).

Animal Treatment

Female CD-1 mice (30 days old, obtained from Charles River Breeding Laboratories (Kingston, NY) were divided into 6 groups of 5 animals each. All test compounds were dissolved in acetone, which served as the solvent (negative) control. A TPA control group was done to indicate the maximum induction of inflammation in mouse ear. Black tea theaflavin derivatives were evaluated in a dose dependent manner for their inhibitory effect on mouse ear inflammation. A positive control using sulindac, a standard NSAID was also used in a few experiments and compared to the anti-inflammatory potential of the theaflavin derivatives.

Both ears of CD-1 mice were treated with 15 µL acetone (solvent control group), TPA (TPA control) or test compound in acetone 20 min before topical application of acetone (solvent control) or 0.4 nmol TPA on the mouse ear. This treatment was continued twice a day for 3.5 days. Alternatively, 0.8 nmol of TPA was applied once a day for 4 days in a few experiments. At the end of the treatment period the mice were sacrificed by cervical dislocation and the ears punched. The ear punches were weighed separately to quantify the level of inflammation in the mouse ear.

The ear punches from each group were combined and homogenized with phosphate buffered saline. The resulting homogenate was centrifuged and the supernatant tested for the levels of various inflammatory biomarkers using the enzyme linked immunosorbent assay (ELISA).

Results

The theaflavin related compounds evaluated in this study significantly inhibited inflammation and inflammatory mediators. Theaflavin mixture inhibited inflammation in the ear as shown by ear punch weight and also inflammatory biomarkers, interleukin-6 (IL-6) and leukotriene B_4 (LTB_4) using ELISA as shown in Table I. However, a dose response effect could only be seen with respect to LTB_4 concentration.

Table I. Effect of theaflavin mixture (TFs) on TPA-induced inflammation, formation of interleukin-6 (IL-6) and leukotriene B_4 (LTB_4) in ears of CD-1 mice

Treatment	Weight of Ear Punch (mg)	IL-6 concentration (pg/mg tissue)	LTB_4 Concentration (pg/mg tissue)
Acetone	7.93 ± 0.40	0.35 ± 0.01	1.90 ± 0.15
TPA (0.4 nmol)	13.77 ± 0.41	4.28 ± 0.34	9.80 ± 1.21
Theaflavins (0.25 μmol) + TPA (0.4 nmol)	9.58 ± 0.50	1.68 ± 0.11	2.53 ± 0.21
Theaflavins (0.5 μmol) + TPA (0.4 nmol)	10.49 ± 0.51	3.76 ± 0.40	2.12 ± 0.17
Theaflavins (1 μmol) + TPA (0.4 nmol)	8.27 ± 0.22	0.47 ± 0.08	1.18 ± 0.04

Female CD-1 mice (30 days old, 5 mice per group) were treated topically with 20 μL acetone or test compound in acetone 20 minutes before application of 20 μL acetone or TPA (0.4 nmol) in acetone twice a day for 3.5 days (total of 7 treatments). The mice were sacrificed by cervical dislocation 6 hours after last dose of TPA. Ear punches (6 mm in diameter) were weighed separately. The ear punches from each group were then combined, homogenized and tested for the levels of inflammatory mediators interleukin-6 (IL-6) and leukotriene B_4 (LTB_4) using ELISA. Data are expressed as the mean ± standard error from 3 separate determinations.

Theaflavin digallate was evaluated at three different concentrations (0.25, 0.5 and 1 μmol) for its anti-inflammatory potential alone and in combination with sulindac. A significant inhibition of inflammation and inflammatory biomarkers, PGE_2 and LTB_4 could be seen. A dose response effect could not be seen for ear punch weight. When theaflavin digallate (0.25 μmol) was used along with sulindac a synergistic inhibition on ear punch weight was observed. The ear punches from each group were then combined, homogenized and tested for the level of PGE_2 and LTB_4 using ELISA. A concentration dependent inhibitory effect was seen on the levels of PGE_2 and LTB_4 as shown in Table II. Theaflavin digallate was a better inhibitor of LTB_4 than sulindac. However, being a COX inhibitor, sulindac inhibited PGE_2 to a greater extent than theaflavin digallate. When used along with sulindac, theaflavin digallate also inhibited PGE_2 concentration to a similar extent, probably due to the effect of sulindac.

The synthetic compounds, EGCCa and EGCGCa exhibited a similar pattern of inhibition on inflammatory biomarkers (Table III). When used together with sulindac, EGCCa showed an additive effect on weight of ear punch. Sulindac when used alone did not reduce IL-6 concentration, but a synergistic effect was observed when administered with the test compounds. EGCCa and EGCGCa showed a significant inhibition of LTB_4 concentration but when used along with sulindac no additive or synergistic effect could be seen.

Discussion

Limited studies conducted with black tea indicate antioxidative, anticarcinogenic and anti-inflammatory properties particularly in *in-vitro* models (*13-16*). Therefore, the present study was conducted to confirm the anti-inflammatory properties of black tea theaflavin derivatives in an *in vivo* TPA-induced mouse ear inflammatory model.

Inflammation is defined as "a localized protective response that serves to destroy, dilute or wall of the injuring agent or the injured tissue" (*17*). This protective mechanism is essential to maintain a physiological balance. During inflammation, inflammatory cells respond to injury by migrating to the site and releasing inflammatory mediators like the cytokines such as interleukin-1 (IL-1) and interleukin-6 (IL-6) which play a vital role in the inflammatory process (*17-19*). In addition, several metabolic lipid inflammatory mediators from membrane bound arachidonic acid, collectively known as eicosanoids, are also produced during inflammation (*20,21*).

Table II. Effect of combination of theaflavin-3,3'-digallate (TF-3) with sulindac (Sul) on TPA-induced inflammation, formation of PGE_2 and LTB_4 in ears of CD-1 mice

Treatment	Weight of ear punch (mg)	PGE_2 Concentration (pg/mg tissue)	LTB_4 Concentration (pg/mg tissue)
Acetone	7.85 ± 0.21	3223.42 ± 550.94	2.11 ± 0.30
TPA (0.8 nmol)	19.63± 0.60	5384.66 ± 366.37	72.47 ± 7.80
TF-3 (0.125 µmol) + TPA (0.8 nmol)	17.23 ± 0.72	4642.38 ± 56.92	49.09 ± 20.06
TF-3 (0.25 µmol) + TPA (0.8 nmol)	19.49 ± 1.00	4140.13 ± 424.85	12.84 ± 0.78
TF-3 (0.5 µmol) + TPA (0.8 nmol)	16.92 ± 0.55	3212.85 ± 445.59	9.96 ± 1.94
Sul (0.3 µmol) + TPA (0.8 nmol)	17.52 ± 0.76	874.94 ± 16.40	20.38 ± 7.13
TF-3 (0.125 µmol) + Sul (0.3 µmol) + TPA (0.8 nmol)	18.39 ± 1.90	1064.96 ± 104.50	15.23 ± 0.91
TF-3 (0.25 µmol) + Sul (0.3 µmol) + TPA (0.8 nmol)	15.44± 1.07	1103.57 ± 4.61	11.68 ± 1.68

Female CD-1 mice (30 days old, 5 mice per group) were treated topically with 20 µL acetone or test compound in acetone 20 minutes before application of 20 µL acetone or TPA (0.8 nmol) in 20 µL acetone once a day for 4 days (total of 4 treatments). The mice were sacrificed by cervical dislocation 6 hours after last dose of TPA. Ear punches (6 mm in diameter) were weighed separately. The ear punches from each group were then combined and homogenized. The tissue homogenate was tested for the levels of inflammatory mediators (PGE_2 and LTB_4) using ELISA. Data are expressed as the mean ± standard error from 3 separate determinations.

Table III. Effect of combination of EGCCa, EGCGCa and Sulindac (Sul) on TPA-induced inflammation, formation of interleukin-6 (IL-6) and leukotriene B_4 (LTB_4) in ears of CD-1 mice

Treatment	Weight of Ear Punch (mg)	IL-6 Concentration (pg/mg tissue)	LTB_4 Concentration (pg/mg tissue)
Acetone	7.74 ± 0.24	0.32 ± 0.05	1.53 ± 0.07
TPA (0.4 nmol)	14.69 + 0.53	4.15 + 0.07	14.00 + 0.51
Sulindac (0.5 μmol) + TPA (0.4 nmol)	13.16 ± 0.80	4.80 ± 0.19	10.12 ± 1.20
EGCCa (0.5 μmol) + TPA (0.4 nmol)	11.04 ± 0.54	3.48 ± 0.08	3.90 ± 0.46
EGCCa (0.5 μmol) + Sul (0.3 μmol) + TPA (0.4 nmol)	9.49 ± 0.28	1.18 ± 0.10	2.55 ± 0.37
EGCGCa (0.5 μmol) + TPA (0.4 nmol)	10.44 ± 1.08	3.15 ± 0.12	3.95 ± 0.19
EGCGCa (0.5 μmol) + Sul (0.3 μmol) + TPA (0.4 nmol)	10.45 ± 0.60	1.12 ± 0.11	2.49 ± 0.41

Female CD-1 mice (30 days old, 5 mice per group) were treated topically with 20 μL acetone or test compound in acetone 20 minutes before application of 20 μL acetone or TPA (0.4 nmol) in 20 μL acetone twice a day for 3.5 days (a total of 7 treatments). The mice were sacrificed by cervical dislocation 6 hours after last dose of TPA. Ear punches (6 mm in diameter) were weighed separately. The ear punches from each group were then combined and homogenized. The tissue homogenate was tested for the levels of inflammatory mediators (LTB_4) using ELISA. Data are expressed as the mean ± standard error from 3 separate determinations.

Arachidonic acid is released from membrane stores during inflammation by phospholipase A_2 and can be metabolized through the cyclooxygenase (COX) or the lipooxygenase (LOX) pathway. The COX enzyme produces prostagladins (PGs) while LOX produces leukotrienes. The cyclooxygenase enzyme exists in at least 2 isoforms – COX-1 and COX-2. COX-1 is the constitutive form and maintains normal epithelial lining of cells while COX-2 is induced during inflammation (*17,22,23*). Each of the black tea derivatives, namely theaflavin mixture, theaflavin digallate, catechol adducts of epigallocatechin and epigallocatechin gallate were found to inhibit inflammatory biomarkers differently. Greater reduction of LTB_4 than PGE_2 by theaflavin digallate suggests that it inhibits LOX to a greater extent than COX.

Synthetic compounds, EGCCa and EGCGCa, are structurally similar to theaflavins due to the presence of benzotropolone ring formed during oxidation reaction. Due to their smaller size they are more polar and may be better absorbed. In addition, the anti-inflammatory potential of our test compounds was also compared with sulindac, a standard nonsteroidal anti-inflammatroy drug (NSAID). In the present study TPA, the active component of croton oil, was used to induce inflammation. Inflammation was quantified by measuring the weight of ear punch as well as important biomarkers of inflammation such as PGE_2, LTB_4, IL-1β and IL-6. During inflammation, cell migration to the site of injury produced localized edema, measured in terms of ear punch weight.

Theaflavin mixture consisting of theaflavin, theaflavin monogallates and theaflavin digallate inhibited all markers of inflammation. The effect was not dose dependent in the case of cytokines, but was dose dependent for eicosanoids (LTB_4). This could be because cytokines and eicosanoids are formed by different pathways. Theaflavin digallate also inhibits all markers of inflammation but to a lesser extent than theaflavin mixture. When administered along with sulindac there was a better inhibition of inflammation than when used alone.

The synthetic compounds, EGCCa and EGCGCa inhibited inflammation. Both these compounds exhibited a synergistic inhibition of IL-6 concentration with sulindac. This may be because IL-6 production is induced by several factors during inflammation. Our test compounds and sulindac could possibly inhibited several of the pathways leading to this strong synergistic inhibition. EGCCa and EGCGCa also inhibit the formation of eicosanoids (LTB_4).

Sulindac, a COX inhibitor, inhibited PGE_2 levels more effectively than the test compounds in this model. When used with the test compounds the inhibition was similar to sulindac, probably due to the effect of sulindac. The commonly used NSAIDs such as sulindac block both isozymes of COX (*23*) and the inhibition of COX-1 is the cause for the side effects of NSAIDs such as stomach ulcers (*24*). Black tea compounds, on the other hand, have been shown to be specific COX-2 inhibitors (*25*), hence show a lower inhibition than sulindac on

PGE$_2$ concentration. Theaflavin related compounds by acting only on COX-2 may not have the side effects of NSAIDs. In addition, while commonly used anti-inflammatory agents inhibit only levels of prostaglandins and COX metabolites, the compounds tested here inhibit other mediators of inflammation as well. Hence these compounds could serve as potential anti-inflammatory agents.

This study showed that the test compounds act through different mechanisms and exhibit a varying of effects on the inflammatory process. More studies are needed in order to evaluate the exact mechanisms of action.

References

1. Yang, C.S.; Yang, G.Y.; Landau, J.M.; Kim, S.; Liao, J. *Exp. Lung Res.* **1998**, *24*, 629-639.
2. Balentine, D.A. In *Phenolic Compounds in Foods and Their Effects on Health I.*; C.-T. Ho, C.Y. Lee and M.T. Huang, Eds.; Americal Chemical Society: Washington, D.C., 1992; pp. 102-117.
3. Robertson, A. In *Tea: Cultivation to Consumption*; K.C. Willson and M.N. Clifford, Eds.; Chapman and Hall: London, 1992; pp 555-601.
4. Mukhtar, H.; Ahmad, N. *Am. J. Clin. Nutr.* **2000**, *71*, 1698S-1702S; discussion 1703S-1694S.
5. Bokuchava, M.A.; Skobeleva, N.I. *Crit. Rev. Food Sci. Nutr.* **1980**, *12*, 303-370.
6. Mahanta, P.K.; Baruah, H.K. *J. Agric. Food Chem.* **1992**, *40*, 860-863.
7. Chen, Y.C.; Liang, Y.C.; Lin-Shiau, S.Y.; Ho, C.-T.; Lin, J. K. *J. Agric. Food Chem.* **1999**, *47*, 1416-1421.
8. Sang, S.; Tian, S.; Meng, X.; Stark, R.E.; Rosen, R.T.; Yang, C.S.; Ho, C.-T. *Tetra. Lett.* **2002**, *43*, 7129-7133.
9. Lambert, J.D.; Yang, C.S. *Mutation Res.* **2003**, *523-524*, 201-208.
10. Yang, C. S.; Maliakal, P.; Meng, X. Inhibition of carcinogenesis by tea. *Ann. Rev. Pharmacol. Toxicol.* **2002**, *42*, 25-54.
11. Miller, N.J.; Castelluccio, C.; Tijburg, L.; Evans, C.R. *FEBS Lett.* **1996**, *392*, 40-44.
12. Sang, S.; Lambert, J.D.; Tian, S.; Hong, J.; Hou, Z.; Rya, J.H.; Stark, R.E.; Rosen, R.T.; Huang, M.T.; Yang, C.S.; Ho, C.-T. *Bioorg. Med. Chem.* **2004**, *12*, 459-467.
13. Conney, A.H.; Lu, Y.; Lou, Y.; Xie, J.; Huang, M. *Proc. Soc. Exp. Biol. Med.* **1999**, *220*, 229-233.
14. Steele, V.E.; Kelloff, G.J.; Balentine, D.; Boone, C.W.; Mehta, R.; Bagheri, D.; Sigman, C.C.; Zhu, S.; Sharma, S. *Carcinogenesis* **2000**, *21*, 63-67.

15. Hong, J.; Smith, T.J.; Ho, C.-T.; August, D.A.; Yang, C.S. *Biochem. Pharmacol.* **2001**, *62*, 1175-1183.
16. Feng, Q.; Torii, Y.; Uchida, K.; Nakamura, Y.; Hara, Y.; Osawa, T. *J. Agric. Food Chem.* **2002**, *50*, 213-220.
17. Gallin, J.; Goldstein, I.; Snyderman, R. *Inflammation: Basic Principles and Clinical Correlates*, Raven Press: New York, NY, 1992.
18. Houck, J.C. *Chemical Messengers of the Inflammatory Process*; Elsevier/North-Holland Biomedical Press.: Washington, D.C., 1979.
19. Higgs, G.A.; Williams, T.J. *Inflammatory Mediators*; VCH Publishers: Deerfield Beach, 1985.
20. Marks, F.; Fürstenberger, G. *Prostaglandins, Leukotrienes, and other Eicosanoids: From Biogenesis to Clinical Applications*; Wiley-VCH: New York, 1999.
21. Smith, W.L.; Dewitt, D.L. *Adv. Immunol.* **1996**, *62*, 167-215.
22. Smith, C.J.; Zhang, Y.; Koboldt, C.M.; Muhammad, J.; Zweifel, B.S.; Shaffer, A.; Talley, J.J.; Masferrer, J.L.; Seibert, K.; Isakson, P.C. *Proceed. Nat. Acad. Sci. USA* **1998**, *95*, 13313-13318.
23. Smith, W.L.; DeWitt, D.L.; Garavito, R M. *Ann. Rev. Biochem.* **2000**, *69*, 145-182.
24. Warner, T.D.; Giuliano, F.; Vajnovic, I.; Bukasa, A.; Mitchell, J. A.; Vane, J.R. *Proceed. Nat. Acad. Sci. USA* **1999**, *96*, 7563-7568.
25. Lu, J.; Ho, C.-T.; Ghai, G.; Chen, K.Y. *Cancer Res.* **2000**, *60*, 6465-6471.

Chapter 21

Antioxidant and Antitumor Promoting Activities of Apple Phenolics

Ki Won Lee[1], Hyong Joo Lee[1], and Chang Yong Lee[2]

[1]Department of Food Science and Technology, School of Agricultural Biotechnology, Seoul National University, Seoul 151–742, South Korea
[2]Department of Food Science and Technology, Cornell University, Geneva, NY 14456

The antioxidant and antitumor promoting activities of major phenolic phytochemicals of apples were investigated. The contribution of each antioxidant to total antioxidant activity of apples was determined using 2,2′-azino-bis(3-ethylbenzothiazoline-6-sulfonic acid) radical scavenging assay. The estimated contribution of major phenolics and vitamin C to total anitoxidant capacity of fresh apples is as follows: quercetin > epicatechin > procyanidin B_2 > phloretin > chlorogenic acid > vitamin C. Recent reports suggest that carcinogenicity of hydrogen peroxide (H_2O_2) is attributable to the inhibition of gap-junctional intercellular communication (GJIC), which is related to tumor promotion process. The inhibition of GJIC by H_2O_2 was recovered effectively by the treatment of apple extracts. Among major antioxidants in apples, quercetin exerted the strongest protective effects on H_2O_2-induced inhibition of GJIC, following epicatechin, procyanidin B2, and vitamin C, while chlorogenic acid and phloretin did not show any effects. The results indicate that flavonoids such as quercetin, epicatechin, and procyanidin B_2 contribute significantly to total antioxidant capacity and antitumor promoting activities of apples, which may contribute to a putative cancer chemopreventive activity.

INTRODUCTION

Reactive oxygen species (ROS) such as hydrogen peroxide (H_2O_2) cause cancer through multiple mechanisms. Antioxidants can protect against oxidative DNA damage, which is implicated in tumor initiation. Although the carcinogenic effect of ROS has primarily been focused on their genotoxicity, ROS have also been known to play a significant role in the promotional stage of carcinogenesis. The involvement of ROS, particularly H_2O_2, in tumor promotion process is supported by both *in vivo* and *in vitro* studies (*1, 2*). Several oxidants and free radical generators are tumor promoters (*2, 3*). A recent theory (*1, 2, 4*) on "epigenetics" also indicates that greater attention must be paid to those processes which do not involve DNA damage in multistage carcinogenesis. The promotional phase of carcinogenesis is a consequence of epigenetic events involving inhibition of gap junction intercellular communication (GJIC), which could be mediated by ROS.

GJIC is essential for maintaining the homeostatic balance through modulating cell proliferation and differentiation in multicellular organisms (*5*). Inhibition of GJIC is strongly related to the carcinogenic process, particularly in tumour promotion. The mechanism by which ROS cause tumor promotion may be associated with the inhibition of GJIC. Recent reports suggest that the mechanism of carcinogenic process of H_2O_2 is involved in the inhibition of GJIC by hyperphosphorylation of connexin43 proteins (Cx43) which modulate GJIC (*5, 6*).

Naturally occurring antioxidants have been reported to play a major role in ameliorating oxidative damage caused by free radicals. Recently, antioxidative vitamins and phenolic substances derived from the daily diet have received considerable attention because of their potential chemopreventive activities. Vitamin C has been considered as one of the most prevalent antioxidative components of fruits and vegetables, which exerts substantial chemopreventive effects without apparent toxicity at a relatively high dosage (*7*). However, the contribution of vitamin C to the total antioxidant activity of fruits has been determined to be generally less than 15% (*8*). It has been suggested that phenolic phytochemicals contribute significantly to the total antioxidant capacity of fruits, vegetables, grains, and tea, and considerable attention has recently been paid to the possible health benefits of dietary phenolics that exhibit stronger antioxidant activity than vitamin C.

Previous studies have shown that the antioxidative and antiproliferative activities of apples are from combined activity of phenolics rather than vitamin C (*9, 10*). In particular, phenolics in apple skin exerted a much higher degree of contribution to the total antioxidant and antiproliferative activities of whole apple than that in apple flesh (*9*). Apples contain various antioxidative phenolics such as chlorogenic acid, epicatechin, procyanidin B2, phloretin, and quercetin as well as vitamin C (*11, 12*). However, the relative contribution of each antioxidant to cancer chemopreventive activity of apples has not been clearly demonstrated.

The contents of total pehnolics or flavonoids in fruits often do not directly relate to total antioxidant and anticarcinogenic capacities. Furthermore, the effects of antioxidants on carcinogenesis may be different depending on their structure and dosage although antioxidants have anti-carcinogenic activity against oxidative DNA damage. Therefore, accurate measurement of antioxidant and anti-carcinogenic capacities of each bioactive compound is warranted. In this study, we identified and quantified major phenolics in various apple cultivars, and investigated their contributions to cancer chemopreventive activity.

MATERALS AND METHODS

Chemicals. Ammonium hydroxide, 2,2'-azino-bis (3-ethylbenzothiazoline-6-sulfonic acid) (ABTS) as diammonium salt, Folin and Ciocalteu's phenol reagent, ammonium phosphate monobasic ($NH_4H_2PO_4$), quercetin, epicatechin, phloretin, and chlorogenic acid were obtained from Sigma Chemical (St. Louis, MO). 2,2'-Azobis (2-amidino-propane) dihydrochloride (AAPH) was obtained from Wako Chemicals (Richmond, VA). Procyanidin B2 was obtained from Shimazu (Kyoto, Japan). Quercetin glycosides (arabinoside, xyloside, gulcoside, galactoside, and rhamnoside) and phloretin glycosides (glucoside and xyloglucoside) were obtained from Extrasynthese (Genay, France). Vitamin C was purchased from Fisher Scientific (Pittsburgh, PA). All other chemicals used were of analytical or HPLC grade.

Apple Cultivars. Six apple cultivars, Golden Delicious, Cortland, Monroe, R. I. Greening, Empire, and NY674, were picked at commercial maturity during the 2000 harvest season at the New York State Agricultural Experiment Station orchard in Geneva, New York. Apples were stored in a 2–5°C cold room. They were carefully cut into slices, the pits were removed and the freeze-dried samples were ground to powder and then stored at -20°C until analyzed.

Extraction of Phenolics. Phenolics were extracted according to the method described previously (*11*). Briefly, phenolics were extracted from 10 g ground freeze-dried sample using 100 mL of 80% aqueous methanol. The mixture was sonicated for 20 min with a stream of nitrogen to prevent possible degradation of phenolics, filtered through Whatman No. 2 filter paper (Whatman International Limited, Kent, England) using a chilled Büchner funnel and rinsed with 50 mL of 100% methanol. Extraction of the residue was repeated under the same conditions. The filtrates were combined and transferred into a 1 L evaporating flask with an additional 50 mL of 80% aqueous methanol. The solvent was evaporated using a rotary evaporator at 40°C. The remaining phenolic concentrate was first dissolved in 50 mL of 100% methanol and diluted to a final volume of 100 mL using distilled deionized water (ddH$_2$O). The samples were then centrifuged in a Sorvall® RC-5B Refrigerated Superspeed Centrifuge (Du Pont, Biomedical Products Department, Wilmington, DE) at 12,000 g using a GSA rotor for 3 min, and the resulting supernatants were used as the final samples.

Identification of Phenolics. HPLC analysis was performed according to the method described previously (11). Extracted sample was filtered through a 0.45-μm poly (tetrafluoroethylene) syringe-tip filter. Using a 20 μL sample loop, the sample was analyzed using an HPLC system (Hewlett Packard Model 1100, Palo Alto, CA) equipped with photodiode array detector, quaternary pump, and vacuum degasser. A C18 reversed-phase Symmetry Analytical column (5-μm × 250-mm × 4.6-mm) was used with a Symmetry Sentry guard column of the same packing material as the analytical column (Waters Corporation, MA). Three mobile phases were used: solvent A, 50 mM ammonium phosphate monobasic (NH$_4$H$_2$PO$_4$), pH 2.6 (pH adjusted with phosphoric acid); solvent B, 80:20 (v/v) acetonitrile/50 mM NH$_4$H$_2$PO$_4$, pH 2.6; and solvent C, 200 mM phosphoric acid (H$_3$PO$_4$), pH 1.5 (pH adjusted with ammonium hydroxide). The gradient for HPLC analysis was linearly changed as follows (total 60 min): 100% A at zero min, 92% A/8% B at 4 min, 14% B/86% C at 10 min, 16.5% B/83.5% C at 22.5 min, 25% B/75% C at 27.5 min, 80% B/20% C at 50 min, 100% A at 55 min, 100% A at 60 min. Flow rate was 1.0 mL/min at constant room temperature (23°C). Phenolic standards were used to generate characteristic UV-Vis spectra and calibration curves. Individual phenolics in the sample were tentatively identified by the comparison of UV-Vis spectra and retention times and with spiked input of polyphenolic standard. Three replicated HPLC analyses were performed for each apple cultivar.

Quantification of Vitamin C. Ascorbic acid was determined by using the 2,6-dichloroindophenol titrimetic method, according to AOAC Method 969.21 (10). Reference material was an ascorbic acid solution (1 mg/mL) prepared from L-ascorbic acid.

ABTS Radical Scavenging Activity. ABTS method described earlier was used with slight modifications (10). Briefly, 1.0 mM of AAPH was mixed with 2.5 mM of ABTS in phosphate buffered saline (PBS) solution (100 mM potassium phosphate buffer containing 150 mM NaCl). The mixture was heated in a 68°C water bath. The resulting blue-green ABTS radical solution was adjusted to an absorbance of 0.30 ± 0.02 at 734 nm. Various doses of antioxidants (each 10 μL) were added to 190 μL of the resulting blue-green ABTS radical solution in a 96 well plate. The control consisted of 10 μL of 99% ethanol and 190 μL of ABTS radical solution. Decrease in absorbance, which resulted from the addition of test compounds, was measured at 734 nm using a micro-plate reader (Emax, Molecular Devices, CA). ABTS radical scavenging activities of test compounds were expressed as % remaining ABTS radicals at each time point. The radical stock solution was prepared fresh each day.

Quantification of Total Antioxidant Capacity. A method developed by Winston et al. (*13*) was empoyed with slight modifications for the quantification of antioxidant value of each compound tested. The area under the kinetic curve was calculated by integration. The total antioxidant capacity (TAC) of each tested compound was then quantified according to following equation.

$$TAC = 100 - (\int SA / \int CA \times 100)$$

Percent increase in integrated area was measured to compare each phenolic and vitamin C. Here, $\int SA$ and $\int CA$ are integrated areas from curves defining the sample and control reactions, respectively. The median effective dose (EC_{50}) of all samples tested was calculated from the dose-response curve. TAC of each phenolic was expressed as vitamin C equivalents antioxidant capacity (VCEAC). All tested samples were replicated six times and presented as mean value ± standard deviation.

Cell Culture. WB-F344 rat liver epithelial cells (WB cells) were kindly provided by Dr. J. E. Trosko at Michigan State University. The cells (passage 8–16) were cultured in D-media supplemented with 10% fetal bovine serum (FBS, GIFCO BRL), 2 mM and penicillin/streptomycin in a 37°C humidified incubator containing 5% CO_2 and 95% air.

Cytotoxicity. The cytotoxicity was measured by the MTT assay as described previously (*14*). Briefly, cells were cultured in 96-well plates at 1000 cells/well in media for 24 h. Each well was filled with fresh D-medium containing various amounts of samples. The cells were then incubated for a further 24 h at 37°C. Each well was then incubated with MTT for 4 h. The liquid was removed and dimethyl sulfoxide was added to dissolve the solid residue. The optical density at 570 nm of each well was then determined by using a micro-plate reader (Emax, Molecular Devices, CA). All data are from at least three replications for each prepared sample.

Bioassay of GJIC. The GJIC was measured by the scrape loading/dye transfer (SL/DT) technique (*15*). Briefly, cells were co-treated with 500 μM H_2O_2 and various concentrations of samples for 1 h. The GJIC assay was conducted at non-cytotoxic dose levels of the samples, as determined by MTT assay. Following incubation, the cells were washed twice with 2 ml of PBS. Lucifer yellow was added to the washed cells and three scrapes were made with a surgical-steel-bladed scalpel at low light intensities. These three scrapes were performed to ensure that the scrape traversed a large group of confluent cells. After 3 min of incubation, the cells were washed with 10 ml of PBS and then fixed with 2 ml of a 4% formalin solution. The number of communicating cells visualized with the dye was counted under an inverted fluorescent microscope

(Olympus Ix70, Okaya, Japan). All data are from at least three replications for each prepared sample.

Statistical Analysis. Data in all figures are presented as mean ± SD. The statistical significance was determined by two-tailed Studant's t-test with the level of significance set at $p < 0.05$.

RESULTS

Composition of Major Antioxidants in Apple

Composition and concentration of the major phenolics of six apple cultivars studied are shown in Table 1. Among the apple cultivars studied, RI Greening showed the highest content of all phenolic phytochemicals analyzed. Average concentrations of the major phenolics were: quercetin glycosides, 13.20 mg; procyanidin B2, 9.35 mg; chlorogenic acid, 9.02 mg; epicatechin, 8.65 mg; and phloretin glycosides, 5.59 mg per 100 g fresh apples. Chlorogenic acid and phloretin glycosides were present in lower amounts compared to quercetin glycosides and procyanidin B2. Several phenolics in apples contained glycosides. In particular, a wide variety of quercetin glycosides were present in apple cultivars. Galactoside was the most abundant form among the glycosides identified in most tested cultivars except NY674, in which rhamnoside was most abundant. In addition, xyloglucoside was an abundant form of phloretin glycosides.

There is increasing evidence that flavonoids can be absorbed into the human body in amounts that are, in principle, sufficient to exert antioxidant or other biological activities *in vivo* (*16-18*). Chlorogenic acid undergoes no structural changes in the small intestine (*18*) and both epicatechin and procyanidin B2 are absorbed as epicatechin (*19*). In general, flavonoids and isoflavones were found to have lower biological activities in gastrointestinal hydrolase. Intestinal conjugation seemed to be an important process for the absorption, because only conjugated forms were detected in the mesenteric vein blood (*20*). Furthermore, when quercetin-glycosides and genistin were fed to rats or humans, quercetin and genistein, only the aglycone form was detected in the urine, respectively (*21, 22*). Therefore, we measured herein the antioxidant and anti-tumor promoting activity of quercetin and phloretin instead of quercetin and phloretin glycosides.

Contribution of Major Antioxidants to Total Antioxidant Capacity of Apple

Vitamin C and phenolics exerted ABTS radical scavenging activity in a dose and time-dependent manner (data not shown). TAC shows the ratio of

Table 1. Composition of major antioxidants of six apple cultivars

Fresh apples with skins (mg/100g)

Antioxidants	Golden Delicious	Cortland	Monroe	R.I. Greening	Empire	NY674	Average
Vitamin C	8.30	6.08	4.50	7.11	6.61	5.81	6.40
Chlorogenic acid	8.48	5.36	10.08	14.28	11.52	4.40	9.02
Epicatechin	7.12	8.32	10.72	19.16	2.28	4.32	8.65
Phloretin glycosides							5.59
glucoside	1.80	1.44	2.40	2.08	2.80	1.84	
xyloglucoside	1.92	3.20	4.92	5.88	1.72	3.56	
Procyanidin B2	6.28	11.32	8.32	21.68	3.44	5.04	9.35
Quercetin glycosides							13.20
Arabinoside	2.16	2.40	4.44	2.88	2.76	1.56	
Xyloside	1.68	1.08	2.28	1.92	2.16	1.20	
Glucoside	2.40	1.56	2.40	1.20	2.40	0.36	
Galactoside	4.20	3.36	4.80	4.32	4.20	1.92	
Rhamnoside	3.84	2.28	3.12	4.08	3.84	2.40	
Total	48.18	46.40	57.98	84.59	43.73	32.41	52.21

integrated value of the area under each data point to the control, representing time- and dose-dependency of each compound. Strong correlations ($r^2 > 0.97$) were observed between the concentrations and the TAC of vitamin C and phenolics in apple. The relative TAC of antioxidants evaluated by ABTS assay compared to vitamin C was as follows: querecetin (3.06) > epicatechin (2.67) > procyanidin B2 (2.36) > phloretin (1.63) > vitamin C (1.00) > chlorogenic acid (0.97) (Table 2). The data show that quercetin has the lowest EC_{50} value among the major phenolics in apple.

Although most phenolics are reported to have antioxidant activity, quercetin has been reported to have structural advantages as an antioxidant because the ortho-dihydroxy moiety in the B ring confers stability to the resulting free radical (23). Since quercetin is mainly present in apple peel (11), it was suggested that consumption of apples with skins is highly desirable in order to maximize apple antioxidant activity (9). In parallel, quercetin showed the highest antioxidant capacity in ABTS radical scavenging assay. Considering the amount of each compound, the estimated contribution of quercetin (36.8%) to the total antioxidant capacity of apples is the highest among major phytochemicals, followed by epicatechin (21.0%) and procyanidin B2 (20.1), whereas chlorogenic acids (8.0) and phloretin glycosides (8.3) provide minimal contribution (Table 2). Moreover, vitamin C contributes only (5.8%) to the total antioxidant capacity of apple (Table 2). These results indicate that flavonoids such as quercetin, epicatechin, and procyanidin B2 rather than vitamin C contribute significantly to total antioxidant activity of apple.

Maximum Non-cytotoxic Concentration of Apple Extracts and Their Major Antioxidants

A dose of 500 µM H_2O_2 maximally inhibited GJIC in WB cells, and is non-cytotoxic as shown in the previous study (24); we therefore also used a dose of 500 µM H_2O_2 in all experiments in the present study. To select the appropriate doses of samples for this study, the maximum non-cytotoxic concentrations of samples were measured using the MTT assay. The maximum non-cytotoxic concentration of apple extracts on WB cells was 50 mg/ml fresh apple equivalent extracts (Table 3) and that of quercetin, epicatechin, procyanidin B2, phloretin, chlorogenic acid, and vitamin C was 40, 100, 100, 100, 200, and 500 µg/ml on WB cells, respectively (Table 3). We performed the experiments described below at non-cytotoxic doses for these samples.

Effects of Apple Extracts and Their Major Antioxidants on Inhibition of GJIC by H_2O_2

The effects of samples on GJIC were investigated using the scrape loading /dye transfer technique. As the positive control, 500 µM of H_2O_2 markedly

Table 2. Contributions of major antioxidants to total antioxidant activity of apple

Antioxidants	Concentration (mg/100g fresh)	EC_{50}	Relative VCEAC values	Total antioxidant activity (mg VCEAC /100g fresh)	Relative Contribution (%)	Relative Contribution (%)
Quercetin glycosides	13.20	0.56	3.06	40.39	36.8	36.8
Epicatechin	8.65	0.64	2.67	23.10	21.0	21.0
Procyanidin B2	9.35	0.72	2.36	22.07	20.1	20.1
Vitamin C	6.40	1.71	1.00	6.40	5.8	5.8
Phloretin glycosides	5.59	1.05	1.63	9.11	8.3	8.3
Chlorogenic acid	9.02	1.76	0.97	8.75	8.0	8.0
Total	52.21			109.82	100.0	100.0

[a]Relative vitamin C equivalent antioxidant capacity (VCEAC) values = VCEAC of each antioxidant/antioxidant capacity of vitamin C

Table 3. Maximum non-cytotoxicity concentration of Golden Delicious apple extracts and their major antioxidants on WB-F344 rat liver epithelial cells (WB cells), as determined by the MTT assay.

	Maximum non-cytotoxic concentration
Apple ext.	50 mg/ml
Quercetin	80 µg/ml
Epicatechin	200 µg/ml
Procyadnidin B2	200 µg/ml
Phloretin	200 µg/ml
Chlorogenic acid	400 µg/ml
Vitamin C	1000 µg/ml

inhibited GJIC (Fig. 1b) compared to the untreated negative control (Fig. 1a). Apple extracts (Golden Delicious), at a concentration of 20 mg/ml fresh apple equivalent extracts, completely prevented the inhibition of GJIC by H_2O_2 (Fig. 1d). Apple extracts effectively protected against the inhibition of GJIC induced by H_2O_2 in a dose-dependent manner (Fig. 1B).

Among major antioxidants in apple, quercetin exerted the strongest effects on the inhibition of GJIC by H_2O_2 (Fig. 2). Epicatechin, procyanidin B2, and vitamin C also recovered H_2O_2-induced inhibition of GJIC (Fig. 2). However, chlorogenic acid and phloretin did not exert any effects on the inhibition of GJIC by H_2O_2 (Fig. 2). These results indicate that the effects of phenolics on the inhibition of GJIC by H_2O_2 depend on their structures. Our results suggest that apple has antitumor promoting activity through modulation of GJIC by the combined activity of several antioxidants.

DISCUSSION

Sun et al. (25) suggested that phytochemicals in fruits including apple show a high correlation with antioxidant capacity ($r^2=0.97$). On the other hand, Imeh et al. (26) observed a weak correlation ($r^2=0.58$) between phenolic content of the fruits and the total antioxidant activity. This was probably due to the unidentified phenolics and synergism among these compounds and major phenolics that may have had some effect on the activity. Apples, like other fruits, vary in chemical composition, even within the same variety, depending on maturity, location produced, and agricultural practices, as well as numerous other environmental factors. Indeed, significant variations in phenolic content and antioxidant activity were observed among cultivars and even among different fruits in the same cultivar (27). In this study, various apple cultivars showed different levels of phenolic content, all phenolics showing different antioxidant activities. Some phenolics such as epicatechin, quercetin, procyanidin B2, chlrogenic acid, and phloretin, have been identified as major antioxidants in apple. The evidences shown herein in terms of the content and the capacity of antioxidants suggest that quercetin may have the highest contribution as an antioxidant in apple. Some flavonoids such as epicatechin and procyanidins also are major antioxidants of apple.

Gap junction channels play an important role in intercellular communication by providing a direct pathway for the movement of molecular information, including ions and polarized and non-polarized molecules up to a molecular mass of 1 kDa between adjacent cells (28). Cell-to-cell communication through gap junction channels is a critical factor in the life and death balance of cells because GJIC has an important function in maintaining tissue homeostasis through the regulation of cell growth, differentiation, apoptosis and adaptive functions of differentiated cells (28). The transfection of

GJIC-deficient cells with connexins suppresses tumor formation, whereas a number of pesticides, pharmaceuticals, dietary additives, polyhalogenated hydrocarbons, and peroxisome proliferators inhibit GJIC through diverse mechanisms (*1, 29*). Furthermore, many tumor promoters including H_2O_2, peroxynitrite, TPA, and pentachlophorbol (PCP) can inhibit GJIC, which is a reversible process (*5, 29*). Thus, substances working against the inhibition of GJIC are anticipated to prevent cancer, because the inhibition of cell-to-cell communication is strongly related to the carcinogenic process, particularly to tumor promotion.

In the human body, ROS such as H_2O_2 cause cancer through diverse cellular processes. The carcinogenic effect of ROS has been primarily due to its genotoxicity, and antioxidants can protect against oxidative DNA damage, which is implicated in tumor initiation. However, ROS have also been known to play a significant role in the promotional stage of carcinogenesis. Several oxidants and free radical generators are tumor promoters. The involvement of ROS, particularly H_2O_2, in the tumor promotion process is supported by both *in vivo* and *in vitro* studies. The mechanism by which ROS cause tumor promotion may be associated with the inhibition of GJIC. Recent reports suggest that the mechanism of carcinogenic process of H_2O_2, reactive non-radical tumor promoter, is involved in the inhibition of GJIC by hyperphosphorylation of connexin43 proteins (Cx43) which mainly modulate GJIC. The apple extracts have protective effect against inhibition of GJIC by H_2O_2 in a dose-dependent manner. Furthermore, among antioxidants in apples, quercetin, epicatechin, procyanidin B2, and vitamin C recovered effectively the inhibition of GJIC by H_2O_2. However, chlorogenic acid and phloretin had no effect on GJIC. Antioxidants such as propylgallate and Trolox did not prevent the inhibition of GJIC by H_2O_2 (*15*). Thus, the effects of antioxidants on inhibition of GJIC by H_2O_2 depend on their structure and dosage.

Recently, considerable efforts have been made to develop chemopreventive agents that could inhibit, retard or reverse multistage carcinogenesis (*2*). Chemopreventive agents or food can act in the initiation, promotion, or progression of carcinogenesis. However, intervention during the promotion stage seems to be the most appropriate and practical strategy since tumor promotion is a reversible event which requires repeated and prolonged exposure to promoting agents (*2*). The promotional phase of carcinogenesis is closely linked to epigenetic events involving inhibition of GJIC, which could be mediated by ROS. Thus, the mechanistic basis for cancer preventive action of apple extracts may be related to the protective effects against inhibition of GJIC as well as free radical scavenging activity.

The nutritional value and health-promoting activity of fruits depend not only on their activity but also on the amount of such foods consumed daily. Apples are one of the major fruits frequentgly consumed by Americans. Among fresh fruits consumed in 1996, apples (8.75 kg/person/year) ranked second after

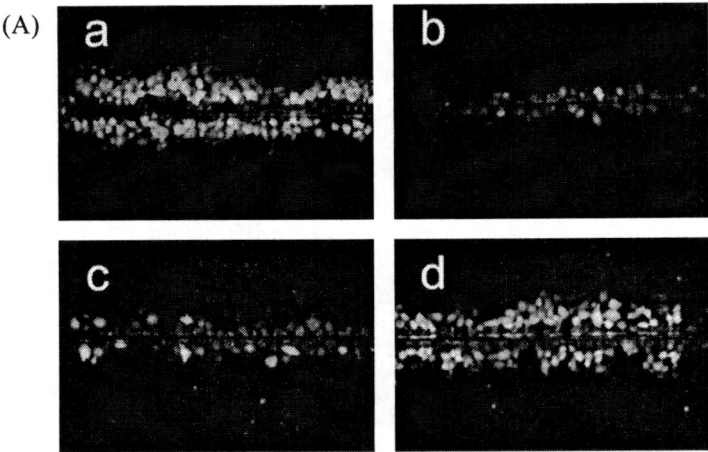

*Figure 1. Effects of Golden Delicious apple extracts on the inhibition of gap junction intercellular communication (GJIC) by H_2O_2 in WB-F344 rat liver epithelial cells (WB cells). The cells were exposed to 500 μM H_2O_2 for 1 h in the absence or presence of different concentrations of apple extracts. GJIC was assessed using the scrape loading/dye transfer method. The data shown are representative of at least three independent experiments. (A) a, control (ddH_2O as vehicle); b, H_2O_2; c, H_2O_2 plus 10 mg/mL fresh apple extract equivalent; d, H_2O_2 plus 20 mg/mL fresh apple extract equivalent. (B) The number of communicating cells visualized with the dye was counted under an inverted fluorescent microscope (Olympus Ix70, Okaya, Japan). Error bar = SD, $n \geq 3$. *Statistically different from control at $p < 0.05$.*

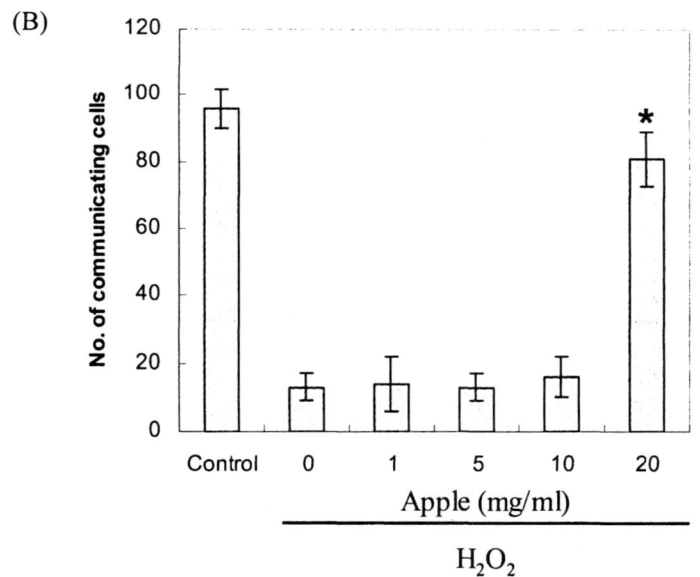

Figure 1. *Continued.*

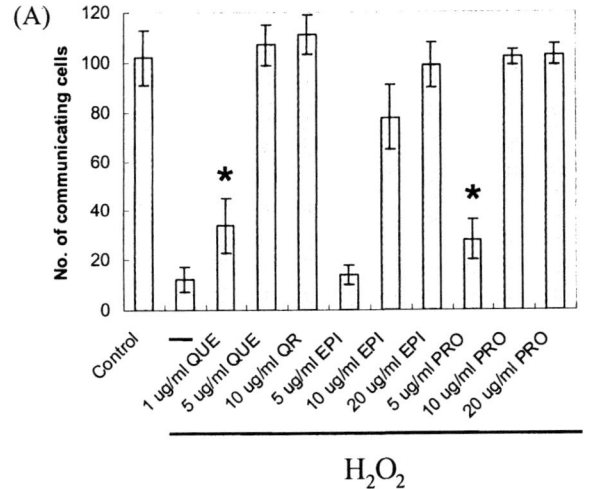

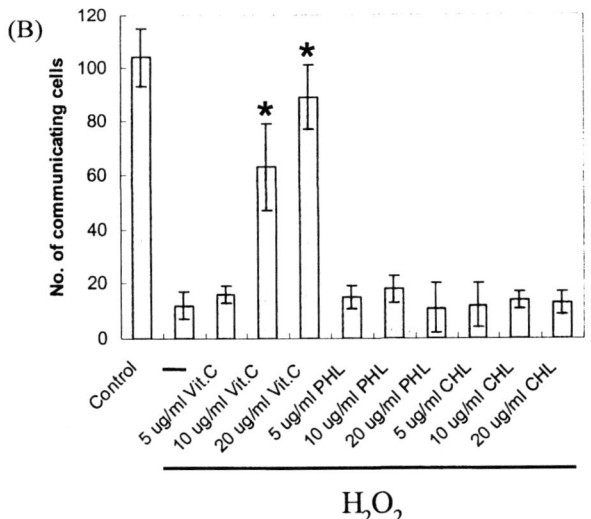

*Figure 2. Effects of quercetin, epicatechin, and procyanidin B2 (A) and vitamin C, phloretin, and chlorogenic acid on the inhibition of GJIC by H_2O_2 in WB cells. QUE, quercetin; EPI, epicatechin; PRO, procyanidin B2; Vit.C, vitamin C; PHE, phloretin; CHL, chlorogenic acid. The cells were exposed to 500 μM H_2O_2 for 1 h in the absence or presence of different concentrations of antioxidants. GJIC was assessed using the scrape loading/dye transfer method and the number of communicating cells visualized with the dye was counted under an inverted fluorescent microscope (Olympus Ix70, Okaya, Japan). Error bar = SD, n ≥ 6. *Statistically different from control at $p < 0.05$.*

bananas (12.7 kg). However, when fresh and processed products were combined, the estimated per capita consumption of apples (21.3 kg) exceeds that of bananas. We suggest that the contribution of each nutrient in daily diet should be carefully considered based on its activity and quantitative data. It is also necessary to study the interactions between active food components to determine the biological activities of each phenolic compound other than antioxidant activity. The results of this study indicate that apples as major antioxidant sources in American diet may provide cancer chemopreventive effects through their antioxidant and antitumor promoting activity. Thus, the consumption of apples including significant amounts of diverse antioxidants may give help to prevent cancer.

ACKNOWLEDGEMENTS

This work was in part supported by a grant from BioGreen 21 Program, Rural Development Administration, Republic of Korea and New York State Apple Research and Development Program. The authors thank Ms. Nancy Smith, a research support specialist, for technical assistance.

REFERENCES

(1) Trosko, J. E.; Chang, C. C. *BioFactors* 2000, *12*, 259-263.
(2) Surh, Y.-J. *Mutat. Res.* 1999, *428*, 305-327.
(3) Cerutti, P. A. *Science* 1985, *227*, 375-381.
(4) Wu, C. T.; Morris, J. R. *Science* 2001, *293*, 1103-1105.
(5) Trosko, J. E. *Mol. Carcinogenesis* 2001, *30*, 131-137.
(6) Huang, R. P.; Peng, A.; Golard, A.; Hossain, M. Z.; Huang, R.; Liu, Y. G.; Boynton, A. L. *Mol. Carcinogenesis* 2001, *30*, 209-217.
(7) Carr, A. C.; Frei, B. *Am. J. Clin. Nutr.* 1999, *69*, 1086-1107.
(8) Wang, H.; Cao, G.; Prior, R. L. *J. Agric. Food Chem.* 1996, *44*, 701-705.
(9) Eberhardt, M. V.; Lee, C. Y.; Liu, R. H. *Nature* 2000, *405*, 903-904.
(10) Kim, D. O.; Lee, K. W.; Lee, H. J.; Lee, C. Y. *J. Agric. Food Chem.* 2002, *50*, 3713-3717.
(11) Burda, S.; Oleszek, W.; Lee, C. Y. *J. Agric. Food Chem.* 1990, *38*, 945-948.
(12) Mayr, U.; Treutter, D.; Santos-Buelga, C.; Bauer, H.; Feucht, W. *Phytochemistry* 1995, *38*, 1151-1155.
(13) Winston, G. W.; Regoli, F.; Dugas, A. J. J.; Fong, J. H.; Blanchard, K. A. A *Free Radic. Biol. Med.* 1998, *24*, 480-493.
(14) Kim, J. Y.; Lee, K. W.; Kim, S. H.; Wee, J. J.; Kim, Y. S.; Lee, H. J. *Planta Medica* 2002, *68*, 119-122.
(15) Upharm, B. L.; Kang, K. S.; Cho, H. Y.; Trosko, J. E. *Carcinogenesis* 1997, *18*, 37-42.
(16) Murota, K.; Terao, J. *Arch. Biochem. Biophys.* 2003, *417*, 12-17.

(17) Heim, K. E.; Tagliaferro, A. R.; Bobilya, D. J. *J. Nutr. Biochem.* 2002, *13*, 572-584.
(18) Azuma, K.; Ippoushi, K.; Nakayama, M.; Ito, H.; Higashio, H.; Terao, J. *J. Agric. Food Chem.* 2000, *48*, 5496-5500.
(19) Spencer, J. P.; Schroeter, H.; Shenoy, B.; Srai, S. K.; Debnam, E. S.; Rice-Evans, C. *Biochem. Biophys. Res. Commun* 2001, *285*, 588-593.
(20) Day, A. J.; Gee, J. M.; DuPont, M. S.; Johnson, I. T.; Williamson, G. *Biochem. Pharmacol.* 2003, *65*, 1199-1206.
(21) Crespy, V.; Morand, C.; Besson, C.; Manach, C.; Demigne, C.; Remesy, C. *J. Nutr.* 2001, *131*, 2109-2114.
(22) Lu, L. J. W.; Grady, J. J.; Marshall, M. V.; Ramanijam, V. M. S.; Anderson, K. E. *Nutr. Cancer* 1995, *24*, 311-323.
(23) Salah, N.; Miller, N. J.; Paganga, G.; Tijburg, L.; Bolwell, G. P.; Rice-Evans, C. *Arch. Biochem. Biophys.* 1995, *322*, 339-346.
(24) Lee, K. W.; Lee, H. J.; Kang, K. S.; Lee, C. Y. *Lancet* 2002, *359*, 172.
(25) Sun, J.; Chu, Y. F.; Wu, X.; Liu, R. H. *J Agric. Food Chem.* 2002, *50*, 7449-7554.
(26) Imeh, U.; Khokhar, S. *J. Agric. Food Chem.* 2002, *50*, 6301-6306.
(27) Lattanzio, V.; Venere, D. D.; Linsalata, V.; Bertolini, P.; Ippolito, A.; Salerno, M. *J. Agric. Food Chem.* 2001, *49*, 5817-5821.
(28) Trosko, J. E.; Ruch, R. J. *Front. Biosci.* 1998, *3*, 208-236.
(29) Trosko, J. E.; Chang, C. C.; Upham, B.; Wilson, M. *Toxicol. Lett.* 1998, *102-103*, 71-78.

Chapter 22

Cranberry Phenolics: Effects on Oxidative Processes, Neuron Cell Death, and Tumor Cell Growth

Catherine C. Neto[1], Marva I. Sweeney-Nixon[2],
Toni L. Lamoureaux[1], Frankie Solomon[2], Miwako Kondo[1],
and Shawna L. MacKinnon[3]

[1]Department of Chemistry and Biochemistry, University of Massachusetts at Dartmouth, North Dartmouth, MA 02747
[2]Department of Biology, University of Price Edward Island, Charlottetown, Prince Edward Island, Canada
[3]Institute for Marine Biosciences, Halifax, Nova Scotia, Canada

The North American cranberry (*Vaccinium macrocarpon*) is rich in natural phenolic compounds which contribute to its biological activity and potential therapeutic effect. Flavonols and anthocyanins from cranberry are potent radical scavengers which effectively protect low-density lipoproteins from oxidation *in vitro*. Whole cranberry extract was recently observed to exert neuroprotective effects *in vitro* under conditions of stroke. The incidence of hydrogen-peroxide-induced necrosis in rat brain neurons treated with whole cranberry extract was 48% lower than control and simulated ischemia-induced necrosis declined by 42%. Similar protective effects were observed for cranberry in rat brain neuron apoptosis assays. The relative contributions of each class of antioxidants is under investigation. Selective inhibition of tumor cell proliferation in a variety of tumor cell lines has been observed for triterpene phenolic esters and proanthocyanidin extracts isolated from whole cranberry. The nature and occurrence of these antitumor principles in cranberries is under further studies.

© 2005 American Chemical Society

The cranberry *(Vaccinium macrocarpon* Ait. Ericaceae) is an important crop in Massachusetts, New Jersey, Wisconsin and parts of Canada. Closely related species include lowbush blueberry (*V. augustifolium* Ait.), highbush blueberry (*V. corymbosum* L.) and bilberry (*V. myrtillus*). Cranberry itself has attracted growing public interest as a functional food due to its significant antioxidant activity (*1*) and therefore potential health benefits. Cranberry can prevent adhesion of *E. coli* in the urinary tract and stomach (*2,3*) and protect against low-density lipoprotein (LDL) cholesterol oxidation in the blood (*4*). There have also been reports in recent years of *in vitro* anticancer activity (*5-8*), including earlier studies by our group. Many of the biological and health-promoting effects of cranberries have been linked to the presence in the fruit of phenolic compounds having diverse molecular structures and bioactivities, including several classes of flavonoids and phenolic acids. In addition, phenolic esters of ursolic acid in cranberries which inhibit tumor growth *in vitro* were recently reported (*9*).

Cranberry Phenolics

Due to its diverse and plentiful phenolic content, cranberry has substantial antioxidant power (*10,11*). Among 20 fruits analyzed, cranberry contained the highest content of phenolics both per serving and by weight and was ranked sixth in overall antioxidant quality (*12*). We reported earlier on the radical-scavenging activities of flavonol glycosides and anthocyanins in whole cranberry fruit and their considerable ability to protect against lipoprotein oxidation *in vitro* (*8*). The bioactivity of flavonoids from other *Vaccinium* have been more thoroughly studied (*13*) than cranberry. Bilberry anthocyanins, for example Myrtocyan® (available in Europe), were observed to inhibit platelet aggregation (*14*), increase vasodilation (*15*), and induce apoptosis in HL-60 leukemia cells and HCT116 colon carcinoma cells (*16*). Although cranberry contains the same basic classes of polyphenolics as other *Vaccinium* fruits and grapes, the composition and structures of individual compounds vary significantly and therefore the bioactivity of cranberry phenolics warrants further studies.

Phenolic antioxidants in cranberry include anthocyanins, flavonols, proanthocyanidins (PACs), phenolic triterpene esters and small phenolic acids. Cranberry flavonol composition includes primarily galactosides and arabinosides of quercetin and myricetin, with a lesser content of quercetin xyloside and rhamnoside. Flavonols can inhibit LDL oxidation and platelet aggregation and adhesion *in vitro* and *in vivo* (*17*). They are also likely to induce endothelium-dependent vasodilation and to increase reverse cholesterol transport (*18*).

Cranberry anthocyanins, which give the berries their deep red color, are primarily cyanidin and peonidin galactosides and arabinosides. Both flavonols and anthocyanins have the potential to prevent oxidative damage caused by reactive oxygen species (ROS) (*19-21*). In an earlier study (*8*), we reported on the free-radical scavenging activity of numerous cranberry flavonol glycosides and cyanidin-3-galactoside. It was found that cyanidin-3-galactoside and the cranberry flavonols had superior radical-scavenging activity to vitamin E. Using an *in vitro* bioassay designed to evaluate the ability of compounds to prevent LDL and VLDL oxidation, we also found that cyanidin-3-galactoside was an effective inhibitor of lipoprotein oxidation and more effective than vitamin E. The flavonol glycosides, though slightly less effective, were comparable to vitamin E in this regard. Therefore, the potential for these compounds to protect against oxidative damage is substantial.

Proanthocyanidins (PACs) in cranberry are somewhat less plentiful than anthocyanins and flavonols, but also have the potential to contribute to prevention of oxidative processes. In addition to their unique antibacterial properties, cranberry PACs have been reported to inhibit copper-induced LDL oxidation (*22*). PACs have been identified as factors responsible for the anti-adhesion effect of cranberry and cranberry juice against *E. coli* bacteria associated with urinary tract infections (*2*). They are polymeric flavan-3-ols composed primarily of epicatechin, and range greatly in size and structure. Some features of PAC structure are fairly unique to cranberries, particularly the prevalence of "A-type" interflavan linkages.

Considering the diversity and abundance of phenolics in cranberry, significant potential exists for cranberry products to prevent oxidative stress related to cardiovascular disease and cancer at the cellular level and *in vivo*. The ability of cranberry antioxidants to provide protection against health conditions related to ROS needs to be more thoroughly investigated. We are focusing our research efforts in two areas:

- Investigating the role of cranberry phenolics in cellular processes related to cardiovascular disease and stroke.
- Investigating the identity and mechanism of cranberry phytochemicals that may play a role in cancer prevention.

Antioxidants and Stroke

In the physiological environment, oxidative stress results from overproduction of ROS which are not adequately inactivated by endogenous or exogenous antioxidant systems. Cardiovascular diseases, such as heart attack and stroke, collectively are the number one cause of mortality worldwide (23). Heart attacks are caused by ischemia (reduced blood flow) to cardiac tissue, while brain ischemia is the cause of 80% of strokes. Brain damage by ischemic stroke is due at least in part to oxidative stress to neurons, induced by ROS produced during reperfusion which are not adequately inactivated by endogenous or exogenous antioxidant systems. Thus, tissue hypoxia followed by reperfusion leads to the cell damage associated with stroke (24,25). Because of this widespread involvement of reactive oxygen species in various pathologies, antioxidants found in food have received much attention as potential preventors against these conditions. There is evidence that dietary intake of flavonoid antioxidants such as quercetin is inversely related to stroke incidence (26). However, little has been done in controlled animal studies to evaluate the effect of dietary phenolics on hypoxic brain damage.

Prior work has shown that rats fed on a diet enriched with extracts of a relative of the cranberry, lowbush blueberry (*Vaccinium angustifolium*), suffered much less neuronal cell death upon induced stroke (27). The possible identity of the compounds responsible for this *in vivo* protection was investigated in isolated rat neurons challenged with a 6 hr incubation of oxygen-glucose deprived medium, to simulate a stroke, or 100 µM hydrogen peroxide, to induce oxidative stress/reperfusion injury. The whole blueberry polyphenol mixture added to cells produced a concentration-dependent reduction in both apoptotic and necrotic cell death (28). Enriched fractions of anthocyanins and proanthocyanidins also provided neuroprotection, although they were less effective than the polyphenolic mixture. We hypothesized that similar effects would be observed with cranberry due to its high content of similar phenolics. A study was then initiated to investigate the effects of cranberry extracts including a whole polyphenolic-enriched extract and subclasses enriched in flavonols and anthocyanins or proanthocyanidins. The effects of these extracts on oxygen-glucose deprivation and oxidative stress in isolated rat cerebellar neurons was studied, the results of which are presented here.

Cranberry Phytochemicals and Cancer

The anticancer properties of cranberry and the nature of compounds providing protection against tumor promotion and proliferation have not been fully investigated. However, published evidence suggests that several groups of cranberry phytochemicals have an impact on cancer-related processes. Classes of compounds with potential anticancer activity in *Vaccinium* species include phenolics such as flavonols, phenolic esters of triterpenes, and proanthocyanidins as well as other triterpenoids and carotenoids. Flavonoid-rich fractions from *Vaccinium* species including bilberry, blueberry and cranberry inhibited the expression of ornithine decarboxylase (ODC) while inducing the xenobiotic detoxification enzyme quinone reductase (QR) *in vitro* (5).

Using bioassay-guided fractionation we screened cranberry extracts for *in vitro* antitumor activity in nine cell lines (8). Further investigation led to the isolation of phenolic esters of ursolic acid which exhibited selective tumor cell growth inhibition in breast, prostate, lung, cervical and leukemia cell lines. These were identified as *cis-* and *trans-* isomers of 3-*O-p*-hydroxycinnamoyl ursolic acid (9). *In vitro* cytotoxicity assays showed that the *cis* isomer selectively inhibited the growth of several tumor cell lines at concentrations of approximately 20 μM. It was particularly effective in MCF-7 (breast), ME180 (cervical), PC3 and DU145 (prostate) cell lines. The *cis* isomer was slightly more active than quercetin, a known inhibitor of tumor growth (29,30) found in cranberries. Cranberry anthocyanins such as cyanidin-3-galactoside, although powerful antioxidants, did not exhibit significant tumor cytotoxicity in our assays.

A recently published study (7) indicated that cranberry PACs may also inhibit carcinogenesis. Extracts of cranberry fruit containing PACs and smaller flavonoids inhibited ODC, an enzyme associated with tumor proliferation, in epithelial cells. These proanthocyanidin extracts were not well-characterized, however. In the course of evaluating cranberry extracts produced in our laboratory for tumor cytotoxicity, we tested the cytotoxicity of the PACs and flavonol/anthocyanin extracts prepared for the neuroprotection study and found that the PAC-rich extract was selectively cytotoxic in several tumor cell lines. Therefore further fractionation and characterization of the cytotoxic proanthocyanidins in this extract are presented here.

Materials and Methods

Materials and Instrumentation

Fresh cranberries (*Vaccinium macrocarpon* cv Stevens) were donated by Decas Cranberry, Wareham, MA. The berries were harvested in October, 2001 in Wareham, MA, and kept frozen at $-20\ ^{\circ}C$ until use. All reagents were of analytical grade. Solvents were purchased from Pharmco Products, Brookfield, CT. Diaion-HP20 was purchased from Supelco, Bellefonte, PA and Sephadex LH-20 was purchased from Fluka Chemical, Buchs, Switzerland.

HPLC analysis was performed on a Waters Millenium HPLC system composed of two Waters 515 pumps with a Waters 996 photodiode array detector. HPLC chromatograms were monitored at 250-600 nm. MALDI-TOF

MS analysis was performed at the University of Wisconsin using a Bruker Reflex II mass spectrometer equipped with N_2 laser.

Preparation of Crude Whole Cranberry Extract

Whole cranberry fruit (200 g) was blended with 300 mL of a 40:40:19:1 (v/v/v/v) mixture of acetone/methanol/water/formic acid designed to maximize extraction of phenolics. After standing one hour, the mixture was filtered and the solids re-extracted twice with 100 mL of solvent mixture. The solutions were dried by a combination of rotary evaporation and lyophilization. The solids were redissolved in 6 mL of methanol and loaded on a Diaion HP-20 column (25 x 290 mm). The column was washed with 300 mL water, then the phenolics eluted with methanol until no further reddish color appeared on the column. The methanol solution was dried to yield 624 mg of crude extract.

Preparation of Flavonol/Anthocyanin and PAC-Enriched Fractions

Crude extract (353 mg) was dissolved in 6 mL of 50:50 (v/v) methanol/water and loaded on a Sephadex LH-20 column (25 x 290 mm). Flavonol/anthocyanin extract was eluted until eluent was colorless (approx. 150 mL) with a 70:25:5 (v/v/v) mixture of methanol/water/formic acid and dried in vacuo to yield 228 mg of flavonol/anthocyanin enriched extract for bioassay. Proanthocyanidins were eluted with 200 mL of 70:30 acetone/water and dried in vacuo to yield 93 mg of PAC-enriched extract for bioassay. The presence of the desired phenolics was confirmed by HPLC analysis on a Waters Symmetry C18 column (4.6 x 250 mm). Solvents used were: (A) water in 2% acetic acid; (B) methanol in 2% acetic acid; program used was: isocratic elution with 100% A from 0 to 5 min, linear gradient to 100% B from 5 to 45 min, isocratic elution with 100% B from 45 to 60 min; flow rate was 0.8 mL/min. PDA detection was used at 525 nm to detect anthocyanins, 350 nm to detect flavonols and 280 nm to detect PACs.

Fractionation and Characterization of Proanthocyanidins

The proanthocyanidin extract prepared as described above (100 mg) was dissolved in 5 mL of 50:50 ethanol/water and loaded onto a Sephadex LH-20 column (20 x 240 mm). The column was eluted first with water/ethanol with ethanol increasing from 50 to 75 to 100% (100 mL each), then with water/acetone with acetone increasing from 50 to 75 to 100% (100 mL each). A final wash of 40:60 acetone/water eluted the remaining colored material. Fractions were collected and pooled into six larger fractions based on TLC analysis on silica gel plates. The fractions were dried in vacuo and used for cytotoxicity assays in tumor cells as described below. Fractions exhibiting cytotoxicity were subjected to MALDI-TOF MS analysis by Christian Krueger at the University of Wisconsin. The samples were deionized and spiked with cesium trifluoroacetate (^{133}Cs), which allowed detection of $[M + Cs]^+$ ions.

Neuroprotection Bioassay

Wistar rats (Charles River Ltd., Quebec) were mated and rat pups removed on post-natal day 6-8 for establishment of primary cultures of cerebellar granule neurons according to published procedures (*31*). Briefly, pups were anaesthetised with halothane, decapitated, and then the cerebellum was removed aseptically. Cells were dissociated at 37°C with trypsin and plated on poly-L-lysine coated plates at a density of 5×10^6 per plate. Neurons were grown at 37°C in a humidified incubator containing 5% CO_2. Two forms of simulated stroke were used in the bioassay. Kreb's Henseleitt solution lacking glucose was gassed with 95% N_2/5% CO_2 (Praxair, Charlottetown, PE, Canada) until pO_2 values were 23-27 mmHg. This solution was designated oxygen-glucose deprived Krebs (OGD) which simulates cerebral ischemia. To simulate oxidative stress seen during reperfusion (REP), 1 mM hydrogen peroxide was made in Kreb's Henseleitt solution containing both oxygen and glucose. After removing growth medium from cells on day 7-10 in culture, OGD or REP Kreb's solution was placed on cells for 6 hr. Control cells received normal glucose- and oxygen-containing Kreb's solutions. The crude, flavonol/ anthocyanin and PAC fractions were tested for inhibition of necrosis and apoptosis. Cranberry extracts were added to cells under control, OGD or REP conditions (n = 6) at three concentrations (30, 100 and 300 µg/mL). After six hours, the solutions were replaced with MEM for 1 hr. One ml of medium from each sample was frozen for analysis of lactate dehydrogenase (LDH) activity (indicative of necrosis) spectrophotometrically using commercially-available kits (Diagnostic Chemicals Ltd., Charlottetown, PE, Canada). The remaining cell contents were collected and frozen on dry ice for caspase-3 analysis (indicative of apoptosis) using colorimetric assay kits (Clontech Ltd., Palo Alto, CA).

Tumor Cell Cytotoxicity Assay

Cytotoxicity of the flavonol/anthocyanin and PAC extracts and the PAC fractions was evaluated in various tumor cell lines using published procedures

(9, 32). Cell lines tested include BALB/c3T3, H460, ME180, M-14, DU145, MCF-7, HT-29, PC3, and K562. Growth inhibition was evaluated at various concentrations of extract by spectrophotometrically quantifying the number of live cells remaining at the end of the incubation period. GI_{50} were calculated as the range of concentrations of extract required to inhibit tumor cell growth by 50% relative to control.

Results and Discussion

Cranberry and Stroke

The effects of cranberry extract on survival of neonatal rat brain cells under conditions of stroke in our *in vitro* bioassay are shown in Table I. Treatment with cranberry extract caused a dose-dependent decrease in the percentage of neurons undergoing necrosis. At the highest dosage level, there was a 49% decrease in necrosis caused by oxygen and glucose deprivation, conditions of ischemia. A 43% decrease in necrosis was observed in the oxidatively-stressed neuron cultures. A similar effect on apoptosis was observed; cranberry extract at 1.0 mg/mL protected 50% of neurons against apoptosis under conditions of ischemia and protected 36% of neurons under conditions of reperfusion. These results suggest that whole cranberry extract may provide some protection against brain damage associated with stroke. The bioavailability of cranberry polyphenolics in the brain tissue is not well-studied. Thus further investigation of the degree of neuroprotection afforded by a cranberry-enriched diet *in vivo* is warranted.

Table I: Effect of Treatment with Cranberry Extract on Necrosis and Apoptosis of Rat Neurons

	% Inhibition of necrosis in neurons treated with cranberry extract vs. control			% Inhibition of apoptosis in neurons treated with cranberry extract vs. control		
Concentration of extract (µg/mL)	30	100	300	30	100	300
Simulated ischemia group[a]	15 %	36 %	49 %	20 %	32 %	50 %
Oxidatively stressed group[b]	11 %	29 %	43 %	8 %	24 %	36 %

[a]Cultures deprived of oxygen and glucose, n = 6; [b]Cultures treated with 1 mM hydrogen peroxide, n = 6

Preliminary results on a flavonol/anthocyanin enriched fraction and a proanthocyanidin fraction indicate that the relative protective effects of these classes of phenolics varies somewhat depending on the conditions. Table II shows the results of treatment of rat brain neurons with each fraction as compared to the whole berry extract.

Table II: Percent Decrease in Rat Neuron Necrosis in Cultures Treated with Various Cranberry Extracts at 300 µg/mL

	Whole phenolics-enriched cranberry extract	*Flavonol/anthocyanin extract*	*Proanthocyanidin extract*
Simulated ischemia group[a]	48.9 %	17.1 %	16.0 %
Oxidatively stressed group[b]	42.7 %	42.4 %	16.5 %

[a]Cultures deprived of oxygen and glucose, n = 6; [b]Cultures treated with 1 mM hydrogen peroxide, n = 6

Under conditions of oxidative stress, the flavonol/anthocyanin fraction was nearly as effective in protecting against necrosis as whole cranberry extract, whereas the PAC-enriched fraction only provided limited protection (16.5% at 300 µg/mL). This is not unexpected, given the superior ability of anthocyanins and flavonols to scavenge free radicals. Under conditions of ischemia, the two fractions were about equally protective against necrosis (17.1% for flavonol /anthocyanin and 16.1% for PACs) and not as effective as the whole cranberry extract. It is possible that the different phenolic classes work synergistically in this case, or that the major contributors to neuroprotection under conditions of oxygen and glucose deprivation are not flavonoids. Whole cranberries also contain an abundance of phenolic acids and triterpene phenolic esters. Further investigation of the relative contributions of the different classes of cranberry phytochemicals is currently underway.

Antitumor Activity of Cranberry Proanthocyanidins

As shown in Table III, the cranberry proanthocyanidin fraction exhibited selective cytotoxicity against several tumor cell lines *in vitro*. Growth was inhibited by 50% at extract concentrations as low as 8 – 31 µg/mL in H460 lung tumor cells and at somewhat higher concentrations (31 – 125 µg/mL) in HT-29 colon carcinoma and K562 leukemia cell lines. By comparison, the anthocyanin/ flavonol fraction was not very effective at inhibiting tumor growth in the cell lines tested.

Table III: *in vitro* Tumor Cytotoxicity of Extracts from Whole Cranberry Fruit

Cell Lines	GI_{50}^{a} ($\mu g/mL$) Anthocyanin/ Flavonol	GI_{50} ($\mu g/mL$) PAC^{b} extract	GI_{50} ($\mu g/mL$) Fraction from PAC extract
3T3 murine fibroblast	> 125	> 125	---
H460 human small cell lung carcinoma	> 125	8 – 31	16 – 63
ME180 human cervical carcinoma	> 125	> 125	63 – 250
DU145 human prostate metastatic carcinoma	> 125	> 125	---
MCF-7 human breast adenocarcinoma	> 125	> 125	---
M-14 human melanoma	> 125	> 125	---
HT-29 human colon adenocarcinoma	> 125	31 – 125	63 – 250
PC3 human prostate adenocarcinoma	> 125	> 125	---
K562 human chronic myelogenous leukemia	> 125	31 – 125	16 – 63

^{a} GI_{50} = concentration range at which 50% of growth is inhibited; ^{b} PAC = proanthocyanidin

Fractionation of the PACs led to isolation of a PAC fraction with cytotoxicity against lung, colon and leukemia cell lines with GI_{50} in the 16 – 63 µg/mL range. MALDI-TOF MS analysis shows that this fraction contains PACs varying in size from two to ten catechin or epicatechin units, and suggests that the predominant oligomers in this fraction are trimers and tetramers with one A-type linkage between epicatechin units. We are currently developing new fractionation methods in an attempt to better understand the effect of proanthocyanidin size and structure on ability to inhibit the growth of tumors.

Given the observed ability of cranberry triterpenes, flavonols and proanthocyanidins to inhibit tumor growth, it is possible that these cranberry phytochemicals could work synergistically to reduce proliferation of some cancers. Further studies are needed to explore the anticancer potential of cranberry phenolics *in vivo*. We are currently investigating the effects of various cranberry fractions on the expression of several genes linked to carcinogenesis in the hopes of better understanding the mechanisms behind the observed tumor cytotoxicity.

Acknowledgements

The authors would like to thank Drs. Christian Krueger and Jess Reed (Univ. of Wisconsin) for MALDI-TOF MS analysis and we also acknowledge the support of the UMD Cranberry Agricultural Research Program and the Cranberry Institute/Wisconsin Cranberry Board.

References

1. Wang, S. Y.; Jiao, H. *J. Agric. Food Chem.* **2000**, *48*, 5677-5684.
2. Howell, A.; Vorsa, N.; Der Marderosian, A.; Foo, L. *N. Engl. J. Med.* **1998**, *339*, 1085-1086.
3. Burger, O.; Ofek, I.; Tabak, M.; Weiss, E. I.; Sharon, N.; Neeman, I. *FEMS Immunol. Med. Microbiol.* **2000**, *29*, 295-301.
4. Wilson, T.; Porcari, J.P.; Harbin, D. *Pharmacology Lett.* **1998**, *62*, 381-386.
5. Bomser, J.; Madhavi, D.L.; Singletary, K.; Smith, M.A. *Planta Med.* **1996**, *62*, 212-216.
6. Guthrie, N. *Proceedings of the Experimental Biology Conference*, April 14-18, **2000**, San Diego, CA, Abstract #531.13.
7. Kandil, F. E.; Smith, M. A. L.; Rogers, R. B.; Pepin, M.-F.; Song, L. L.; Pezzuto, J. M.; Seigler, D. S. *J. Agric. Food Chem.* **2002**, *50*, 1063-1069.
8. Yan, X.; Murphy, B. T.; Hammond, G. B.; Vinson, J. A.; Neto, C. C. *J. Agric. Food Chem.* **2002**, *20*, 5844-5849.
9. Murphy, B. T.; MacKinnon, S. L.; Yan, X.; Hammond, G. B.; Vaisberg, A. J.; Neto, C. C. *J. Agric. Food Chem.* **2003**, *51*, 3541-3545.
10. Heinonen, I. M.; Lehtonen, P. J.; Hopia, A. I. *J. Agric. Food Chem.* **1998**, *46*, 24-31.
11. Wang, S. Y.; Stretch, A. W. *J. Agric. Food Chem.* **2001**, *49*, 69-974.
12. Vinson, J. A.; Su, X.; Zubik, L.; Bose, P. *J. Agric. Food Chem.* **2001**, *49*, 5315-5321.
13. Kalt, W. *Hort. Rev.* **2001**, *27*, 269-315.
14. Morazzoni, P.; Magistretti, M. J. *Fitoterapia* **1990**, *61*, 13-21.

15. Bettini, V.; Aragno, R.; Bettini, G.; Braggion, G.; Calore, L.; Concolato, M. T.; Favaro, P.; Penada, G. *Fitoterapia* **1991**, *62*, 15-28.
16. Katsube, N.; Iwashita, K.; Tsushida, T.; Yamaki, K.; Kobori, M. *J. Agric. Food Chem.* **2003**, *51*, 68-75.
17. Folts, J. D.; Osman, H.; Shanganayagam, D.; Reed, J. *Haemostasis* **1996**, *95*, 573.
18. Reed, J. *Critical Reviews in Food Science and Nutrition* **2002**, *42*, 301-316.
19. Ioku, K.; Tsushida, T.; Takei, Y.; Nakatani, N.; Terao, J. *Biochim Biophys Acta* **1995**, *1234*, 99-104.
20. Vinson, J. A.; Dabbagh, Y. A.; Serry, M. M.; Jang, J. *J. Agric. Food Chem.* **1995**, *43*, 2800-2802.
21. Pietta, P-G. *J. Nat. Prod.* **2000**, *63*, 1035-1042.
22. Porter, M. L.; Krueger, C. G.; Wiebe, D.A.; Cunningham, D. G.; Reed, J. D. *J. Sci. Food Agric.* **2001**, *81*, 1306-1313.
23. Murray C.J.; Lopez A.D. *Lancet* **1997**, *349*, 1436-1442.
24. Sweeney, M.I.; Yager, J.Y.; Walz, W.; Juurlink, B.H.J. *Can. J. Physiol. Pharmacol.* **1995**, *73*, 1525-1535.
25. Chan, P.H. *J. Cereb. Blood Flow Metab.* **2001**, *21*, 2-14.
26. Knekt, P., Jarvinen, R., Reunanen, A.; Maatela, J. *Br. Med. J.* **1996**, *312*, 478-481.
27. Sweeney, M.I.; Kalt, W.; MacKinnon, S.L.; Ashby, J.; Gottschall-Pass, K.T. *Nutr. Neurosci.* **2002**, *5*, 427-431.
28. MacKinnon, S.L.; Craft, C.; Clark, K.J.; Gottschall-Pass, K.J.; Kalt, W.; Sweeney, M.I. *50th Annual Congress of the Society for Medicinal Plant Research*, Barcelona, Spain. September 8-12, **2002**.
29. Kawaii, S.; Tomono, Y.; Katase, E.; Ogawa, K.; Yano, M. *Biosci. Biotechnol. Biochem.* **1999**, *63*, 896-899.
30. Manthey, J. A.; Guthrie, N. *J. Agric. Food Chem.* **2002**, *50*, 5837-5843.
31. Logan, M.; Sweeney, M. I. *Mol. Chem. Neuropathol.* **1997**, *31*, 119-133.
32. Skehan, P.; Storeng, R.; Scudiero, D.; Monks, A.; McMahon, J.; Vistica, D.; Warren, J.; Bokesch, H.; Kenney, S.; Boyd, M. *J. Natl. Cancer Inst.* **1990**, *82*, 1107-1112.

Indexes

Author Index

Ae, Shutaro, 176
Amarowicz, Ryszard, 57, 67, 94
Bode, Ann M., 225
Dong, Zigang, 225
Dykes, Gary A., 94
Ekanem, Albert, 118
Fu, Hui-Yin, 46
Ghai, Geetha, 242
Goodwin, Douglas C., 161
Hertwig, Kristen M., 161
Ho, Chi-Tang, 1, 46, 118, 129, 197, 242
Hong, Jungil, 212
Hori, Hitoshi, 176
Hsu, Hseng-Kuang, 46
Huang, Mou-Tuan, 242
Huang, Tzou-Chi, 46
Juliani, Rodolfo, 118
Kondo, Miwako, 271
Kouno, Isao, 188
Laband, Kimberley A., 161
Lambert, Joshua D., 212
Lamoureaux, Toni L., 271
Lee, Chang Yong, 254
Lee, David Y-W, 19
Lee, Hyong Joo, 254
Lee, Ki Won, 254
Lee, Mao-Jung, 212
Liang, Chia-Pei, 118
Liang, Yu-Chih, 197
Lin, Jen-Kun, 197
Lin-Shiau, Shoui-Yn, 197
Liu, Yanze, 19
Liu, Yue, 242

Liyana-Pathirana, Chandrika M., 33
Lu, Hong, 212
MacKinnon, Shawna L., 271
Madhujith, Terrence, 83
Masuda, Hideki, 129
Matsuo, Yosuke, 188
Meng, Xiaofeng, 212
Mine, Chie, 188
Naczk, Marian, 57, 67
Nagasawa, Hideko, 176
Neto, Catherine C., 271
Pan, Min-Hsiung, 197
Parry, John W., 10, 107
Pegg, Ronald B., 67, 94
Peterson, Devin G., 143
Ramji, Divya, 242
Rosen, Robert T., 242
Sang, Shengmin, 212, 242
Shahidi, Fereidoon, 1, 33, 57, 83, 94
Simon, James E., 118
Solomon, Frankie, 271
Sweeney-Nixon, Marva I., 271
Tanaka, Takashi, 188
Totlani, Vandana M., 143
Troszynska, Agnieszka, 94
Ueno, Toshio, 129
Uto, Yoshihiro, 176
Wang, Mingfu, 118
Watarumi, Sayaka, 188
Yang, Chung S., 212, 242
Yu, Liangli (Lucy), 10, 107
Zadernowski, R., 57
Zhou, Kequan, 10, 107

Subject Index

A

ABTS. *See* 2,2'-Azino-bis(3-ethylbenzothiazoline-6-sulfonic acid) (ABTS)
Acetic acid, concentration in solid-fermented *Toona sinensis*, 47–48, 49*t*
Acetone. *See* Bearberry-leaf
Active efflux, tea polyphenols, 217–218
Alanine, model Maillard reaction, 147
Amino acids, model Maillard reaction, 147
Amylase inhibitors, glucose lowering effect, 84*t*
Anthocyanidin
  name, substitution, and dietary source, 4*t*
  structure, 3*f*
Anthocyanins
  bean seed coats, 85
  cranberry, 273
  structures of, and phenolic acids in bean seed coats, 86*f*
Antibacterial properties, green tea extracts, 96, 103, 104*t*
Anticancer effects
  bioactive compounds in pulses, 84*t*
  black and green teas, 202
  modulation of tumor promotion signaling proteins by theaflavin and catechin, 203–204
  phenolics, 6–7
  See also Cancer prevention
Anti-inflammatory properties. *See* Theaflavins
Antimicrobial assays, green tea extract preparations, 98–99

Antimutagenic effects, black and green tea extracts, 200–202
Antioxidant activity
  apples, 255–256
  bean extracts, 87–89
  canola hull phenolics in β-carotene-linoleate model system, 64*f*, 65
  changes of extracts from fermented *Toona sinensis* leaves, 51*t*
  dechlorophyllized bearberry-leaf and fractions in cooked pork, 79, 80*f*
  edible beans, 85, 86*t*
  honeybush tea, 126–127
  phenolics, 6
  rapeseed meal and cakes in β-carotene-linoleate model system, 58–59
  sesame fractions, 43
  See also Sesame seeds; Total antioxidant activity (TAA)
Antioxidants
  benefits, 10
  cold-pressed seed oils, 108
  composition of, in apple, 259, 260*t*
  dietary, 34
  natural, 68, 69*f*, 255
  phenolic, in cranberry, 272–273
  stroke, 273–274
  synthetic, 68
  wheat, 11
  See also Apple phenolics; Artepillin C; Cranberry phenolics
Antitumor promoting activities. *See* Apple phenolics
Apple phenolics
  ABTS (2,2'-azino-bis(3-ethylbenzothiazoline-6-sulfonic

acid) radical scavenging activity, 257–258
apple cultivars, 256
bioassay of gap junction intercellular communication (GJIC), 258–259
cell culture, 258
cell-to-cell communication, 264–265, 269
composition of major antioxidants in apple, 259, 260t
content and antioxidant activity, 264
contribution of major antioxidants to total antioxidant capacity (TAC) of apple, 259, 261, 262t
cytotoxicity, 258
effect of epicatechin, procyanidin B2, and vitamin C on inhibition of GJIC by H_2O_2, 264, 268f
effect of quercetin on inhibition of GJIC by H_2O_2, 264, 268f
effects of extracts and major antioxidants on inhibition of GJIC by H_2O_2, 261, 264
extraction of phenolics, 256–257
fruit content, 256
gap junction channels, 264–265, 269
GJIC, 255
Golden Delicious extracts inhibiting GJIC, 266f, 267f
identification of phenolics, 257
materials and methods, 256–259
maximum non-cytotoxic concentration of extracts and major antioxidants, 261, 263t
previous studies, 255–256
quantification of TAC, 258
quantification of vitamin C, 257
statistical analysis, 259
Arctostaphylos uva-ursi L. Sprengel. See Bearberry-leaf
Aroma compounds, thermal generation
choice of flavonoid compound and magnitude of inhibitory effect, 154, 157f
composition of model Maillard reaction mixture, 147t
effect of epicatechin (EC) addition on, for model food systems, 154, 158f, 159
effect of reaction time on EC inhibition on aroma generation, 149, 153f, 154
formation in Maillard reactions, 146
gas chromatography/mass spectrometry (GC/MS) analysis method, 149
granola bar model systems, 147–148
influence of EC on, at different pH values, 149, 150f, 151f, 152f
materials and methods, 146–149
model food systems, 147–148
model Maillard reaction system, 147
reaction temperature and EC effects on aroma generation, 154, 155f, 156f
roasted cocoa sample, 148
See also Maillard chemistry
Artepillin C
chemical structure, 178f, 185
DPPH radical scavenging activity, 179–180, 182–184
first total synthesis, 177
highest occupied molecular orbital (HOMO) energy calculation, 183t, 185
inhibition of rat liver mitochondria (RLM) lipid peroxidation, 180
inhibitory activity of, on RLM lipid peroxidation, 184, 185–186
materials and methods, 177–181
molecular orbital (MO) calculation, 180

octanol-water partition coefficient calculation, 181
prenylation, 185
prenylation of p-iodophenol in toluene, 181
preparation of rat liver mitochrondria (RLM), 180
product distributions in prenylation of p-iodophenol, 181t
synthesis, 178–179, 181–182
time-course of DPPH radical scavenging reaction with antioxidants, 183f
Ascorbate, kinetic evaluation of peroxidases, 170
Ascorbate-dependent chronometric method, oxidation of capsaicinoids, 167–169
Ascorbic acid
inhibition of gap junction intercellular communication (GJIC), 264, 268f
natural antioxidant, 68, 69f, 255
quantification method, 257
2,2'-Azino-bis(3-ethylbenzothiazoline-6-sulfonic acid) (ABTS)
apple phenolics, 259, 261
cold-pressed seed oils, 113, 114f
radical scavenging activity, 109–110, 257–258

B

Beans
anthocyanins and phenolic acids in bean seed coats, 85, 86f
anticarcinogenic effect, 84t
antioxidative activity of edible beans, 85, 86t
bioactive compounds in pulses and their health effects, 84t
Fe^{2+} concentration and percentages of chelated ferrous ion, 89, 90t
glucose lowering effect, 84t
health benefits, 83–84
hypolipaemic effect, 84t
inhibition of oxidation of human low density lipoprotein (LDL), 90–91
metal chelation capacity, 89, 90t
polyphenolic compounds, 84–85
prevention of deoxyribonucleic acid (DNA) strand breakage by, 91
structures of anthocyanins and phenolic acids in bean seed coats, 86f
total phenolic content, 87–88
Trolox equivalent antioxidant capacity (TEAC), 87, 88, 89f
Bearberry-leaf
antioxidant activity by β-carotene-linoleic acid model, 70
antioxidant capacity as percent inhibition of lipid oxidation, 74
biological activity of tannins by dye-labeled protein assay, 73–74
block diagram reporting percent inhibition of lipid oxidation of cooked pork, 79, 80f
dye-labeled bovine serum albumin assay, 79
ethanol and acetate fractions, 79, 81
flow diagram for preparation of crude extract, 76f
fractionation of crude extract, 71
liquid-liquid fractionation, 75, 76f
liquid-liquid fractionation of ethanol and acetone fractions, 81
materials and methods, 71–74
phenolics and condensed tannins, 75, 76t
preparation of extract, 71
protein precipitating capacity of various fractions of phenolics, 80t

protein precipitating potential of
tannins, 77–78
protein-tannin complex formation
determinations, 72–74
selection of quantification standard,
77
tannin contents in extract and
fractions, 72
2-thiobarbituric acid reactive
substances (TBARS)
determination, 74
total phenolics in extract and
fractions, 72
vanillin and proanthocyanidin
assays, 75, 77
Benzotropolone ring
theaflavins and synthetic
compounds, 251
See also Theaflavins
Beverages
scientific research publications,
189f
See also Black tea; Coffee; Green
tea; Oolong tea; Tea; Tea
fermentation
BHT. See 2,6-Di-*t*-butyl-4-
methylphenol (BHT)
Bioavailability
pharmacokinetics of tea catechins,
218–220
proposed model for factors
affecting tea polyphenol, 221f
theaflavins, 214
See also Tea polyphenols
Biological activity, tannins in
bearberry-leaf extract and fractions,
73–74
Biotransformation
tea catechins, 215–217
See also Tea polyphenols
Black caraway seed oil
comparison of radical DPPH
scavenging activity, 111–112
oxygen radical absorbing capacities
(ORAC), 114, 115f

radical cation scavenging capacity,
113, 114f
total phenolic content of cold-
pressed, 115, 116f
See also Edible seed oils
Black hull extract
total phenolic content, 88t
See also Beans
Black raspberry seed oil
comparison of radical DPPH
scavenging activity, 111–112
total phenolic content of cold-
pressed, 115, 116f
See also Edible seed oils
Black tea
anti-carcinogenic effects, 202
antimutagenic effects of extracts,
200–202
antioxidative effects of, and
catechin, 198–200
cancer prevention, 229–231
concentrations of tea polyphenols,
207t
health effects, 189
health promotion, 207
inhibition of xanthine oxidase and
reactive oxygen species (ROS)
by tea polyphenols, 200
production, 198
production and composition, 213
production of polyphenols, 189–
190
protection against lipid
peroxidation by catechin and,
199t
published papers, 199t
scientific research publications,
189f
theasinensin production, 192–195
See also Green tea; Oolong tea;
Tea; Tea fermentation
Black whole bean extract
Fe^{2+} concentration and percentages
of chelated ferrous ion, 90t
total phenolic content, 88t

See also Beans
Bran. See Wheat bran
Brown hull extract
  total phenolic content, 88t
  See also Beans
Brown whole bean extract
  Fe^{2+} concentration and percentages of chelated ferrous ion, 90t
  total phenolic content, 88t
  See also Beans

C

Camellia sinenesis
  benefits from tea consumption, 213
  tea from leaves, 198
  See also Tea; Tea polyphenols
Cancer, cranberry phytochemicals and, 274–275
Cancer prevention
  [6]-gingerol, 234–236
  inositol hexaphosphate (IP_6), 232, 233f
  mechanism of [6]-gingerol anticancer effects, 236, 237f
  mitogen-activated protein (MAP) kinases, 226–227
  phenolics, 6–7
  potential of dietary supplements, 225–226
  resveratrol, 232–234
  signal transduction and carcinogenesis, 226–229
  tea and constituents with models, 213
  tea polyphenols, 220, 222, 229–231
  tumor promoters inducing activation and phosphorylation of MAP kinase cascades, 227, 228f
  tumor promoters inducing activation of transcription factor NF-κB, 228–229
  See also Apple phenolics

Canola hulls
  antioxidant activity determination method, 59
  antioxidant activity of phenolics in β-carotene-linoleate model system, 64f, 65
  chemical composition of seeds, cotyledons, and hulls, 58t
  estimating total phenolic compound content, 59
  materials and methods, 59–60
  protein precipitating capacity of extracts, 62t
  protein precipitating potential, 60–61
  scavenging effect on α,α-diphenyl-β-picrylhydrazyl (DPPH) radical, 59, 61, 63f
  statistical analysis, 59–60
  total content of condensed tannins in extracts, 62t
  total phenolic content by Folin–Denis assay, 60
  total phenolic content in extracts, 62t
  yield of phenolic extracts from, 60t
Capsaicin
  effect of ascorbate on accumulation of oxidation products, 168f
  oxidation by peroxidase under steady-state conditions, 167f
  reduction by o-methoxyphenols, 171, 172t
  reduction of peroxidase compounds, 165f, 166f
Capsaicinoids
  AH reducing substrates for peroxidase compounds I and II, 163f
  ascorbate-dependent chronometric method, 167–169
  compound II reduction, 165, 166f
  compound I reduction, 164, 165f
  effect of ascorbate on accumulation of oxidation products, 168f

o-methoxyphenols, 162, 163f
peroxidase-catalyzed oxidation, 162–164
physiological considerations of oxidation and ascorbate, 170
proposed interaction with peroxidase catalytic cycle, 163f
rate constants for peroxidase compound I and II reduction by o-methoxyphenols, 172t
standard spectrophotometric methods, 166–167
steady-state kinetics of oxidation, 166–169
structures, 163f
substituent structure and, oxidation, 170–171
transient-state kinetics of oxidation, 164–165
Capsicum fruits. *See* Capsaicinoids
Carcinogenesis, signal transduction and, 226–229
β-Carotene-linoleate assays
determination of antioxidant activity, 61, 70
phenolic extracts of canola hulls, 64f, 65
Carrot seed oil
comparison of DPPH radical scavenging activity, 111–112
dose and time effects of DPPH scavenging activity, 112f
oxygen radical absorbing capacities (ORAC), 114, 115f
radical cation scavenging capacity, 113, 114f
See also Edible seed oils
Cartenoid, natural antioxidant, 68, 69f
Catechins
antioxidative effects of black tea and, 198–200
bioavailability, 214
biotransformation of tea, 215–217
citral with added, 130

concentration in green tea extracts, 97t
p-cymen-8-ol formation with added, 136–137
Fe^{2+} concentration and percentages of chelated ferrous ion, 90t
ferrous ion chelation capacity by black and white sesame seeds and hulls and, 40, 41f
green tea, 95
green tea preparations by reversed-phase high performance liquid chromatography (HPLC), 99, 100f
inhibition of low density lipoprotein (LDL) oxidation, 42–43
name, substitution, and dietary source, 4t
oxidation products from citral with added, 133, 135–136
pharmacokinetics in rats, mice, and humans, 218–220
possible mechanisms for trapping citral-derived radical intermediates with phenoxy radicals from, 138f
protection against lipid peroxidation by, and black tea, 199t
structures, 3f, 214f
tea composition, 213, 243
See also Citral
Cedrela sinensis
chemical composition, 47
See also Toona sinensis
Cholesterol. *See* Low density lipoprotein (LDL)
Citral
p-cymen-8-ol formation with added catechins, 136–137
degradation products, 130
effects of added catechins on formation of oxidation products

in absence of Fe^{3+}, 133, 135–136
effects of epigallocatechin (EGC) and epigallocatechin gallate (EGCg) on formation of oxidation products from, in presence of Fe^{3+}, 137, 139–140
experimental, 130–132
formation behavior of oxidation products, 135*f*
formation behavior of oxidation products from, with added catechins and Fe^{3+}, 140*f*
formation pathways of oxidation products from, under acidic aqueous conditions, 132–133
gas chromatography/mass spectrometry (GC/MS) method, 131
GC analysis of acidic buffer solution of, before and after storage, 133*f*
GC method, 131–132
hypothetical formation inhibition of off-odorants from, with added catechins and Fe^{3+}, 139*f*
model reactions, 130–131
possible mechanisms for formation behavior of oxidation products with added catechins, 136*f*
possible mechanisms for trapping of radical intermediates with phenoxy radicals from catechins, 137, 138*f*
preparation of analytical samples, 131
proposed formation pathways of oxidation products, 134*f*
stability under acidic aqueous conditions, 130
Cocoa, roasted
aroma compounds with added epicatechin, 154, 158*f*, 159
model food system, 148

Coffee, scientific research publications, 189*f*
Cold-pressed seed oils. *See* Edible seed oils
Colorado. *See* Wheat bran
Coumaric acid
winter red and white wheats, 12–15, 16*f*, 17*f*
See also Wheat bran
Coumestans, honeybush tea, 122
Coupled oxidation mechanism, theaflavin synthesis, 190–191
Cranberry phenolics
anthocyanins, 273
antioxidant power, 272
antioxidants and stroke, 273–274
antitumor activity of cranberry proanthocyanidins (PAC), 279–281
cranberry and stroke, 278–279
diversity and abundance, 273
effect of treatment with cranberry extract on necrosis and apoptosis of rat neurons, 278*t*
fractionation and characterization of PAC, 276–277
in vitro tumor cytotoxicity of extracts from whole fruit, 280*t*
materials and instrumentation, 275–276
materials and methods, 275–278
neuroprotection bioassay, 277
percent decrease in rat neuron necrosis in treated cultures, 279*t*
phenolic antioxidants, 272–273
phytochemicals and cancer, 274–275
preparation of crude whole cranberry extract, 276
preparation of flavonol/anthocyanin and PAC-enriched fractions, 276
proanthocyanidins, 273
tumor cell cytotoxicity assay, 277–278
Cranberry seed oil

comparison of radical DPPH
    scavenging activity, 111–112
radical cation scavenging capacity,
    113, 114f
total phenolic content of cold-
    pressed, 115, 116f
See also Edible seed oils
Curcumin
    cancer preventive potential, 6
    o-methoxyphenols, 162
Cyclopia species
    brewing for African tea, 119
    See also Honeybush tea
Cytotoxicity
    assay for measuring, 258
    cranberry proanthocyanidins, 279–281
    in vitro tumor, of extracts from cranberry fruit, 280t
    maximum non-cytotoxic concentration of apple extracts and antioxidants, 261, 263t
    tumor cell, assay, 277–278

D

Deoxyribonucleic acid (DNA), preventing strand breakage by bean extracts, 91
2,6-Di-*t*-butyl-4-methylphenol (BHT)
    DPPH radical scavenging reaction with, 183f
    food additive, 10
    inhibitory activity on rat liver mitochondria (RLM) lipid peroxidation, 184, 185
    structure, 178f
Dietary source, flavonoids, 4t
Dietary supplements
    potential effects, 225–226
    See also Cancer prevention
Dihydroquercetin
    characterization of isomeric, 23t
    structures, 22

See also Phenolic flavonolignans
Dioscorides
    milk thistle and medicinal use, 20
    See also Milk thistle
1,1-Diphenyl-2-picrylhydrazyl (DPPH) radical
    antioxidative properties of artepillin C, 182–184
    canola hull extracts, 61, 63f
    comparing scavenging activity for cold-pressed seed oils, 111–112
    dose and time effects on activity of carrot seed oil, 111–112
    estimating scavenging capacity, 109
    radical scavenging mechanism for black and white sesame seeds and hulls, 39–40
    reaction kinetics of cold-pressed seed oil extracts with, 112, 113f
    scavenging activity of *Toona sinensis* leaf, 50–51
    scavenging capacity, 40f
    scavenging method, 37, 179–180
Disease prevention, phenolics and, 6–7
DPPH. *See* 1,1-Diphenyl-2-picrylhydrazyl (DPPH) radical
Dye-labeled protein assay
    biological activity of tannins in bearberry-leaf extract, 73–74
    condensed tannins with bearberry-leaf, 79

E

Edible seed oils
    ABTS$^{·+}$ scavenging capacities of cold-pressed seed oils, 113, 114f
    cold-pressing procedure, 108
    comparison of radical 1,1-diphenyl-2-picrylhydrazyl (DPPH) scavenging activity, 111f

dose and time effects of DPPH
scavenging activity, 112f
free radical scavenging properties
of cold-pressed seed oils, 110–
112
materials and methods, 108–110
natural antioxidants, 108, 110
ORAC assay (oxygen radical
absorbing capacities), 110
ORAC values of cold-pressed seed
oils, 114, 115f
preparation of antioxidant extract,
109
radical cation ABTS·+ scavenging
activity, 109–110
radical DPPH scavenging activity,
109
reaction kinetics of cold-pressed
seed oils extracts with DPPH
radicals, 113f
statistical analysis, 110
total phenolic content by Folin–
Ciocalteu reagent, 110
total phenolic content of cold-
pressed seed oils, 115, 116f
Enhancer wheat
effect of growing condition on
phenolic acid composition, 15,
17f
phenolic acid composition, 14f, 15,
16f
See also Wheat bran
Enterococcus, lab acid bacteria
strains, 48–49
Enzyme systems, effects of theaflavin
and epigallocatechin gallate, 203t
Epicatechin (EC)
active efflux, 217–218
antioxidants in apples, 264, 265
aroma compounds in model granola
bar during baking, 154, 158f
aroma compounds in roasting of
cocoa nibs with, 154, 158f, 159
aroma generation in Maillard
reaction model, 154, 157f

effect of reaction time on EC
inhibition of aroma generation,
149, 153f, 154
influence of, and pH on aroma
compound generation, 149,
150f, 151f, 152f
influence of temperature on aroma
generation, 154, 155f, 156f
inhibition of gap junction
intercellular communication
(GJIC), 264, 268f
pharmacokinetics, 218–220
phytonutrient, 144
structure, 145f, 214f
See also Aroma compounds,
thermal generation
Epicatechin gallate (ECG)
active efflux, 217–218
biotransformation, 215–217
structure, 214f
See also Catechins; Tea
polyphenols
Epidermal growth factor receptor
(EGFR), overexpression in cancers,
206
Epigallocatechin (EGC)
anti-inflammatory properties of
catechol adduct, 248, 250t
biotransformation, 215–217
catechol adduct, 243, 245f
formation behavior of oxidation
products from citral with added, 140f
oxidation products, 192
oxidation products formation from
citral with, 137, 139–140
pharmacokinetics, 218–220
structure, 214f
See also Catechins; Citral
Epigallocatechin gallate (EGCg)
active efflux, 217–218
anticancer effects, 230–231
anti-inflammatory properties of
catechol adduct, 248, 250t
aroma generation in Maillard
reaction model, 154, 157f

biotransformation, 215–217
catechol adduct, 243, 245f
disease prevention, 6
effects on signal transduction, 231
formation behavior of oxidation products from citral with added, 140f
modulation of tumor promotion signaling proteins, 203–204
oxidation products, 192
oxidation products formation from citral with, 137, 139–140
pharmacokinetics, 218–220
phytonutrient, 144
structure, 145f, 214f
See also Aroma compounds, thermal generation; Catechins; Citral
Ethanol. See Bearberry-leaf

F

Fatty acid synthase (FAS), theaflavin and epigallocatechin gallate, 204–206
Fermentation
honeybush tea, 119–120
See also Tea fermentation
Ferulic acid
reduction by o-methoxyphenols, 171, 172t
winter red and white wheats, 12–15, 16f, 17f
See also Wheat bran
Flavanol
green tea, 95–96
name, substitution, and dietary source, 4t
structure, 3f
Flavanones
honeybush tea, 120–121
name, substitution, and dietary source, 4t
structure, 3f
Flavones
honeybush tea, 121
name, substitution, and dietary source, 4t
structure, 3f
Flavonoids
antioxidants, 6
basic structure, 145f
chemical properties affecting flavor attributes, 144
classes, 4t
dietary sources, 4t
effects on flavor properties of foods, 144
influence on Maillard chemistry, 145–146
inhibitory effect on aroma generation, 154, 157f
lignans, 5
natural antioxidant, 68, 69f
occurrence, 2, 4–5
potentially chemically reactive forms for Maillard reaction, 145f
stilbenes, 5
structures, 3f
substitution patterns, 4t
tannins, 5
See also Aroma compounds, thermal generation
Flavonolignans. See Milk thistle; Phenolic flavonolignans
Flavonols
chelating metal ions, 6
honeybush tea, 121
name, substitution, and dietary source, 4t
structure, 3f
Flavononol
name, substitution, and dietary source, 4t
structure, 3f
Flavor properties, functional foods, 144
Food models. See Model food systems
Food therapy

disease prevention, 225–226
See also Cancer prevention
Fructose, model Maillard reaction, 147
Fruit
  antioxidative components, 255–256
  consumption and health promotion, 265, 269
  See also Apple phenolics
Functional foods
  beans, 83–84
  flavor for consumption, 144

G

Gallic acid, natural antioxidant, 68, 69f
Gap junction channels
  bioassay of gap junction intercellular communication (GJIC), 258–259
  inhibition of communication, 255
  intercellular communication, 264–265, 269
  See also Apple phenolics
Gas chromatography (GC), citral analysis before and after storage, 132, 133f
Gas chromatography/mass spectrometry (GC/MS)
  analysis for aroma compounds, 149
  citral reactions, 131
  honeybush tea extract, 124–126
[6]-Gingerol
  anticancer effects, 6–7, 234–236
  mechanism of anticancer effects, 236, 237f
  o-methoxyphenols, 162
Glucose, model Maillard reaction, 147
Glucose lowering effect, bioactive compounds in pulses, 84t
Glycine, model Maillard reaction, 147
Golden Delicious apple extracts
  inhibition of gap junction intercellular communication (GJIC), 264, 266f, 267f
  See also Apple phenolics
Granola bar model
  aroma compounds with added epicatechin, 154, 158f
  model food system, 147–148
Green tea
  antibacterial activity of green tea extracts (GTE), 103, 104t
  antibacterial properties, 96
  anti-carcinogenic effects, 202
  antimicrobial assays, 98–99
  antimutagenic effects of extracts, 200–202
  antioxidant assay, 97–98
  cancer prevention, 229–231
  catechins, 95–96
  concentrations of tea polyphenols, 207t
  effects of GTE on food-related bacterial strains, 103, 104t
  experimental, 96–99
  flavanols, 95–96
  health promotion, 207
  high performance liquid chromatography (HPLC) of catechins from extract, 99, 100f
  inhibition of TBARS (2-thiobarbituric reactive species) formation by, extract preparations, 99, 101f
  inhibition of TBARS formation by GTE, reconstituted GTE, and proanthocyanidins fraction, 99, 102f
  medicinal use, 95
  percent inhibition of lipid oxidation, 98
  percent inhibition of meat lipid oxidation by, extract, 99, 101t, 103t
  production, 198
  production and composition, 213

published papers, 199t
sample preparation and chromatography, 96–97
scientific research publications, 189f
tea consumption, 95
TBARS, 97–98
variation in concentration of catechins from extract preparations, 97t
See also Black tea; Oolong tea; Tea; Tea fermentation
Guaiacol, reduction by o-methoxyphenols, 171, 172t

H

Health benefits
  antioxidants, 10
  beans, 83–84
  phytochemicals, 34
  tea, 189, 207, 213, 243
Hemp seed oil
  comparison of radical DPPH scavenging activity, 111–112
  oxygen radical absorbing capacities (ORAC), 114, 115f
  radical cation scavenging capacity, 113, 114f
  total phenolic content of cold-pressed, 115, 116f
  See also Edible seed oils
High performance liquid chromatography (HPLC)
  green tea preparations by reversed-phase, 99, 100f
  honeybush tea extract analysis, 122, 125f
  identification of phenolics, 257
  phenolic acids and hydrolysate of bran extract, 12, 13f
  phenolic flavonolignan analysis, 28
Honeybush tea
  analysis of volatile components, 124–126
  analytical methods, 122–124
  bio-activities, 126–127
  C6-C1, C6-C2, and C6-C3 metabolites, 121
  chemical components, 120–122
  chemical composition of volatile components, 126t
  coumestans, 122
  extraction of phenolic compounds, 122–123
  fermentation, 119–120
  flavones, 121
  flavonols, 121
  flavonones, 120–121
  high performance liquid chromatography (HPLC) method, 122
  HPLC of, sample, 125f
  isoflavones, 121
  liquid chromatography/mass spectrometry (LC/MS) method, 123–124
  production, 119
  total ion chromatogram of, sample, 125f
  volatile components, 126
  xanthones, 120
Horse gram
  antioxidative activity, 86t
  See also Beans
Hot peppers. *See* Capsaicinoids
Hulls. *See* Beans; Canola hulls; Sesame seeds
Human diet, flavonoids, 4–5
Human low density lipoprotein (LDL)
  inhibition of copper-mediated, by sesame seeds, 37
  inhibition of oxidation by bean extracts, 90–91
  See also Low density lipoprotein (LDL)
Hyacinth bean
  antioxidative activity, 86t

See also Beans
Hydrogen peroxide
  carcinogenic effect, 265
  effect of antioxidants on inhibition of gap junction intercellular communication (GJIC) by, 261, 264
3-{4-Hydroxy-3,5-di(3-methyl-2-butenyl)phenyl}-2(E)-propenoic acid. *See* Artepillin C
4-Hydroxy-3-methoxymandelic acid (HMMA), reduction by o-methoxyphenols, 171, 172t
Hydroxybenzoic acids
  occurrence, 2
  winter red and white wheats, 12–15, 16f, 17f
  See also Wheat bran
Hydroxycinnamic acids, occurrence, 2
Hypolipaemic effect, bioactive compounds in pulses, 84t
Hypolipidemic effects, theaflavin and epigallocatechin gallate, 204–206

I

Inflammation
  anti-inflammatory properties of black tea theaflavin derivatives, 248, 251–252
  definition, 248
  See also Theaflavins
Inhibition. *See* Lipid oxidation; Rat liver mitochondria (RLM)
Inositol hexaphosphate (IP_6)
  effects on signal transduction, 232, 233f
  interfering with cancer cell cycle, 232
Intercellular communication
  gap junction channels, 264–265, 269
  See also Apple phenolics
Iron(II) chelating capacity
  determination method, 37
  extracts of black and white sesame seeds and hulls, 40, 41f
Iron(III) ion. *See* Citral
Isoflavones
  honeybush tea, 120–121
  hypolipaemic effect, 84t
  name, substitution, and dietary source, 4t
  structure, 3f
Isoleucine, model Maillard reaction, 147

K

Kidney bean
  antioxidative activity, 86t
  See also Beans
Kinetics
  cold-pressed seed oil extracts with DPPH radicals, 112, 113f
  steady-state, of capsaicinoid oxidation, 166–169
  transient-state, of capsaicinoid oxidation, 164–165

L

Lab acid bacteria strains, solid-fermented *Toona sinensis*, 48–50
Lactic acid
  changes in bacteria counts in fermented *Toona sinensis* leaf, 51f
  concentration in solid-fermented *Toona sinensis*, 47–48, 49t
  See also Toona sinensis
Lactobacillus, lab acid bacteria strains, 48–50
Lactococcus, lab acid bacteria strains, 48–49
Lakin wheat

effect of growing condition on phenolic acid composition, 15, 17f
phenolic acid composition, 14f, 15
See also Wheat bran
Lectins, glucose lowering effect, 84t
Leucine, model Maillard reaction, 147
Leuconostoc, lab acid bacteria strains, 48–50
Lignans
  anti-carcinogenic, 84t
  occurrence, 5
Lipid oxidation
  antioxidants, 10
  concentration dependence of percent inhibition by green tea extract (GTE), 103t
  percent inhibition of, for antioxidant capacities, 74, 98
  percent inhibition of meat, by GTE, reconstituted GTE, and proanthocyanidins fraction, 101t
  See also Low density lipoprotein (LDL)
Lipid peroxidation
  protection against, by catechin and black tea, 199t
  See also Rat liver mitochrondria (RLM)
Liquid chromatography/mass spectrometry (LC/MS), honeybush tea extract, 123–125
Liquid-liquid fractionation
  bearberry leaves using ethyl acetate:water, 75, 76f
  ethanol and acetone fractions of bearberry-leaf, 79, 81
Low density lipoprotein (LDL)
  inhibition of copper-mediated human, 37
  inhibition of human LDL oxidation by bean extracts, 90–91
  inhibition of human LDL oxidation by sesame extracts, 41–43
Lutein, natural antioxidant, 68, 69f

M

Maillard chemistry
  aroma formation pathways, 146
  effect of reaction time on epicatechin (EC) inhibition of aroma generation, 149, 153f, 154
  influence of EC and pH on aroma compound generation, 149, 150f, 151f, 152f
  influence of flavonoids, 145–146
  mechanisms, 146
  model system, 147
  potentially chemically reactive forms of flavonoids, 145f
  See also Aroma compounds, thermal generation
Meat systems. See Bearberry-leaf
Mechanisms
  Maillard chemistry, 146
  radical scavenging, 108
  theaflavin synthesis, 190–191
  theasinensin production, 192–195
  tumor production and reactive oxygen species, 265
Medicinal use, milk thistle, 20
Metal chelation capacity, ferrous iron by bean extracts, 89, 90t
o-Methoxyphenols
  capsaicinoids, 162, 163f
  gingerol and curcumin, 162
  rate constants for peroxidase compound I and II reduction by, 170–171, 172t
  See also Capsaicinoids
Micellar electrokinetic capillary chromatography (MECC), phenolic flavonolignan analysis, 29
Milk thistle
  analytical methods for phenolic flavonolignans, 26, 28–31
  biomimetic synthesis of diastereoisomeric silybin and isosilybin, 24f

characterization of isomeric dihydroquercetin, 23t
dihydroquercitin stereoisomers, 21, 22
high performance liquid chromatography (HPLC), 28
medicinal use, 20
micellular electrokinetic capillary chromatography (MECC), 29
nuclear magnetic resonance (NMR), 29, 30f
optical rotation data and stereochemistry of compounds, 26t
primary components, 20–21
seeds of *Silybum marianum*, 21f
silybin, isosilybin, silychristin, and silydianin, 22, 23
single crystal of isosilybin A under microscope, 25f
stereochemistry at C-2 and C-3 of phenolic flavonolignans, 21, 25
stereochemistry at C-7' and C-8' of phenolic flavonolignans, 25–26
structures of flavonolignans, 22–24
X-ray crystallography, 31
X-ray structure of isosilybin A bis-methanolate, 27f
Mitogen-activated protein (MAP) kinases, signal transduction, 226–227, 228f
Model food systems
granola bar model, 147–148
roasted cocoa, 148
See also Aroma compounds, thermal generation
Model Maillard reaction
epicatechin inhibition on aroma generation, 149, 154t
system, 147
See also Aroma compounds, thermal generation; Maillard chemistry
Molecular orbital (MO) calculation, artepillin C, 180, 183t, 185

Multidrug resistance-associated proteins (MRP), active efflux, 217–218

N

Natural antioxidants
beans, 83–84
edible seed oils, 108, 110–111
typical, 69f
use in food systems, 68, 70
See also Antioxidants
Neuroprotection bioassay
cranberry and stroke, 278–279
method using rats, 277
Nonivamide
rate constant determination for oxidation by peroxidase, 169f
reduction by o-methoxyphenols, 171, 172t
North American cranberry. *See* Cranberry phenolics
Nuclear magnetic resonance (NMR)
correlation for identifying silybin and isosilybin, 26, 27f
phenolic flavonolignan analysis, 29, 30f

O

Oil. *See* Edible seed oils
Oolong tea
concentrations of tea polyphenols, 207t
production, 198
production and composition, 213
published papers, 199t
theasinensin production, 192–195
See also Black tea; Green tea; Tea; Tea fermentation
Organic acids, concentration in solid-fermented *Toona sinensis*, 47–48, 49t

Oxidation
  capsaicinoids and peroxidase-
    catalyzed, 162–164
  See also Capsaicinoids; Lipid
    oxidation
Oxidation products
  effects of added catechins on, from
    citral, 133, 135–136
  epigallocatechin and its gallate,
    192
  formation pathways from citral
    under acidic aqueous conditions,
    132–133
  theaflavin, 191
  See also Citral
Oxygen radical absorbing capacities
  (ORAC)
  assay, 110
  cold-pressed seed oils, 114, 115f
  See also Edible seed oils

P

Peroxidases
  capsaicinoid catabolism, 162–164
  catalytic cycle, 163f
  See also Capsaicinoids
pH
  aroma compound generation in
    Maillard-type reactions, 149,
    150f, 151f, 152f
  fermentation of fresh *Toona
    sinensis* leaves, 47–48
Pharmacokinetics, tea polyphenols,
  218–220
Phaseolus vulgaris L.. See Beans
Phenolic acids
  beans, 86f
  classes, 2
  See also Wheat bran
Phenolic flavonolignans
  analytical methods, 26, 28–31
  chemical structures, 22–24
  components in milk thistle, 20–21

  high performance liquid
    chromatography (HPLC), 28
  micellular electrokinetic capillary
    chromatography (MECC), 29
  nuclear magnetic resonance
    (NMR), 29, 30f
  optical rotation data and
    stereochemistry, 26t
  stereochemistry at C-2 and C-3 of
    phenolic, 21, 25
  stereochemistry at C-7' and C-8' of
    phenolic, 25–26
  X-ray crystallography, 27f, 31
  See also Milk thistle
Phenolics
  antioxidant properties, 6
  bearberry-leaf extract, 72, 75, 76t
  definition, 1
  disease prevention, 6–7
  extraction method, 256–257
  identification method, 257
  occurrence, 2–5
  yield of, extracts from canola hulls,
    60t
  See also Apple phenolics; Canola
    hulls; Cranberry phenolics
Phenols, simple, 2
Phytic acid, health effects, 84t
Phytochemicals
  beneficial properties, 34
  cranberry, and cancer, 274–275
  fruits, 264
  See also Cancer prevention;
    Sesame seeds
Phytoesterogen, anticarcinogenic, 84t
Phytosterols, hypolipaemic effect, 84t
Polyphenol, definition, 1
Polyphenolic compounds
  beans, 84–85
  biological effects, 40–41
  health benefits, 189
  mechanism of production of black
    tea polyphenols, 189–190
  published research papers, 198,
    199t

See also Bearberry-leaf; Green tea; Tea polyphenols; Theaflavins
Pork
  antioxidant activity of bearberry-leaf extract in cooked, 79, 80*f*
  antioxidant assay for green tea extracts in pork model system, 97–98, 99, 101*f*, 102*f*
  See also Bearberry-leaf
Proanthocyanidins (PAC)
  antitumor activity of cranberry, 279–281
  bearberry-leaf extracts, 75, 77
  cranberry, 273
  fractionation and characterization, 276–277
  green tea extracts, 100, 101*t*, 102*f*
Procyanidin B2
  antioxidants in apples, 264, 265
  inhibition of gap junction intercellular communication (GJIC), 264, 268*f*
Protease inhibitors, anticarcinogenic, 84*t*
Protein precipitation potential
  bearberry phenolics, 80*t*
  canola hulls extracts, 60–61, 62*t*
  tannins from bearberry-leaf extract, 77–78
Protein-tannin complex, formation determination, 72–74
Pulses
  bioactive compounds and health effects, 84*t*
  See also Beans

Q

Quercetin
  antioxidants in apples, 264, 265
  extent of antioxidant potential for derivatives, 53
  inhibition of gap junction intercellular communication (GJIC), 264, 268*f*
  rat ingestion, 53–55
  solid-fermented *Toona sinensis*, 52–53
  See also Toona sinensis

R

Radical cation scavenging activities
  cold-pressed seed oils, 113, 114*f*
  evaluation, 109–110
  See also Edible seed oils
Radical scavenging activity
  honeybush tea, 126–127
  mechanisms, 108
  See also 1,1-Diphenyl-2-picrylhydrazyl (DPPH) radical; Scavenging activity
Rapeseed
  alcoholic extracts, 58
  seeds of rape and turnip rape, 57
  See also Canola hulls
Rat liver mitochondria (RLM)
  inhibition of artepillin C on RLM lipid peroxidation, 180, 184, 185–186
  preparation, 180
Rats
  cranberry and stroke, 278–279
  ingestion of quercetin, 53–55
  neuroprotection bioassay, 277
  See also Cranberry phenolics
Reaction kinetics, cold-pressed seed oil extracts with DPPH radicals, 112, 113*f*
Reactive oxygen species (ROS)
  carcinogenic effect, 265
  inhibition by tea polyphenols, 200
  oxidative damage, 255
Red hull extract
  total phenolic content, 88*t*
  See also Beans

Red whole bean extract
    Fe^{2+} concentration and percentages of chelated ferrous ion, 90t
    total phenolic content, 88t
    See also Beans
Red winter wheat
    phenolic acid composition, 14f, 16f
    See also Wheat bran
Resveratrol
    anti-cancer effects, 232–234
    effects on signal transduction, 234, 235f
Roasted cocoa
    aroma compounds with added epicatechin, 154, 158f, 159
    model food system, 148
Rutin, changes in fermented *Toona sinensis* leaves, 52–53

S

Saponins
    green tea, 95
    health effects, 84t
Scarlet runner bean
    antioxidative activity, 86t
    See also Beans
Scavenging activity
    black and white sesame seeds and hulls, 39–40
    canola hull extracts, 61, 63f
    Toona sinensis leaf extracts, 50–51
    See also 1,1-Diphenyl-2-picrylhydrazyl (DPPH) radical; Radical scavenging activity
Seeds. See Beans; Edible seed oils; Sesame seeds
Sesame seeds
    chelating effect of sesame extracts vs. catechin, 40, 41f
    determination of iron(II) chelating capacity, 37
    determination of total phenolic content of extracts, 36
    effect of catechin and black sesame hull extracts against low density lipoprotein (LDL) oxidation, 42–43
    ferrous ion chelation capacity, 41f
    free radical scavenging mechanism for antioxidant action, 39–40
    inhibition of copper-induced oxidation of human LDL, 42f
    inhibition of copper-mediated human LDL cholesterol, 37
    inhibition of human LDL oxidation, 41–43
    in vivo and in vitro effects of polyphenolic compounds, 40–41
    materials and methods, 35–38
    measurement of total antioxidant activity by Trolox equivalent antioxidant capacity (TEAC) assay, 36
    overall order of antioxidant activity, 43
    preparation of samples, 35
    scavenging capacity of 1,1-diphenyl-2-picrylydrazyl (DPPH) radical by, 40f
    scavenging of DPPH radical, 37
    statistical analysis, 38
    total antioxidant activity (TAA) of black and white seeds and hulls, 38–39
    total phenolic content (TPC) of black and white seeds and hulls, 38, 39f
    worldwide production, 34–35
Signal transduction
    carcinogenesis, 226–229
    effects of [6]-gingerol, 236, 237f
    effects of inositol hexaphosphate, 232, 233f
    effects of resveratrol, 234, 235f
    effects of tea polyphenols, 231f
    mitogen-activated protein (MAP) kinases, 226–227, 228f
Silybum marianum

medicinal use, 20
primary components, 20–21
See also Milk thistle
Spectrophotometric methods, peroxidase-catalyzed oxidation of capsaicinoids, 166–167
Stability, citral, 130
Steady-state kinetics, capsaicinoid oxidation, 166–169
Stereochemistry. See Milk thistle; Phenolic flavonolignans
Stilbenes, occurrence, 5
Streptococcus, lab acid bacteria strains, 48–49
Stroke
  antioxidants and, 273–274
  effects of cranberry extract, 278–279
Superoxide anion radical scavenging, honeybush tea extract, 126–127
Synthetic antioxidants
  examples, 68
  See also Antioxidants
Syringic acid
  winter red and white wheats, 12–15, 16f, 17f
  See also Wheat bran

T

Tannins
  bearberry-leaf extract, 72
  biological activity in bearberry-extract, 73–74
  condensed, in bearberry-leaf extracts and fractions, 75, 76t
  glucose lowering effect, 84t
  occurrence, 5
  properties, 70
  protein-tannin complex formation determination, 72–74
Tea
  Camellia sinenesis leaves, 198, 243
  catechins, 243
  concentrations of polyphenols in different teas, 207t
  consumption, 95
  health benefits of polyphenols, 189
  health promotion, 207, 243
  published research papers, 198, 199t
  theaflavins, 243, 244f
  See also Black tea; Green tea; Honeybush tea; Oolong tea
Tea fermentation
  coupled oxidation mechanism evidence for theaflavin synthesis, 190–191
  mechanism for production of dehydrotheasinensin A, 193f
  mechanism for theasinensin production, 192–195
  oxidation products of epigallocatechin and its gallate, 192
  oxidation products of theaflavin, 191
  oxidation-reduction dismutation of dehydrotheasinensin A, 194f
  possible pathways for theasinensin production, 195f
Tea polyphenols
  active efflux, 217–218
  bioavailability, 214
  bioavailability of green tea and catechins in human volunteers, 219–220
  biotransformation, 215–217
  biotransformation of tea catechins, 215f
  cancer prevention, 220, 222, 229–231
  chemical composition of teas, 213
  effects on signal transduction, 231f
  pharmacokinetics, 218–220
  proposed model for factors affecting bioavailability, 221f
  structures, 214f
Temperature

influence on aroma generation in Maillard reaction, 154, 155f, 156f
See also Aroma compounds, thermal generation
Theaflavins
  animals and chemicals, 244
  animal treatment, 246
  benzotropolone ring, 251
  bioavailability, 214
  black tea, 243
  black tea theaflavin-3,3'-digallate (TF3), 198
  confirming anti-inflammatory properties, 248, 251–252
  evaluation of theaflavin digallate with sulindac, 248, 249t
  evidence for coupled oxidation mechanism, 190–191
  inhibiting inflammation in ears of mice, 247
  inhibition of xanthine oxidase and reactive oxygen species (ROS) by tea polyphenols, 200
  materials and methods, 244, 246
  modulation of tumor promotion signaling proteins by TF3 and epigallocatechin gallate, 203–204
  oxidation products, 191
  oxidation products during fermentation, 189
  preparation of theaflavin digallate and related compounds, 246
  published papers, 199t
  structures, 214f, 244f
  structures of experimental compounds, 245f
  sulindac inhibiting cyclooxygenase (COX), 251–252
  See also Tea fermentation
Thearubigins, published papers, 199t
Theasinensin
  mechanism for production, 192–195
  mechanism for production of dehydrotheasinensin A, 193f
  oxidation-reduction dismutation of dehydrotheasinensin A, 194f
  possible pathways for production, 195f
  See also Tea fermentation
Thermal generation. *See* Aroma compounds, thermal generation
2-Thiobarbituric acid reactive substances (TBARS)
  antioxidant assay of tea extracts, 97–98, 99, 101f, 102f
  cooked pork model with bearberry-leaf, 79, 80f
  determination method, 74
Thistle. *See* Milk thistle
α-Tocopherol
  chemical structure, 178f
  comparison of 2,2-diphenyl-1-picrylhydrazyl radical (DPPH) scavenging activity, 111–112
  DPPH radical scavenging reaction with, 183f
  inhibitory activity on rat liver mitochondria (RLM) lipid peroxidation, 184, 185
  natural antioxidant, 68, 69f
Toona sinensis
  changes in lactic acid bacteria (LAB) counts and total bacterial counts in naturally fermented, leaf, 51f
  changes of antioxidative activity of extracts from fermented leaves, 51t
  changes of pH in naturally fermented, 48f
  chemical composition, 47
  2,2-diphenyl-1-picrylhydrazyl radical (DPPH) method for scavenging activity, 50–51
  extent of antioxidant potential in vivo for quercetin derivatives, 53

fermentation process, 47, 48f
identification and characterization of major LAB strains from solid-fermented, 50t
LAB strains, 48–50
leaves and stems, 47
organic acids in solid–fermented, 47–48, 49t
quercetin and derivative concentrations in rat plasma after ingestion, 54f
quercetins concentration and distribution, 52–53
rutin (quercetin-3-O-β-rutinoside), 52–53
tests on rats, 53–55
total plasma quercetin level in rats, 55
Total antioxidant activity (TAA)
extracts of black and white sesame seeds and hulls, 38–39
Trolox equivalent antioxidant capacity (TEAC) assay, 36
Total antioxidant capacity (TAC)
contribution of major antioxidants to TAC of apple, 259, 261
quantification, 258
See also Apple phenolics
Total phenolic content (TPC)
bean extracts, 87–88
cold-pressed seed oils, 115, 116f
extracts of black and white sesame seeds and hulls, 38, 39f
honeybush tea extract, 122–123
procedure for determination, 36
Transient-state kinetics, capsaicinoid oxidation, 164–165
Trolox equivalent antioxidant capacity (TEAC)
assay for measurement of total antioxidant activity, 36
whole seed and hull extracts of beans, 88, 89f

Tumor promoters
activation and phosphorylation of mitogen-activated protein (MAP) kinases, 226–227, 228f
inducing activation of transcription factor, 228–229
reactive oxygen species (ROS), 255
See also Cancer prevention

V

Vaccinium macrocarpon. See Cranberry phenolics
Vanillic acid
winter red and white wheats, 12–15, 16f, 17f
See also Wheat bran
Vanillin, reduction by o-methoxyphenols, 171, 172t
Vanillin assay, condensed tannins in bearberry-leaf extracts, 75, 77
Vanillylamine, reduction by o-methoxyphenols, 171, 172t
Venago wheat
phenolic acid composition, 15, 16f
See also Wheat bran
Vitamin C
antioxidants in apples, 265
inhibition of gap junction intercellular communication (GJIC), 264, 268f
natural antioxidant, 68, 69f, 255
quantification method, 257
Vitamin E
comparison of radical DPPH scavenging activity, 111–112
natural antioxidant, 68, 69f
Volatile components
analysis of honeybush tea extract, 124–125
honeybush tea, 126

W

Wheat bran
- antioxidants, 11
- comparing white and red wheat for phenolic acid content, 14f
- effect of growing condition on phenolic acid composition, 15, 17f
- extraction and testing sample preparation, 12
- high performance liquid chromatography (HPLC) analysis method, 12
- HPLC chromatograms of standard phenolic acids and hydrolysate of bran extract, 13f
- Lakin and Enhancer wheat grown at Burlington and Walsh Colorado, 14f, 15
- materials and methods, 11–12
- phenolic acid analysis of Enhancer and Venago red winter wheats, 15, 16f
- phenolic acid compositions, 12–13, 15
- phenolic acid compositions of two red wheat bran extracts, 16f
- phenolic acids, 11
- statistic analysis, 12

White hull extract
- total phenolic content, 88t
- See also Beans

White whole bean extract
- total phenolic content, 88t
- See also Beans

White winter wheat
- phenolic acid composition, 14f, 17f
- See also Wheat bran

Wine, scientific research publications, 189f

Winged bean
- antioxidative activity, 86t
- See also Beans

Winter wheat. See Wheat bran

X

Xanthine oxidase, inhibition of, and reactive oxygen species (ROS) by tea polyphenols, 200

Xanthones, honeybush tea, 120

X-ray crystallography, phenolic flavonolignan analysis, 27f, 31